# C–H Bond Activation in Organic Synthesis

# C–H Bond Activation in Organic Synthesis

Edited by
## Jie Jack Li
University of San Francisco
California, USA

CRC Press
Taylor & Francis Group
Boca Raton London New York

CRC Press is an imprint of the
Taylor & Francis Group, an **informa** business

CRC Press
Taylor & Francis Group
6000 Broken Sound Parkway NW, Suite 300
Boca Raton, FL 33487-2742

First issued in paperback 2017

ISBN-13: 978-1-4822-3310-0 (hbk)
ISBN-13: 978-1-138-89411-2 (pbk)

---

**Library of Congress Cataloging-in-Publication Data**

---

C-H bond activation in organic synthesis / edited by Jie Jack Li.
    pages cm
    "A CRC title."
    Includes bibliographical references and index.
    ISBN 978-1-4822-3310-0 (hardcover : alk. paper) 1. Organic compounds--Synthesis.
    2. Activation (Chemistry) 3. Chemistry, Analytic. 4. Carbon. 5. Hydrogen. I. Li, Jie
    Jack, editor. II. Title: Carbon-hydrogen bond activation in organic synthesis.

QD271.C27 2015
547'.27--dc23                                          2014047996

---

**Visit the Taylor & Francis Web site at**
**http://www.taylorandfrancis.com**

**and the CRC Press Web site at**
**http://www.crcpress.com**

# Contents

**Chapter 10** C—H Activation of Heteroaromatics

*Donna A. A. Wilton*

# Preface

The field of C—H activation in organic chemistry has flourished during the last decade. The approach is especially attractive to organic chemists because C—H activation is "greener" than conventional methods. New methodology is being published every day. This volume has attempted to summarize the state of the art of C—H activation for functionalization. It is to serve as the first stop for the reader who would like to gain an overview of this exciting playground of chemistry.

I am indebted to the contributing authors for their hard work and expertise in writing each chapter. I have learned enormously from this book and I hope you will too.

**Jie Jack Li**
*August 2014*
*San Francisco*

# Editor

**Jie Jack Li** earned his PhD in organic chemistry in 1995 at Indiana University. After a stint as a postdoctoral fellow MIT, he worked as a medicinal chemist at Pfizer and Bristol-Myers Squibb from 1997 to 2012. Since then he has been an associate professor of chemistry at the University of San Francisco teaching organic and medicinal chemistry. He has published 23 books ranging from organic and medicinal chemistry, to the history of drug discovery.

# Contributors

**Nadia M. Ahmad**
Vertex Pharmaceuticals (Europe)
    Limited
Oxfordshire, United Kingdom

**Narendra B. Ambhaikar**
Piramal Enterprises Limited
R&D, Pharma Solutions
Chennai, Tamilnadu, India

**Adam M. Azman**
Department of Chemistry
Butler University
Indianapolis, Indiana

**Jesse D. Carrick**
Department of Chemistry
Tennessee Tech University
Cookeville, Tennessee

**Timothy T. Curran**
Chemical Development
Vertex Pharmaceuticals
Cambridge, Massachusetts

**Marion H. Emmert**
Department of Chemistry and
    Biochemistry
and
Department of Mechanical Engineering
Worcester Polytechnic Institute
Worcester, Massachusetts

**Micheal Fultz**
Department of Chemistry
West Virginia State University
Institute, West Virginia

**Eric J. Medici**
Department of Chemistry
Davidson College
Davidson, North Carolina

**Nicole L. Snyder**
Department of Chemistry
Davidson College
Davidson, North Carolina

**Andrew C. Williams**
Eli Lilly and Company
Surrey, United Kingdom

**Donna A. A. Wilton**
Theravance Biopharma
South San Francisco, California

**Ji Zhang**
HEC R&D Center
Process Research and Development
Guang-Dong, People's Republic
    of China

# 1 Introduction

*Marion H. Emmert*

## CONTENTS

## 1.1   C—H FUNCTIONALIZATION AND GREEN CHEMISTRY

The vast majority of organic molecules and materials is derived from nonrenewable feedstocks such as crude oil and natural gas.[1,2] From a chemist's point of view, making these feedstocks usable involves breaking and making new C—C bonds and transforming C—H bonds into targeted functional groups. Most traditional pathways have achieved this objective by initial C—H bond functionalizations such as radical functionalizations[3-5] or partial aerobic oxidations,[6] which are then followed by a modifiable sequence of steps to introduce desired functional groups or C—C bonds constructing the desired frameworks. This approach, known as the functional group interconversion strategy,[7,8] is so fundamentally important to the way organic chemists think about synthesis, that the introductory chapters of organic chemistry textbooks and lectures focus on exactly this sequence: Radical C—H functionalizations (halogenations) are followed by substitution and elimination reactions, carboxyl and carbonyl chemistry, as well as oxidations and reductions.

Despite its well-deserved place in the toolbox of organic chemists, the functional group interconversion strategy suffers from a conceptual drawback: it typically leads to highly wasteful processes by requiring several steps from unfunctionalized feedstocks to functionalized products. As a result, high E factors (kg waste/kg product; see Table 1.1) are the rule and not the exceptions when synthesizing complex or highly functionalized organics, such as pharmaceuticals and fine chemicals.

C—H bond activations with subsequent functionalizations *in situ* provide a new approach to introducing functional groups. In principle, multistep syntheses can thus be transformed into one-step syntheses, thereby preventing waste generation

**TABLE 1.1**

**E Factors: Waste Produced as a Proportion of Product**

| Industry Segment | Annual Production ($t$) | E Factor (kg Waste/kg Product) |
|---|---|---|
| Oil refining | $10^6$–$10^8$ | <0.1 |
| Bulk chemicals | $10^4$–$10^6$ | <1–5 |
| Fine chemicals | $10^2$–$10^4$ | 5–50 |
| Pharmaceuticals | $10$–$10^3$ | 25–100 |

*Source:* Sheldon, R. A. *Green Chem.* **2007**, *9*, 1273–1283.

and enabling direct access to desired target structures even at late stages of a synthesis.[7,10,11]

R–H ——— C—H Functionalization ———→ R–FG$^1$

R–FG$^2$ ——→ R–FG$^3$

**Functional group interconversion**

## 1.2   INSPIRATION FROM MOTHER NATURE

Nature is the greatest and most efficient user of C—H bond functionalization strategies, and is able to install new functional groups at any stage in a biosynthetic process.[12] Enzymatic C—H bond functionalizations are not limited to early or late stages of a synthesis and proceed regardless of other, more reactive groups in the molecule. A classic example for this principle is the biosynthesis of biotin (**2**), in which two C—H bonds are cleaved to form a cyclic thioether. Remarkably, the traditionally more reactive urea and carboxylic acid functional groups in the substrate **1** remain unmodified, while two strong C—H bonds are broken.[13]

Biotin (**2**)

Characterization of the reaction pathways in biological C—H functionalizations is often challenging due to the fleeting nature and low concentrations of the reactive intermediates. As in other fields of chemistry, density functional theory (DFT) calculations have greatly contributed to the description and understanding of transient high-energy states in enzymes. This approach provides insights into the fundamental reactivity of many active centers and is remarkably useful for estimating the likelihood of a proposed mechanism.[14–16]

In reactions that include C—H bond activations, the fate of the reactive intermediates (cations, radical cations, or radicals) formed by an enzyme varies greatly depending on its identity. Reactions after the initial C—H cleavage step may include radical recombinations, radical or cationic rearrangements, and many others. However, the selectivity of C—H oxidation in enzymes is typically determined by the arrangement of the substrate in the enzymatic active site. As such, substrate over-oxidations or nonselective C—H oxidations are rarely observed.[17]

## 1.3  SELECTIVITY: THE HOLY GRAIL OF C—H BOND FUNCTIONALIZATIONS

In contrast to the superior chemoselectivity of enzymatic C—H oxidations, synthetic C—H activation catalysts still struggle to achieve similar selectivities. The magnitude of the challenge for catalyst development is illustrated well by considering the significant energies that are required to directly cleave C—H bonds (see Table 1.2). With bond dissociation energies (BDEs) between 96 and 105 kcal/mol, C—H bonds of simple hydrocarbons are among the bonds in organic chemistry and are most difficult to cleave. For comparison, the BDEs for Me—Cl and Me—Br, which are easily functionalized, are 83.7 and 72.1 kcal/mol, respectively. In addition, C—H bonds do not possess suitable lone pairs to coordinate to a catalyst, which makes C—H bond activation also kinetically challenging in comparison to other C—X bond cleaving reactions.[18]

To be broadly synthetically useful, C—H bond functionalization methodologies do not only need to cleave very strong, weakly coordinating bonds. Suitable catalysts for these transformations further need to be (i) stable in the presence of the

**TABLE 1.2**

**BDEs of C—H Bonds (kcal mol$^{-1}$)**

| 133 | 113 | 111 | 105 | 101 | 99 | 98 | 97 |

| 94 | 90 | 89 | 79 | 70 | 94 | 70 |

*Source:*  Blanksby, S. J.; Ellison, G. B. *Acc. Chem. Res.* **2003**, *36*, 255–263.

required oxidants, (ii) not deactivated by coordination of other functional groups in the product or solvent, (iii) selective for a specific type of C—H bond in a molecule, and (iv) slow to catalyze the further functionalization (over-oxidation) of the products. Considering all these critical challenges, it is not surprising that site-selective and product-selective C—H bond functionalization has been called the holy grail of catalysis research.[19]

## 1.4 CLASSIFICATIONS OF C—H BOND FUNCTIONALIZATIONS

Due to the high BDE values for alkanes, C—H bond functionalization methods for these substrates typically proceed under rather harsh conditions. One famous protocol, using the Periana catalyst system (3),[20] is able to functionalize methane, which is the alkane with the highest BDE (105 kcal/mol). The reaction proceeds in fuming sulfuric acid using $SO_3$ as oxidant and produces methylbisulfate (4). Fuming sulfuric acid is used as a suitable solvent without C—H bonds and further acts as a catalyst activator by protonating the ligand backbone of Pt complex 3 to render it to be electron deficient at the metal center and thus catalytically active.[21,22] Product 4 is protected against over-oxidation due to the electron-withdrawing characteristics of the bisulfate functional group; additionally, an excess of methane is employed to favor methane C—H bond activation instead of over-oxidation. The Periana protocol is one of the few examples of successful and product-selective functionalization of alkanes.[23–26] However, complex molecules require the use of much milder conditions than alkanes and, as such, offer their own challenges and opportunities.

$$H_3C-H \xrightarrow[\substack{102\% \; H_2SO_4 \\ 220°C, \; 81\%}]{3} H_3C-OSO_3H$$

Active catalyst
X = Cl⁻, HSO₄⁻

One such opportunity to address the thermodynamic challenge caused by large BDEs is the selective functionalization of "activated" C—H bonds. Benzylic C—H bonds (BDE 89.8 kcal/mol) or C—H bonds in α-positions to heteroatoms (BDE 92.1 kcal/mol for THF α-C—H) exhibit much lower BDEs than alkanes. Therefore, site selectivity can be achieved by cleaving the weakest C—H bonds among all others in a complex molecule. This strategy has been implemented for diverse substrates and has shown to be particularly successful when no organometallic intermediates with M—C bonds are formed along the reaction pathways. Reactions that fall into this category are many C—H bond functionalizations through radical intermediates,[27,28] metal nitrenoids,[29,30] metal carbenoids,[31,32] metal-oxo species,[33–35] and with organic reagents in the absence of metal catalysts.[36,37] Using these strategies, the site of C—H bond oxidation can be predicted for several complex molecules (6–8), based on electronic, steric, and stereoelectronic factors, providing powerful tools for late-stage functionalizations.[27,38]

Alternatively, C—H bond activation of otherwise unreactive bonds with high BDE values can be achieved kinetically through the use of catalyst directing groups.

This proximity-based strategy is often successful, when the reaction intermediate contains an M—C bond. New functional groups can be installed in simple structures bearing substituents with a pendant heteroatoms such as in **9**.[39,40] Diversifying functionalizations are also possible—for example, with pyrazole substrate **10**.[41] Overall, chelate-assisted C—H functionalizations[42] employing catalyst directing groups have been found to be highly versatile and applicable for a large variety of C—H functionalizations.[43]

## 1.5   MECHANISMS OF C—H ACTIVATION AND THEIR INFLUENCE ON SELECTIVITY

As indicated in the sections above, C—H bond activations can proceed through mechanistically distinct manifolds: (i) H-atom abstraction followed by radical functionalization (radical rebound mechanism), (ii) direct insertion into the C—H bond, and (iii) organometallic C—H activation proceeding through an intermediate with an M—C bond. The different mechanisms determine the types of C—H bonds that can be broken and can thus be used to achieve selectivity.

Radical rebound mechanisms are frequently invoked for reactions of metal-oxo catalysts,[44] but also for some functionalizations proceeding through metal nitrenoids.[45] Since the formation of radical intermediates (11) is involved, selectivity for the functionalization of weak C—H bonds is typical in such protocols. Stereoselective C—H functionalizations are often difficult to achieve with this mechanistic manifold due to the stereochemical liability of 11,[41] but can be overcome if radical recombination is fast (e.g., in enzymes).[46]

$$[M]{=}\mathbf{X} + H{-}\mathbf{C}R_3 \longrightarrow [M]{-}\mathbf{X}H \ \cdot CR_3 \longrightarrow [M] + H\mathbf{X}{-}CR_3$$

$$\overset{+n}{\phantom{.}} \qquad\qquad\qquad \overset{+(n-1)}{\phantom{.}} \qquad \overset{+(n-2)}{\phantom{.}}$$

**11**

Direct insertions into C—H bonds have been postulated for reactions that proceed through metal carbenes and nitrenes (12/13).[47,48] In these mechanisms, the unsaturated M=X intermediates 12 or 13 directly attack the C—H bond and form the product in one concerted step. Often, a preference for cleaving weak C—H bonds is observed; however, C—H functionalizations of alkanes are known that have been proposed to follow this so-called C—H insertion pathway. Stereoselective and even enantioselective methodologies have been developed based on this strategy using suitable enantiomerically pure ancillary ligands.[49,50]

$$N_2{=}\mathbf{CR^1R^2} \xrightarrow{\ [M]\ } [M]{=}\mathbf{CR^1R^2} \xrightarrow{\ H{-}CR_3\ } R_3C{-}\overset{H}{\underset{\big|}{\mathbf{CR^1R^2}}} + [M]$$

**12**

$$\begin{matrix} N_2{=}\mathbf{NR} \\ \text{or} \\ \mathbf{X}{-}\mathbf{NHR} \\ \text{or} \\ \mathbf{NH_2R}/\text{oxidant} \end{matrix} \xrightarrow{\ [M]\ } [M]{=}\mathbf{NR} \xrightarrow{\ H{-}CR_3\ } R_3C{-}\overset{H}{\underset{\big|}{\mathbf{NR}}} + [M]$$

**13**

Finally, C—H bond activations that proceed through organometallic intermediates with M—C bonds are common mechanistic pathways toward C—H functionalizations. Several distinct mechanisms for C—H cleavage (σ-bond metathesis; oxidative addition; concerted metalation deprotonation through 4- or 6-membered transition states) have been elucidated for this step. The exact pathway of C—H bond activation typically depends on the identity of the metal, its oxidation state, and the ancillary ligands.

**Concerted metalation deprotonation (CMD)**

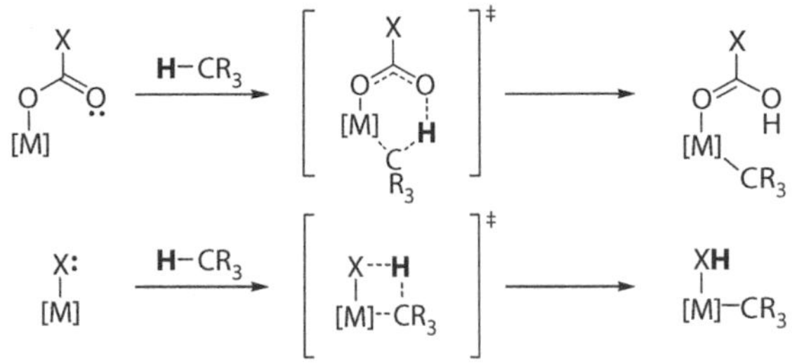

$$\left[\; X\cdots H \;\right]^{\ddagger}$$

**Sigma bond metathesis (SBM)**

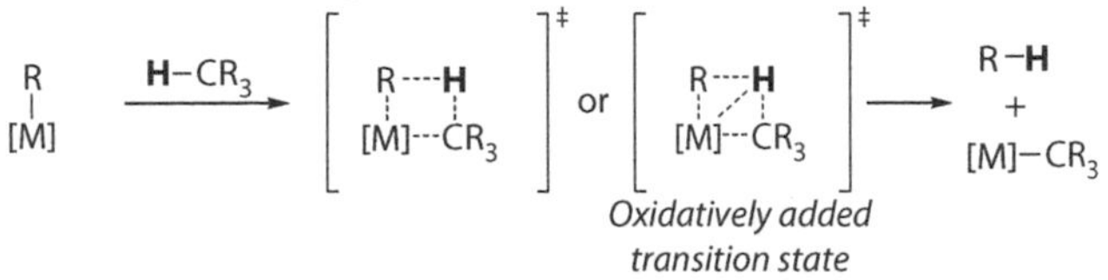

*Oxidatively added transition state*

**Oxidative addition**

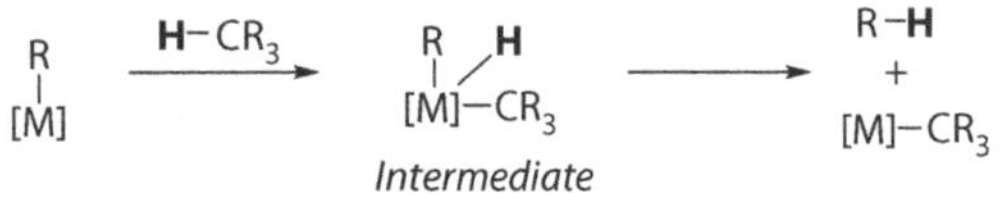

*Intermediate*

To arrive at a C—H functionalized product, the organometallic intermediates **14** that are formed through C—H activation typically react with an oxidant and form the product through reductive elimination pathways; the sequence of the last two steps of the catalytic cycle depends on the nature of the oxidant, product, and metal catalyst.

**Product formation from organometallic intermediate**

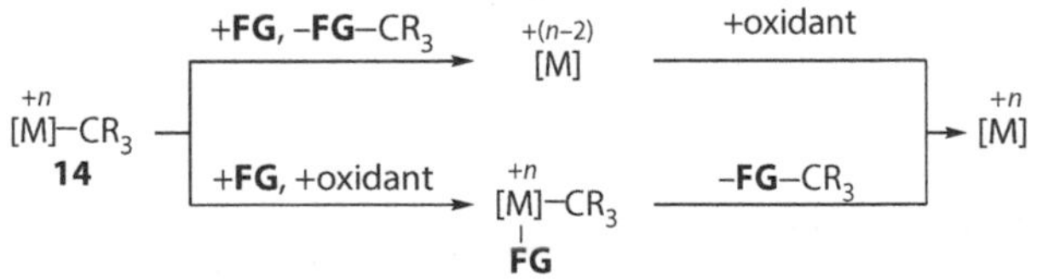

Interestingly, C—H functionalizations through organometallic intermediates preferentially cleave strong C—H bonds in a molecule. Recent mechanistic insights through experiments and computations suggest that the reason for this selectivity lies in the strength of the formed M—C bonds, as the values of M—C bond energies are often higher than the corresponding C—H bond energies.[18,51,52]

## 1.6 Pd: VERSATILE AND OMNIPRESENT

Pd catalyzed C—H bond functionalization protocols are particularly versatile with regard to their substrate and product scope.[37,53,54] Furthermore, Pd catalysts can be used in both simple and complex molecule settings.

Interestingly, several Pd catalysts (e.g., **15/16**) can be found among the few examples of efficient C—H functionalizations of challenging substrates such as methane and simple arenes.[25,55,56]

A wide array of experimental and computational mechanistic studies provides a good insight into reactivity patterns of Pd complexes in C—H functionalizations.[57–65] The relevant oxidation states Pd$^0$, Pd$^{II}$, Pd$^{III}$, and Pd$^{IV}$ have all been invoked in catalytic cycles, and catalytic mechanisms are often characterized as Pd$^{0/II}$ or Pd$^{II/IV}$ pathways, reflecting the oxidation states of the respective catalytic intermediates. Catalytically relevant organometallic Pd complexes have been isolated for most of these oxidation states.[62,66–68] While C—H bond activation during catalysis is believed to occur most often at Pd$^{II}$ centers,[69] recent experimental findings suggest that C—H bond activations can indeed occur at Pd$^{IV}$ complexes.[70] With regard to the functionalization step after C—H activation, C—C bonds can be formed from Pd$^{II}$,[71,72] Pd$^{III}$ (**17**),[73] or Pd$^{IV}$ (**19**),[74] while the formation of polarized C—X bonds typically requires the presence of a strong oxidant and has been shown to be facile from high-oxidation state Pd complexes.[75,76]

One interesting use of Pd compounds is in Catellani-type reactions, which result in functionalizing several positions on an aromatic substrate in one pot. Catellani reactions are initiated by oxidative insertion of the catalyst into a $C_{aryl}$–halogen bond. The resulting $Pd^{II}$–aryl complex **20** then reacts with norbornene to form the Pd–alkyl intermediate **21**, which undergoes C—H activation in *ortho*-position to the original insertion site. Subsequent functionalization of the new Pd—$C_{aryl}$ bond and catalyst regeneration by norbornene elimination leads to the formation of highly functionalized products such as **22**.[77–79]

Another interesting feature of Pd complexes is their ability to use oxygen as terminal oxidant for oxidative functionalizations. Several examples of oxidative $C_{aryl}$—H olefinations (oxidative Heck reactions) proceed with $O_2$ as a reagent.[80,81] Furthermore, aerobic dehydrogenative aryl couplings with high site selectivity have been developed.[82]

Remarkably, C—H oxygenations can also be performed with $O_2$ as terminal oxidant, with and without the use of co-catalysts.[80,83–87] Variations of the catalytic conditions result in efficient protocols for introducing C—O bonds in allylic (**24**), aromatic (**25**), benzylic (**27**), and aliphatic positions (**28**) with high atom economy.

Instead of oxygen, a wide variety of two electron oxidants and one electron oxidant can be applied in Pd catalyzed C—H functionalizations, including metals salts and organic oxidants. This compatibility with many conditions, together with a broad functional group compatibility, makes these catalysts probably the best studied C—H bond functionalization catalysts in organic chemistry. Recent developments in the field aim at combining C—H bond activation chemistry with other types of reactivity in order to achieve C—H functionalizations under milder and more atom-economic conditions.[88] One example for these efforts is a C—H arylation reaction reported by Sanford and coworkers that uses visible-light photocatalysis in order to form the biaryl **29**, whose synthesis in the absence of light typically requires elevated temperatures.[89]

## 1.7 Rh: HALF-SANDWICH COMPLEXES, C—H INSERTIONS, AND BEYOND

Together with the homologous Cp*Ir complexes (Cp* = $\eta^5$-C$_5$Me$_5$), Cp*Rh complexes are among the first catalyst systems that have been used for studying fundamental aspects of organometallic C—H bond activations. Despite their broad use in this context,[19,90,91] Rh catalysts have only recently been established as versatile C—H functionalization catalysts. One reason for the initial hesitation of the organic community to adapt Rh catalyzed protocols might be the high price of Rh ($3812/mol), which is among the highest prices for precious metals and only surpassed by the prices for Au ($8707/mol) and Pt ($9933/mol).[92]

Rh catalysts have achieved impressive transformations that are without precedent with other catalysts, in particular with regard to stereoselective C—C and C—N bond formations. Du Bois and coworkers have developed a series of $Rh^{II}$ complexes bearing carboxylate and carboxamidate ligands. The resulting catalysts show remarkably high reactivities and selectivities in intramolecular (**30**) and intermolecular C—H aminations (**31**).[93,94]

Structurally similar, dimeric $Rh^{II}$ catalysts with four carboxylate or carboxamidate ligands efficiently catalyze carbene insertions into C—H bonds. In this field, highly enantioselective catalysts have been developed by Davies and coworkers, together with mechanistic models that help predict selectivities in these transformations. The resulting protocols have been widely applied, in particular for the synthesis of complex molecules such as R-(–)-rolipram (**34**).[46,95]

Proceeding through distinctly different mechanisms than the C—H insertions with dimeric Rh catalysts above, organometallic C—H bond functionalizations with Rh catalysts have also found tremendous use in organic synthesis. Interestingly, the different oxidation states of Rh have been shown to activate C—H bonds through different mechanisms: Rh$^I$ complexes prefer C—H activation through an oxidative addition mechanism,[96] while Rh$^{III}$-catalyzed C—H bond activations are proposed to proceed through electrophilic pathways.[97,98]

Work with Rh$^I$ catalyst precursors in the Ellman group has resulted in the development of addition reactions of C—H bonds to unsaturated bonds in imines,[97] olefins,[99] isocyanates,[100] or aldehydes[101–103] through Rh-aryl intermediates **35**. The resulting protocols form new C—C bonds in the process and many interesting heterocyclic structures such as **37**, **38**, and **39** can be synthesized.[104,105]

Rh catalyzed C—N bond formations, such as in the synthesis of **40**, have been developed by Glorius, Fagnou, and others. As a result, a large variety of protocols exists that achieve C—H aminations presumably through organometallic intermediates.[106,107] Most of the substrates used require the presence of a directing group. One interesting approach to functionalize these required substituents in one step is the use of oxidizing directing groups, which are transformed under the reaction conditions (e.g., in product **41**). These directing groups act as internal oxidants; thus, no external oxidants are needed to promote the reaction.[108–112]

Notably, Rh catalysts have also been employed by Hartwig for efficient C—H borylations, which enable even C—H functionalizations of aliphatic C—H bonds in alkanes (**42**).[113,114] C—H borylations of alkanes are remarkably selective for functionalization at primary C—H bonds.

## 1.8 Ru: MURAI COUPLINGS AND MORE

Despite the practical incentive of low cost ($203/mol) compared to Pd ($2780/mol) and Rh ($3812/mol),[89] Ru complexes are less developed as catalysts for C—H bond functionalizations. Most Ru catalyzed protocols focus on C—C bond formations, while other C—H functionalizations are underdeveloped with Ru catalyst systems.[115]

One remarkable reaction catalyzed by Ru complexes is the Murai coupling, in which alkyl arenes **44** are formed by hydroarylations of olefins. Interestingly, β-hydride elimination is not a problem in these reactions, as would be expected with other transition metal catalysts.[116,117] Murai-type protocols are especially useful as they can employ electron-rich olefins, which are typically low-yielding substrates in Pd-catalyzed, oxidative Heck (Fujiwara–Moritani) reactions.[116–118]

Another striking feature of Ru catalysts is that the usable directing groups in their chemistry are not restricted to strongly electron-donating heterocycles. Many reaction protocols are known that employ ketones as directing groups in order to functionalize *ortho*-C—H bonds such as in **43**.[119] Taking into account that ketones can be directly used for further syntheses, this capability is advantageous for syntheses of complex molecules.

Overall, Ru catalyzed C—H arylations,[120] allylations,[121,122] alkenylations,[115] alkylations,[123–125] and acylations[126] are known. Furthermore, Ru catalysts have the remarkable ability to enable C—H functionalizations in *meta*-position to a catalyst directing group. In these reactions, the organometallic intermediate **47** reacts with electrophiles at the *para*-position of the M—C bond instead of reacting at the M—C bond. Such reactions are likely analogous to electrophilic substitutions with the metal playing the role of a directing substituent. Overall, sulfonylations (**45**) and alkylations (**46**) can thus be performed.[127,128]

## 1.9   RECENT APPLICATIONS IN COMPLEX MOLECULE SYNTHESIS

After highlighting particularly useful metal catalyst systems in the previous sections, the following subchapter focuses on applications of C—H bond functionalizations in the synthesis of complex molecules. Many of the latest catalysts have been used to enable shorter syntheses and fundamentally new approaches, thus steamlining synthetic processes. Additional examples of such protocols can be found in the following chapters of this book and have also been reviewed elsewhere.[129,130]

The first highly enantioselective catalytic reaction proceeding through aromatic C—H bond activation was reported by Ellman and Bergman for the synthesis of PKCβ inhibitor JTT-010 (**51**).[100,101]

Employing a chiral ligand **49** for the key C—C bond forming step allowed isolation of intermediate **50** in 61% yield and 90% *ee*, which subsequently was converted into the target structure **51** in several steps.

Moreover, White and coworkers have developed a remarkable late-stage C—H oxidative macrocyclization, which enables the direct access to ring structures found in erythromycins (**54**).[131] The ring-closing reaction affording the 14-membered macrocycle **53** is a Pd-catalyzed allylic C—H bond oxygenation in which the intramolecular carboxylic acid as a nucleophile. Through adjustment of the reaction conditions, both possible diastereomeric products are available independently, which can be rationalized by a catalytic chelate-control model. Thus, stereoselective C—H oxygenation at step 19 of 22 of the overall synthesis provides flexibility for late-stage modifications, which is highly valuable for drug discovery. Furthermore, the C—H functionalization strategy eliminates the necessity to install another functional group at C13 (**52**) early in the synthesis and, therefore, prevents unwanted side reactions such as eliminations or isomerizations.

Another significant strategy involving late-stage C—H bond oxidations has been developed by Baran and coworkers for the total synthesis of terpenes.[132] Conceptually, this approach is inspired by the proposed biosynthesis of these natural products, which is divided into a cyclase and an oxidase phase. During the cyclase phase, the carbon framework is assembled, while the introduction of functional groups takes place later in the oxidase phase. This strategy depends on the availability of reliable C—H oxidation protocols and exhibits considerable advantages that stem from the lack of functionalization in the framework building blocks.

Following the cyclase/oxidase strategy, (+)-ingenol (**59**) was synthesized in 14 steps and 1.2% overall yield. The total synthesis makes a class of promising biologically active polyoxygenated terpenoids *and their derivatives* available in large enough quantities to rival bioengineered pathways for their production. Two late-stage, allylic C—H oxidations were performed, both using SeO₂ as an oxidant to synthesize intermediate **57** as well as the final product **59**. Besides the savings in step count, this strategy has the additional advantage that the use of protecting groups can be kept to a minimum and that nonstrategic redox manipulations are absent, advancing the overall "greenness" of the synthetic sequence.[133]

(+)-Ingenol

The three examples detailed above all show remarkable creativity in the use of C–H functionalizations in order to arrive at greener syntheses. Thus, direct C–H functionalization methods are expected to continue to greatly contribute to the mission of green chemistry: Low-energy, waste-free, and atom-economic transformations for the synthesis of organic materials and biologically active molecules in the twenty-first century.

## REFERENCES

1. Hess, J.; Bednarz, D.; Bae, J.; Pierce, J. *Am. J. Public Health* **2011**, *101*, 1568–1579.
2. Schwartz, B. S.; Parker, C. L.; Hess, J.; Frumkin, H. *Am. J. Public Health* **2011**, *101*, 1560–1567.
3. Curran, D. P. *Synthesis* **1988**, 417–439.
4. Walling, C.; Jacknow, B. B. *J. Am. Chem. Soc.* **1960**, *82*, 6108–6112.
5. Recupero, F.; Punta, C. *Chem. Rev.* **2007**, *107*, 3800–3842.
6. Brühne, F.; Wright, E., Benzaldehyde. In *Ullmann's Encyclopedia of Industrial Chemistry*, Wiley-VCH: 2000, pp 5–6.
7. Davies, H. M. L.; Du Bois, J.; Yu, J.-Q. *Chem. Soc. Rev.* **2011**, *40*, 1855–1856.
8. Das, P.; Dutta, A.; Bhaumik, A.; Mukhopadhyay, C. *Green Chem.* **2014**, *16*, 1426–1435.
9. Sheldon, R. A. *Green Chem.* **2007**, *9*, 1273–1283.
10. Potavathri, S.; Pereira, K. C.; Gorelsky, S. I.; Pike, A.; LeBris, A. P.; DeBoef, B. *J. Am. Chem. Soc.* **2010**, *132*, 14676–14681.
11. Ackermann, L.; Fenner, S. *Org. Lett.* **2011**, *13*, 6548–6551.
12. Bollinger Jr, J. M.; Broderick, J. B. *Curr. Opin. Chem. Biol.* **2009**, *13*, 51–57.

13. Booker, S. J., *Curr. Opin. Chem. Biol.* **2009**, *13*, 58–73.
14. Siegbahn, P. E. M.; Crabtree, R. H. *J. Am. Chem. Soc.* **1997**, *119*, 3103–3113.
15. Groves, J. T. *Nat. Chem.* **2014**, *6*, 89–91.
16. Siegbahn, P. E. M.; Borowski, T. *Acc. Chem. Res.* **2006**, *39*, 729–738.
17. Lewis, J. C.; Coelho, P. S.; Arnold, F. H. *Chem. Soc. Rev.* **2011**, *40*, 2003–2021.
18. Blanksby, S. J.; Ellison, G. B. *Acc. Chem. Res.* **2003**, *36*, 255–263.
19. Arndtsen, B. A.; Bergman, R. G.; Mobley, T. A.; Peterson, T. H. *Acc. Chem. Res.* **1995**, *28*, 154–162.
20. Periana, R. A.; Taube, D. J.; Gamble, S.; Taube, H.; Satoh, T.; Fujii, H. *Science* **1998**, *280*, 560–564.
21. Xu, X.; Kua, J.; Periana, R. A.; Goddard, W. A. *Organometallics* **2003**, *22*, 2057–2068.
22. Mironov, O. A.; Bischof, S. M.; Konnick, M. M.; Hashiguchi, B. G.; Ziatdinov, V. R.; Goddard, W. A.; Ahlquist, M.; Periana, R. A. *J. Am. Chem. Soc.* **2013**, *135*, 14644–14658.
23. Caballero, A.; Despagnet-Ayoub, E.; Diaz-Requejo, M. M.; Diaz-Rodriguez, A.; Gonzalez-Nunez, M. E.; Mello, R.; Munoz, B. K. et al. *Science* **2011**, *332*, 835–838.
24. Goldman, A. S.; Roy, A. H.; Huang, Z.; Ahuja, R.; Schinski, W.; Brookhart, M. *Science* **2006**, *312*, 257–261.
25. Meyer, D.; Taige, M. A.; Zeller, A.; Hohlfeld, K.; Ahrens, S.; Strassner, T. *Organometallics* **2009**, *28*, 2142–2149.
26. Munz, D.; Strassner, T. *Angew. Chem. Int. Ed.* **2014**, *53*, 2485–2488.
27. Bigi, M. A.; Reed, S. A.; White, M. C. *Nat. Chem.* **2011**, *3*, 216–222.
28. Qvortrup, K.; Rankic, D. A.; MacMillan, D. W. C. *J. Am. Chem. Soc.* **2013**, *136*, 626–629.
29. Gephart, R. T.; Warren, T. H. *Organometallics* **2012**, *31*, 7728–7752.
30. Dauban, P.; Lescot, C.; Diaz-Requejo, M. M.; Perez, P. J. *Rh-, Ag-, and Cu-Catalyzed C−N Bond Formation.* In *Innovative Catalysis in Organic Synthesis*, Wiley-VCH: Weinheim, Germany: 2012; pp 257–277.
31. Diaz-Requejo, M. M.; Perez, P. J. *Chem. Rev.* **2008**, *108*, 3379–3394.
32. Lian, Y.; Hardcastle, K. I.; Davies, H. M. L. *Angew. Chem. Int. Ed.* **2011**, *50*, 9370–9373.
33. Que, L.; Tolman, W. B. *Nature* **2008**, *455*, 333–340.
34. Hitomi, Y.; Arakawa, K.; Funabiki, T.; Kodera, M. *Angew. Chem. Int. Ed.* **2012**, *51*, 3448–3452.
35. Nam, W. *Acc. Chem. Res.* **2007**, *40*, 522–531.
36. Foo, K.; Sella, E.; Thomé, I.; Eastgate, M. D.; Baran, P. S. *J. Am. Chem. Soc.* **2014**, *136*, 5279–5282.
37. Gui, J.; Zhou, Q.; Pan, C.-M.; Yabe, Y.; Burns, A. C.; Collins, M. R.; Ornelas, M. A.; Ishihara, Y.; Baran, P. S. *J. Am. Chem. Soc.* **2014**, *136*, 4853–4856.
38. Chen, M. S.; White, M. C. *Science* **2010**, *327*, 566–571.
39. Lyons, T. W.; Sanford, M. S. *Chem. Rev.* **2010**, *110*, 1147–1169.
40. Zhang, Q.; Yang, F.; Wu, Y. *Chem. Commun.* **2013**, *49*, 6837–6839.
41. Dai, H. X.; Stepan, A. F.; Plummer, M. S.; Zhang, Y. H.; Yu, J. Q. *J. Am. Chem. Soc.* **2011**, *133*, 7222–7228.
42. Kuhl, N.; Hopkinson, M. N.; Wencel-Delord, J.; Glorius, F. *Angew. Chem. Int. Ed.* **2012**, *51*, 10236–10254.
43. Godula, K.; Sames, D. *Science* **2006**, *312*, 67–72.
44. Borovik, A. S. *Chem. Soc. Rev.* **2011**, *40*, 1870–1874.
45. Barman, D. N.; Liu, P.; Houk, K. N.; Nicholas, K. M. *Organometallics* **2010**, *29*, 3404–3412.
46. Fokin, A. A.; Schreiner, P. R. *Chem. Rev.* **2002**, *102*, 1551–1594.
47. Doyle, M. P.; Duffy, R.; Ratnikov, M.; Zhou, L. *Chem. Rev.* **2009**, *110*, 704–724.
48. Díaz-Requejo, M. M.; Caballero, A.; Fructos, M.; Pérez, P. *Alkane Catalytic Functionalization by Carbene or Nitrene Insertion Reactions.* In *Alkane C−H*

*Activation by Single-Site Metal Catalysis*, Pérez, P. J., Ed.; Springer: Netherlands: 2012; Vol. 38, pp 229–264.

49. Fraile, J. M.; García, J. I.; Mayoral, J. A.; Roldán, M. *Org. Lett.* **2007**, *9*, 731–733.
50. Davies, H. M.; Manning, J. R. *Nature* **2008**, *451*, 417–424.
51. Balcells, D.; Clot, E.; Eisenstein, O. *Chem. Rev.* **2010**, *110*, 749–823.
52. Clot, E.; Me´gret, C.; Eisenstein, O.; Perutz, R. N. *J. Am. Chem. Soc.* **2009**, *131*, 7817–7827.
53. Dick, A. R.; Sanford, M. S. *Tetrahedron* **2006**, *62*, 2439–2463.
54. Chen, X.; Engle, K. M.; Wang, D.-H.; Yu, J.-Q. *Angew. Chem. Int. Ed.* **2009**, *48*, 5094–5115.
55. Emmert, M. H.; Gary, J. B.; Villalobos, J. M.; Sanford, M. S. *Angew. Chem. Int. Ed.* **2010**, *49*, 5884–5886.
56. Emmert, M. H.; Cook, A. K.; Xie, Y. J.; Sanford, M. S. *Angew. Chem. Int. Ed.* **2011**, *50*, 9409–9412.
57. Boutadla, Y.; Davies, D. L.; Macgregor, S. A.; Poblador-Bahamonde, A. I. *Dalton Trans.* **2009**, 5820–5831.
58. Ke, Z.; Cundari, T. R. *Organometallics* **2010**, *29*, 821–834.
59. Zhang, S.; Shi, L.; Ding, Y. *J. Am. Chem. Soc.* **2011**, *133*, 20218–20229.
60. Chaumontet, M.; Piccardi, R.; Audic, N.; Hitce, J.; Peglion, J. L.; Clot, E.; Baudoin, O. *J. Am. Chem. Soc.* **2008**, *130*, 15157–15166.
61. Gorelsky, S. I.; Lapointe, D.; Fagnou, K. *J. Org. Chem.* **2012**, *77*, 658–668.
62. Guo, L.; Xu, Y.; Wang, X.; Liu, W.; Lu, D. *Organometallics* **2013**, *32*, 3780–3783.
63. Powers, D. C.; Xiao, D. Y.; Geibel, M. A.; Ritter, T. *J. Am. Chem. Soc.* **2010**, *132*, 14530–14536.
64. Deprez, N. R.; Sanford, M. S. *J. Am. Chem. Soc.* **2009**, *131*, 11234–11241.
65. Bercaw, J. E.; Hazari, N.; Labinger, J. A.; Oblad, P. F. *Angew. Chem. Int. Ed.* **2008**, *47*, 9941–9943.
66. Ye, Y.; Ball, N. D.; Kampf, J. W.; Sanford, M. S. *J. Am. Chem. Soc.* **2010**, *132*, 14682–14687.
67. Tang, F.; Zhang, Y.; Rath, N. P.; Mirica, L. M. *Organometallics* **2012**, *31*, 6690–6696.
68. Mirica, L. M.; Khusnutdinova, J. R. *Coord. Chem. Rev.* **2013**, *257*, 299–314.
69. Jia, C.; Kitamura, T.; Fujiwara, Y. *Acc. Chem. Res.* **2001**, *34*, 633–639.
70. Racowski, J. M.; Ball, N. D.; Sanford, M. S. *J. Am. Chem. Soc.* **2011**, *133*, 18022–18025.
71. Lafrance, M.; Rowley, C. N.; Woo, T. K.; Fagnou, K. *J. Am. Chem. Soc.* **2006**, *128*, 8754–8756.
72. Yagyu, T.; Hamada, M.; Osakada, K.; Yamamoto, T. *Organometallics* **2001**, *20*, 1087–1101.
73. Khusnutdinova, J. R.; Rath, N. P.; Mirica, L. M. K. *J. Am. Chem. Soc.* **2010**, *132*, 7303–7305.
74. Canty, A. J. *Acc. Chem. Res.* **1992**, *25*, 83–90.
75. Racowski, J. M.; Dick, A. R.; Sanford, M. S. *J. Am. Chem. Soc.* **2009**, *131*, 10974–10983.
76. Pérez-Temprano, M. H.; Racowski, J. M.; Kampf, J. W.; Sanford, M. S. *J. Am. Chem. Soc.* **2014**, *136*, 4097–4100.
77. Martins, A.; Mariampillai, B.; Lautens, M. *Synthesis in the Key of Catellani: Norbornene-Mediated ortho C—H Functionalization.* In *C—H Activation*, Yu, J.-Q.; Shi, Z., Eds.; Springer: Berlin, Heidelberg: 2010; Vol. 292, pp 1–33.
78. Catellani, M. *Novel Methods of Aromatic Functionalization Using Palladium and Norbornene as a Unique Catalytic System.* In *Palladium in Organic Synthesis*, Tsuji, J., Ed. Springer: Berlin, Heidelberg: 2005; Vol. 14, pp 21–53.
79. Dong, Z.; Dong, G. *J. Am. Chem. Soc.* **2013**, *135*, 18350–18353.
80. Zhang, Y. H.; Shi, B. F.; Yu, J. Q. *J. Am. Chem. Soc.* **2009**, *131*, 5072–5074.
81. Campbell, A. N.; Stahl, S. S. *Acc. Chem. Res.* **2012**, *45*, 851–863.

82. Izawa, Y.; Stahl, S. S. *Adv. Synth. Catal.* **2010,** *352,* 3223–3229.
83. Campbell, A. N.; White, P. B.; Guzei, I. A.; Stahl, S. S. *J. Am. Chem. Soc.* **2010,** *132,* 15116–15119.
84. Stowers, K. J.; Kubota, A.; Sanford, M. S. *Chem. Sci.* **2012,** *3,* 3192–3195.
85. Vedernikov, A. N. *Acc. Chem. Res.* **2011,** *45,* 803–813.
86. Zhang, Y.-H.; Yu, J.-Q. *J. Am. Chem. Soc.* **2009,** *131,* 14654–14655.
87. Stahl, S. S. *Angew. Chem. Int. Ed.* **2004,** *43,* 3400–3420.
88. Wencel-Delord, J.; Droge, T.; Liu, F.; Glorius, F. *Chem. Soc. Rev.* **2011,** *40,* 4740–4761.
89. Kalyani, D.; McMurtrey, K. B.; Neufeldt, S. R.; Sanford, M. S. *J. Am. Chem. Soc.* **2011,** *133,* 18566–18569.
90. Jiao, Y.; Evans, M. E.; Morris, J.; Brennessel, W. W.; Jones, W. D. *J. Am. Chem. Soc.* **2013,** *135,* 6994–7004.
91. Corkey, B. K.; Taw, F. L.; Bergman, R. G.; Brookhart, M. *Polyhedron* **2004,** *23,* 2943–2954.
92. Based on market prices (Rh: $1050/oz; Pd: $740.50/oz; Ru: $57/oz) retrieved 01/13/14 from http://www.infomine.com/investment/metal-prices/
93. Du Bois, J. *Org. Process Res. Dev.* **2011,** *15,* 758–762.
94. Roizen, J. L.; Harvey, M. E.; Du Bois, J. *Acc. Chem. Res.* **2012,** *45,* 911–922.
95. Davies, H. M. L.; Lian, Y. *Acc. Chem. Res.* **2012,** *45,* 923–935.
96. Astruc, D. *C−H Activation and Functionalization of Alkanes and Arenes.* In *Organometallic Chemistry and Catalysis,* Springer: Berlin, Heidelberg: 2007; pp 409–429.
97. Park, S. H.; Kwak, J.; Shin, K.; Ryu, J.; Park, Y.; Chang, S. *J. Am. Chem. Soc.* **2014,** *136,* 2492–2502.
98. Tauchert, M. E.; Incarvito, C. D.; Rheingold, A. L.; Bergman, R. G.; Ellman, J. A. *J. Am. Chem. Soc.* **2012,** *134,* 1482–1485.
99. Thalji, R. K.; Ellman, J. A.; Bergman, R. G. *J. Am. Chem. Soc.* **2004,** *126,* 7192–7193.
100. Hesp, K. D.; Bergman, R. G.; Ellman, J. A. *J. Am. Chem. Soc.* **2011,** *133,* 11430–11433.
101. Li, Y.; Zhang, X.-S.; Chen, K.; He, K.-H.; Pan, F.; Li, B.-J.; Shi, Z.-J. *Org. Lett.* **2012,** *14,* 636–639.
102. Lian, Y.; Bergman, R. G.; Ellman, J. A. *Chem. Sci.* **2012,** *3,* 3088–3092.
103. Lian, Y.; Bergman, R. G.; Lavis, L. D.; Ellman, J. A. *J. Am. Chem. Soc.* **2013,** *135,* 7122–7125.
104. Colby, D. A.; Tsai, A. S.; Bergman, R. G.; Ellman, J. A. *Acc. Chem. Res.* **2011,** *45,* 814–825.
105. Colby, D. A.; Bergman, R. G.; Ellman, J. A. *Chem. Rev.* **2010,** *110,* 624–655.
106. Grohmann, C.; Wang, H.; Glorius, F. *Org. Lett.* **2013,** *15,* 3014–3017.
107. Guimond, N.; Gouliaras, C.; Fagnou, K. *J. Am. Chem. Soc.* **2010,** *132,* 6908–6909.
108. Wang, H.; Glorius, F. *Angew. Chem. Int. Ed.* **2012,** *51,* 7318–7322.
109. Guimond, N.; Gorelsky, S. I.; Fagnou, K. *J. Am. Chem. Soc.* **2011,** *133,* 6449–6457.
110. Rakshit, S.; Grohmann, C.; Besset, T.; Glorius, F. *J. Am. Chem. Soc.* **2011,** *133,* 2350–2353.
111. Too, P. C.; Wang, Y.-F.; Chiba, S. *Org. Lett.* **2010,** *12,* 5688–5691.
112. Li, B.-J.; Wang, H.-Y.; Zhu, Q.-L.; Shi, Z.-J. *Angew. Chem. Int. Ed.* **2012,** *51,* 3948–3952.
113. Chen, H.; Schlecht, S.; Semple, T. C.; Hartwig, J. F. *Science* **2000,** *287,* 1995–1997.
114. Hartwig, J. F. *Acc. Chem. Res.* **2011,** *45,* 864–873.
115. Arockiam, P. B.; Bruneau, C.; Dixneuf, P. H. *Chem. Rev.* **2012,** *112,* 5879–5918.
116. Murai, S.; Kakiuchi, F.; Sekine, S.; Tanaka, Y.; Kamatani, A.; Sonoda, M.; Chatani, N. *Nature* **1993,** *366,* 529–531.
117. Helmstedt, U.; Clot, E. *Chem. Eur. J.* **2012,** *18,* 11449–11458.
118. Le Bras, J.; Muzart, J. *Chem. Rev.* **2011,** *111,* 1170–1214.
119. Kakiuchi, F.; Sato, T.; Tsujimoto, T.; Yamauchi, M.; Chatani, N.; Murai, S. *Chem. Lett.* **1998,** *27,* 1053–1054.

120. Ackermann, L.; Vicente, R. *Ruthenium-Catalyzed Direct Arylations Through C−H Bond Cleavages*. In *C−H Activation*, Yu, J.-Q.; Shi, Z., Eds. Springer: Berlin, Heidelberg: 2010; Vol. 292, pp 211–229.

121. Oi, S.; Tanaka, Y.; Inoue, Y. Ortho-selective allylation of 2-pyridylarenes with allyl acetates catalyzed by ruthenium complexes. *Organometallics* **2006**, *25*, 4773–4778.

122. Onodera, G.; Imajima, H.; Yamanashi, M.; Nishibayashi, Y.; Hidai, M.; Uemura, S. *Organometallics* **2004**, *23*, 5841–5848.

123. Foley, N. A.; Lail, M.; Lee, J. P.; Gunnoe, T. B.; Cundari, T. R.; Petersen, J. L. *J. Am. Chem. Soc.* **2007**, *129*, 6765–6781.

124. Lee, D.-H.; Kwon, K.-H.; Yi, C. S. *Science* **2011**, *333*, 1613–1616.

125. Ackermann, L.; Novák, P.; Vicente, R.; Hofmann, N. *Angew. Chem. Int. Ed.* **2009**, *48*, 6045–6048.

126. Kochi, T.; Urano, S.; Seki, H.; Mizushima, E.; Sato, M.; Kakiuchi, F. *J. Am. Chem. Soc.* **2009**, *131*, 2792–2793.

127. Saidi, O.; Marafie, J.; Ledger, A. E. W.; Liu, P. M.; Mahon, M. F.; Kociok-Köhn, G.; Whittlesey, M. K.; Frost, C. G. *J. Am. Chem. Soc.* **2011**, *133*, 19298–19301.

128. Hofmann, N.; Ackermann, L. *J. Am. Chem. Soc.* **2013**, *135*, 5877–5884.

129. McMurray, L.; O'Hara, F.; Gaunt, M. J. *Chem. Soc. Rev.* **2011**, *40*, 1885–1898.

130. Gutekunst, W. R.; Baran, P. S. *Chem. Soc. Rev.* **2011**, *40*, 1976–1991.

131. Stang, E. M.; Christina White, M. *Nat. Chem.* **2009**, *1*, 547–551.

132. Chen, K.; Baran, P. S. *Nature* **2009**, *459*, 824–828.

133. Jørgensen, L.; McKerrall, S. J.; Kuttruff, C. A.; Ungeheuer, F.; Felding, J.; Baran, P. S. *Science* **2013**, *341*, 878–882.

# 2 Radical-Mediated C—H Bond Activation

*Adam M. Azman*

## CONTENTS

## 2.1  INTRODUCTION

This chapter begins with an acknowledgement of the difficulties in writing this chapter. The problem is a matter of definition: what, exactly, constitutes a "radical-mediated C—H bond activation"? First, what qualifies as "C—H activation"? If a C—H bond has bond dissociation energy (BDE) less than, arbitrarily, 95 kcal mol$^{-1}$ (see Chapter 1, "Introduction"), is it already "activated" and, thus, unable to undergo "C—H activation." What if the C—H bond is cleaved via an acid/base step? Should the Friedel–Crafts reaction or the Snieckus directed *ortho*-lithiation reaction be considered "C—H activation"? Is it necessary for the C—H bond to be cleaved before or during the rate-determining step?

These questions may be pertinent to every chapter in this book. For this chapter specifically, some additional questions arise: What does "radical-mediated" mean? Can a radical hydrogen-atom abstraction meet the requirements of C—H activation? Does it depend on the particular mechanism of the reaction, the C—H BDE, or the timing of the radical abstraction step? How can a researcher be sure that absolutely no transition metals are present and active during the reaction? What if a transition metal is an additive, but only serves as an electron shuttle and does not specifically cleave to or react with the C—H bond?

Can a transition-metal-free reaction even be labeled as C—H activation? Some chemists have argued that true C—H activation requires oxidative addition or other metal C—H bond insertion step;[1] therefore, some "radical-mediated C—H bond activation" reactions might be better classified as homolytic aromatic substitution (HAS), the radical variant of the better known electrophilic aromatic substitution (EAS).[2]

Further complicating the development of new "transition-metal-free" reactions, subsequent investigations sometimes show "advantageous" metal contamination, even at parts per billion level, is truly responsible for the reactivity being described.[3–7] The burden of proof is clearly on the corresponding authors to ensure the absence of trace transition-metal in the system.[8,9]

This chapter will not attempt to provide the definitive answer to the questions posed above. Instead, this chapter will take the broad view of C—H activation (otherwise the chapter would end after this paragraph). Recognizing that different scientists may have good-faith disagreements about the suitability of some reactions for inclusion, this chapter will survey reactions under the following conditions: Substitution of a C—H bond for some other functional group, where a radical intermediate is responsible for the key bond-forming step, and where a transition-metal is not intimately involved in binding or cleaving the C—H bond (though some radical reactions involving transition-metals for other purposes will be highlighted).[10]

From a green chemistry perspective, transition-metal-free radical-mediated reactions are superior to transition-metal-based C—H activations, as they avoid potential metal contamination of the desired product and the production of unwanted metal waste. They can sometimes be more atom-efficient than similar transition-metal-based reactions and decrease the reliance on potentially expensive or toxic transition-metal reagents and catalysts.

## 2.2 RADICAL-MEDIATED C(sp³)—H FUNCTIONALIZATION

### 2.2.1 RADICAL HALOGENATION

Radical alkane halogenation is ubiquitous and often the first reaction learned in undergraduate organic chemistry courses. The reaction is quite atom-efficient since every atom in every reagent, except those caught in non-productive radical termination steps, is incorporated into the product. Hydrogen-atom abstraction by some previously generated halogen-centered radical forms a carbon-centered radical. Propagation occurs when the carbon-centered radical abstracts a halogen atom to form product and regenerate the halogen-centered radical.[11,12] The initial halogen-centered radical is prototypically generated by the homolytic bond cleavage of an elemental halogen, typically $Br_2$ or $Cl_2$, though other initiators are known.

Reaction with $Cl_2$ is known to be less regioselective than $Br_2$ due to the decreasing bond strength from HCl to HBr, the byproduct of the rate-determining propagation step. Along with the diminished regioselectivity, polychlorination is common. Selectivities of radical chlorination reactions can be improved with the irradiation of N-chloroamines and sulfuric acid, with regioselectivities favoring the ω-1 position of straight chain alkanes.[13]

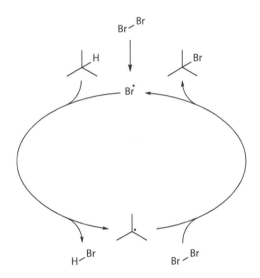

Non-functionalized alkanes are preferentially brominated at tertiary carbon atoms in preference to primary or secondary positions, with selectivity decreasing with increasing temperature. Allylic halogenation may be difficult to achieve due to competing ionic addition of $Br_2$ across the alkene (see Section 2.2.2), but benzylic positions are brominated faster than tertiary positions.[14] Electron withdrawing groups, such as carbonyls, deactivate the α-position as halogen-centered radicals can be thought of as somewhat electrophilic, favoring abstraction of more electron dense hydrogen atoms.[15] Hydrogen-atom abstraction occurs readily at carbon atoms α to an ether. Neighboring group participation of an adjacent halogen can direct subsequent halogenation to the β-carbon with respect to the original halogen.[16]

## 2.2.2   Radical Allylic and Benzylic Halogenation (Wohl–Ziegler Bromination)

As mentioned above, attempts to affect radical halogenation using elemental halo-gen is complicated when the substrate contains an alkene, as electrophilic addition of $X_2$ across the alkene predominates. N-bromosuccinimide (NBS is the most common reagent of choice to successfully accomplish allylic halogenation. In the presence of an initiator (peroxides or AIBN), NBS liberates a bromine radical to begin the propaga-tion sequence. Hydrogen-atom abstraction occurs to produce a resonance-stabilized, allylic carbon-centered radical and hydrogen bromide. The hydrogen bromide reacts with unreacted NBS through an ionic pathway to generate $Br_2$, from which a halogen atom can be abstracted to yield the product and regenerate the halogen-centered radical necessary to propagate the chain reaction. Since the initial bromine-centered radical is present only in low concentration, HBr is generated only in low concentration, and thus $Br_2$ is generated only in low concentration. This low concentration of $Br_2$, and lack of build-up of HBr is key to avoiding electrophilic addition across the alkene.[12]

## 2.2.3 INTRA-MOLECULAR δ-CHLORINATION/PYRROLIDINE FORMATION (HOFMANN–LÖFFLER–FREYTAG REACTION)

Perhaps the first directed C—H functionalization reaction, reported in the 1880s, creates pyrrolidines from N-chloroamines.[17] The original reaction conditions call for protonation of an N-chloroamine 1 under strongly acidic condition. Homolytic bond cleavage of the N—Cl bond by heat, light or radical initiator provides a nitrogen-centered radical cation, which intra-molecularly abstracts a hydrogen atom from the δ-position (1,5-hydrogen atom abstraction). The alkyl radical abstracts a halogen from another N-chloroamine to complete the catalytic cycle. Intra-molecular nucleophilic substitution of the newly formed alkyl halide 2 by the tethered amine under basic conditions yields the final pyrrolidine 3.

The Hoffmann–Löffler–Freytag reaction can be improved if the δ-position is itself α to a heteroatom, as oxygen and nitrogen can stabilize the carbon-centered radical intermediate, allowing the reaction to occur under milder conditions. Oishi took advantage of this modification in the synthesis of (±)-dihydrodeoxyepiallocernuine 4.[18,19] The Togo group has utilized an iodine-based modification to the Hofmann–Löffler–Freytag reaction in the lactonization of ortho-alkylbenzoic acid derivatives 5.[20] The Togo group has also carried out periodination of a methyl group ortho to a sulfonamide utilizing this technology followed by ring closure and hydrolysis of the resulting diiodo group to gain efficient access to saccharin derivatives 6.[21]

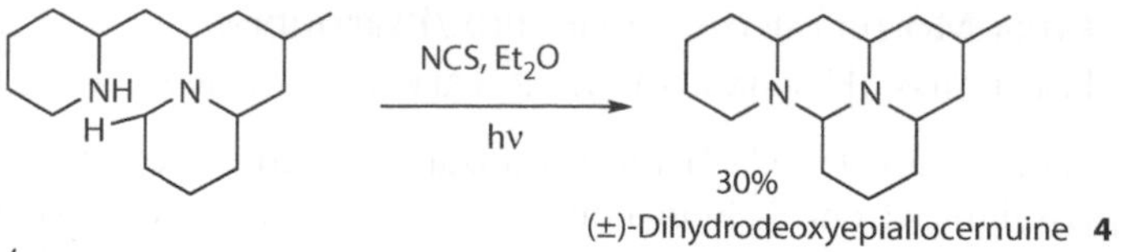

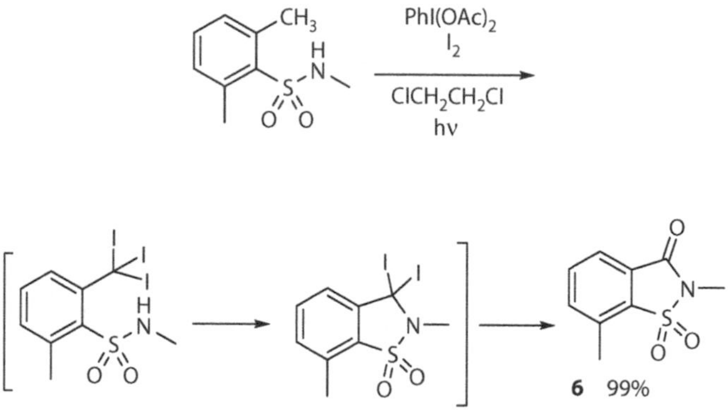

30%
(±)-Dihydrodeoxyepiallocernuine **4**

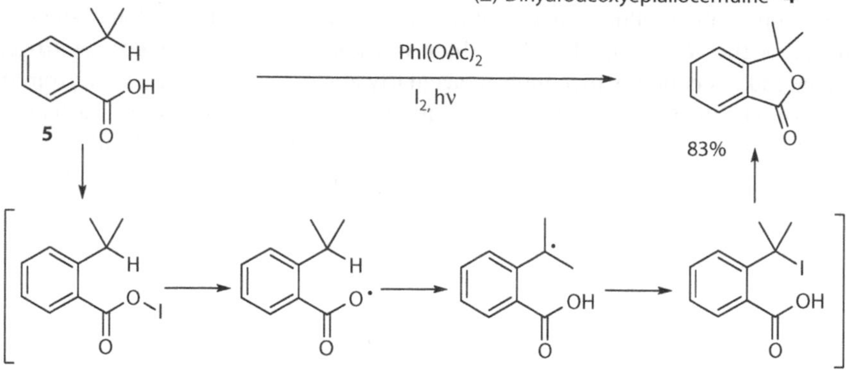

**5**

83%

**6** 99%

## 2.2.4 INTRA-MOLECULAR δ-BROMINATION/1,3-DIOL FORMATION

Expanding on the Hofmann–Löffler–Freytag reaction (see Section 2.2.3), the Baran group developed a conversion of an alcohol into an *N*-bromocarbamate **7**, irradiation of which, in the presence of silver carbonate, selectively functionalizes the most substituted C—H bond beta to the alcohol, ultimately generating a 1,3-diol **8** after cyclization and subsequent hydrolysis.[22]

Two main challenges needed to be overcome to find success in this reaction. First, targeting a 1,3-diol required an intra-molecular 1,6-hydrogen-atom abstraction through a seven-membered transition state, rather than the 1,5-hydrogen atom abstraction typical in the Hofmann–Löffler–Freytag reaction. Furthermore, assuming the desired hydrogen-atom abstraction occurs, cyclization can occur at either the oxygen or nitrogen atom of the carbamate.

To solve the first challenge, the structure of the carbamate was tuned to optimal reactivity of the nitrogen-centered radical. Ultimately, a trifluoroethyl carbamate **7** proved to affect the desired intra-molecular C—H functionalization most productively. Investigations into the second challenge established that silver carbonate ensured the carbamate cyclized through the carbonyl oxygen atom, rather than the carbamate nitrogen atom.

While olefins, free carboxylic acids, amines, amides, unprotected alcohols and azides are not tolerated, esters and epoxides are, even though the reaction conditions might appear to render these functional groups incompatible. The reaction is only feasible for tertiary or benzylic C—H bonds; secondary C—H bonds are not productively functionalized.

## 2.2.5  INTRA-MOLECULAR γ-HYDROXY OXIME FORMATION (BARTON NITRITE ESTER REACTION)

Going through a similar functionalization mechanism as the Hofmann–Löffler–Freytag reaction, the Barton nitrite ester reaction generates a γ-hydroxy oxime product. The reaction works well in rigid systems such as steroid skeletons. Formation and homolysis of a nitrite-, hypchlorite-, or hypoiodite ester generate an oxygen-centered radical. Abstraction of a hydrogen atom from the δ-position (1,5-hydrogen atom abstraction) forms a carbon-centered radical. Combination with, commonly, a nitrogen-centered NO radical and subsequent tautomerization yields the γ-hydroxy oxime **9**.[23]

Addition of radical trapping agents can extend the functionality of the Barton nitrite ester reaction. The Čekovič group added Michael acceptors to the reaction, and the carbon-centered radical adds to the Michael acceptor to form a chain elongated product **10**.[24] A large excess of Michael acceptor is necessary for yields to be practical.

## 2.2.6  C(sp³)−H FUNCTIONALIZATION BY HYPERVALENT IODINE

Hypervalent iodine has been thoroughly investigated as a catalytic, non-metal oxidant.[25,26] Reagents involving hypervalent iodine are typically safe, green, and their reactions simple to employ.[27] Its utility will be explored throughout the rest of this chapter.

### 2.2.6.1  Of Unfunctionalized Alkanes

While phenyliodine(III) diacetate and TMSN₃ are known to form $PhI(N_3)_2$ in solution and homolytically cleave to the azide radical (see this section, below), the combination is sensitive to decomposition at elevated temperatures.[28] A more stable azide source was developed by Zhdankin and coworkers. Azidoiodinane **11** is thermally stable and in fact needs a radical initiator to affect the radical reaction. Following radical initiation, hydrogen-atom abstraction leads to a carbon-centered radical which can abstract azide from the azidoiodinane to form an iodine-centered radical in equilibrium with the oxygen-centered radical. The chain sequence is propagated by hydrogen atom abstraction from another alkane.[28]

Site-selective functionalization of alkanes is an ambitious goal, as finely tuned discriminating factors are required. Maruoka,[29] and later Kita,[27] used hypervalent iodine to selectively oxidize remote methylene groups of unfunctionalized alkanes. Kita's system involved an iodine(III) species with an electron-withdrawing ligand and a good leaving group to allow for a more electrophilic iodine-centered radical. Phenyliodine(III) with a (phosphinoyl)oxy and triflate ligand 12 served this purpose well. Secondary carbons are oxidized in the presence of primary, tertiary, or even tertiary benzylic. Site selectivity can be achieved for cyclohexane derivatives, with preferential oxidation at different positions around the carbocycle depending on the electronic and steric influence of the original substitution of the ring. Expectedly, acyclic alkanes showed little selectivity, except to oxidize secondary carbons over primary or tertiary.

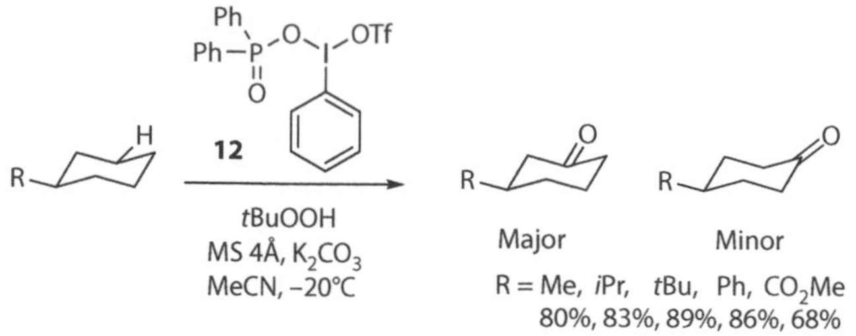

Major      Minor

R = Me, $i$Pr, $t$Bu, Ph, $CO_2$Me
80%, 83%, 89%, 86%, 68%

## 2.2.6.2   At the Benzylic/Allylic Position

A number of papers showcase a variety of hypervalent iodine reagents for the goal of oxidizing a benzylic C—H bond. While some might not consider this C—H activation, as the argument could be made that the benzylic position is already activated, these reactions will nonetheless be covered here.

Under radical initiation conditions, typically peroxides, hypervalent iodine reagents can be homolytically cleaved to iodine-centered radicals. These iodine centered radicals abstract a hydrogen atom from a labile benzylic C—H bond to yield a resonance-stabilized benzylic radical. At this point in the mechanism, researchers seem divided on the next step. Some propose a second single electron transfer (SET) to form a benzylic carbocation,[30] which undergoes ionic reactions to form product. Others suggest radical combination to form an alkyl halide or organic peroxide which reacts further under the reaction conditions to form product.

The Fan group[31] and Nicholas group[32] independently propose the radical mechanism in the amination reaction they developed. While the source of the iodine-centered radical differs, the mechanistic concept is the same. An $N$-iodo species can homolytically cleave to a nitrogen-centered radical. Hydrogen atom abstraction from the benzylic C—H bond and iodine atom abstraction from the $N$-iodo species form a benzylic iodide. Substitution of the iodide with the amine yields the product.

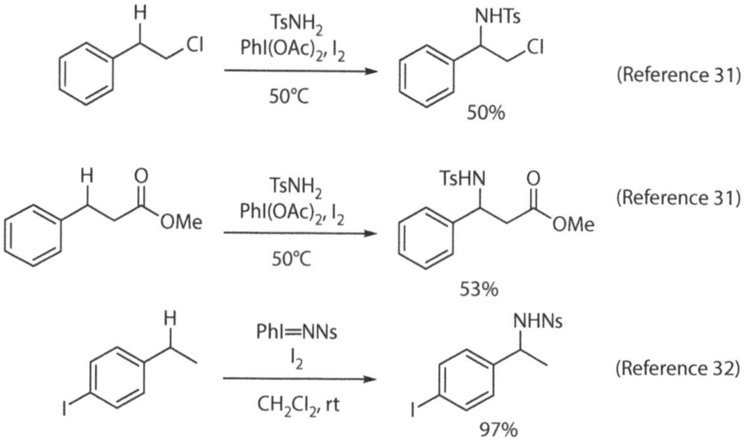

(Reference 31)

(Reference 31)

(Reference 32)

While not iodine-based, Inoue and coworkers also report an interesting amination of the benzylic position using *N*-hydroxyphthalimide **13** and dialkyl azodicarboxylates **14**. An oxygen-centered radical from the homolytic cleavage of the N—O bond of the hydroxyphthalimide abstracts the benzylic hydrogen atom. Addition into the N=N pi bond of the azodicarboxylate forms the C—N bond. Treatment with zinc and acetic acid reveals the benzylic amine **15**.[33]

Benzylic ethers can also easily have a hydrogen atom abstracted to form a benzylic radical. Through ligand exchange and homolysis, phenyliodine(III) diacetate and TMSN$_3$ generate azide radicals. The azide radical abstracts the benzylic hydrogen atom to form the benzylic radical. Azide abstraction from a molecule of PhI(N$_3$)$_2$ forms the product **16**.[34]

The stable azide source **11** developed by Zhdankin discussed at the beginning of this section can also be used to add an azide to the allylic position of alkenes.[28]

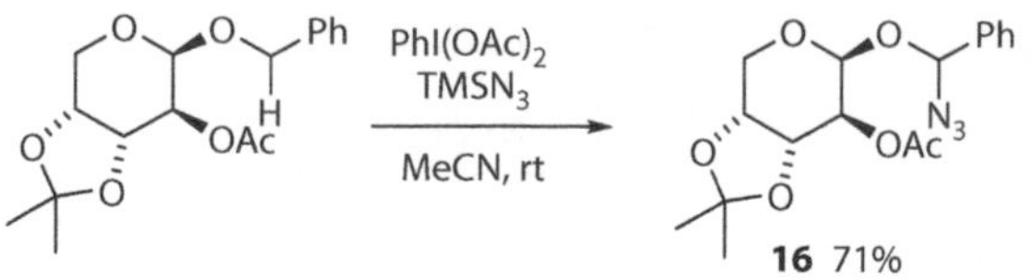

**16** 71%

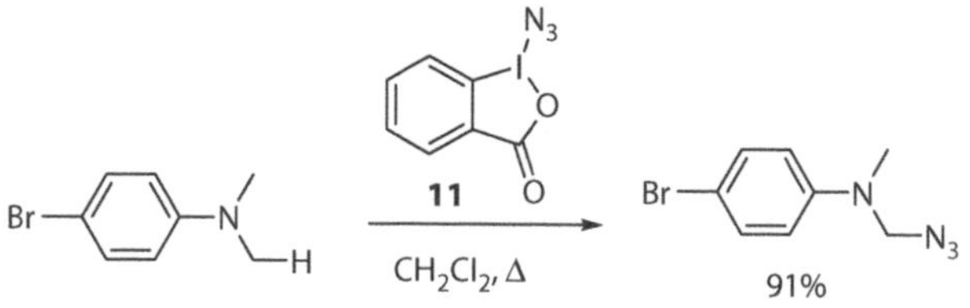

91%

Kita and coworkers use potassium bromide to depolymerize iodosobenzene **17** by generating an I—Br bond.[35] Homolytic cleavage of the I—Br bond to generate the iodine-centered radical allows formation of the benzylic radical **18**. The authors propose the benzylic radical is trapped by bromine radical to generate a benzylic bromide **19**, a known byproduct of the reaction conditions. Conversion to the alcohol and subsequent oxidation yields the aryl ketone **20**. While the ketone products are useful themselves, Ochiai utilizes this mechanism, though with a different hypervalent iodine reagent, to deprotect benzylic ethers such as **21** to alcohols **2**.[36]

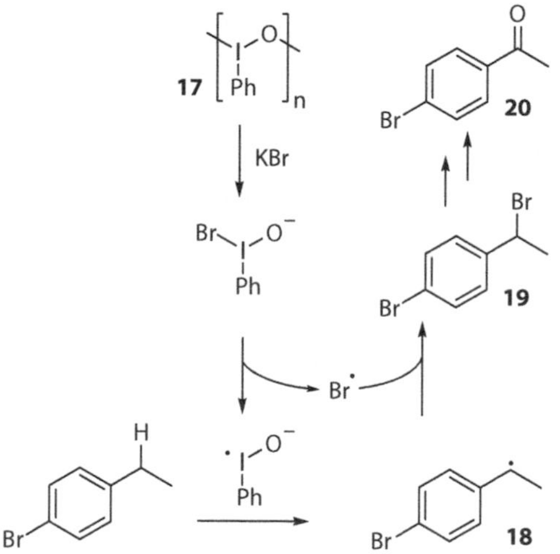

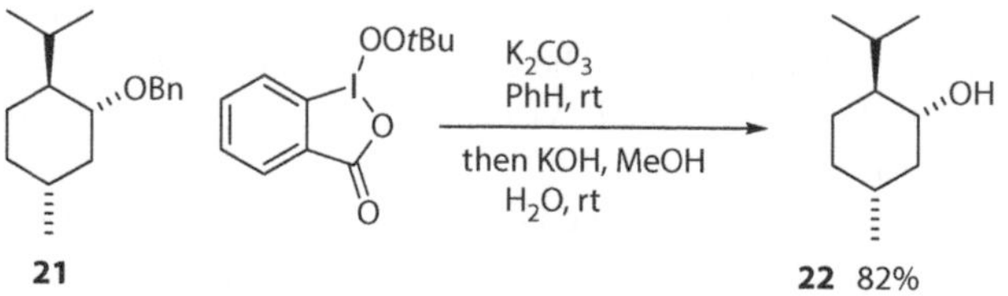

**21**                                              **22** 82%

Yeung and coworkers oxidize the allylic position of alkenes using a combination of phenyliodine(III) diacetate and *t*-butyl hydroperoxide in an ester solvent.[37] A variety of substituted cyclic alkenes, including steroid frameworks, can be oxidized to α,β-unsaturated ketones. Ester solvents proved crucial for the success of the reaction, leading to the supposition that coordination of the carbonyl of the ester to the iodine center is necessary for the reaction to proceed.

Following up on the allylic oxidation and the interesting requirement of an ester solvent, the Yeung group attempted the allylic oxidation reaction in the absence of the alkene, and noticed some oxidation of the ester solvent. Development of this as its own reaction demonstrated the successful oxidation of a number of *O*-alkyl and *N*-alkyl esters such as **23** and amides **24**. The oxidation typically occurred on a methylene carbon on the alkoxy- or amino-carbon chain several carbons away from the ester center.[38] Attempts to oxidize *n*-octane resulted in no reaction, further corroborating the necessity of carbonyl coordination to the iodine center.

## 2.2.7   C(sp³)—H FUNCTIONALIZATION α TO A HETEROATOM

### 2.2.7.1   α to an Oxygen Atom

The Fuchs group, in preparing acetylenic triflones **25** for an unrelated purpose, noted an unexpected exotherm upon mixing with tetrahydrofuran **26**. Investigation of the remnants uncovered coupling of the alkyne to the carbon atom α to the oxygen **27**. Reaction with ultra-purified tetrahydrofuran gave no reaction, suggesting a radical process. This discovery was expanded into a general methodology for coupling alkanes to alkynes.[39,40]

Similar to the Fuchs group's acetylenic triflones **25**, acetylenic bromides **28** can couple with carbon-centered radicals. The Li group used peroxides to abstract a hydrogen atom α to an oxygen in cyclic ethers **29**.[41] Homolytic substitution at the sp-hybridized carbon liberates a bromine-centered radical which can abstract another hydrogen atom from the cyclic ether to propagate the chain reaction.

### 2.2.7.2  α to a Nitrogen Atom

Zhdankin's azide source **11** can affect azidation α to a nitrogen atom through the same mechanism discussed in the previous section (see Section 2.2.6.1).[28]

Xiang, Li, and coworkers used *t*-butyl hydroperoxide to form a carbon-centered radical α to the nitrogen of an amide **30**. The carbon-centered radical can abstract a sulfur atom from a disulfide **31** to form a new C—S bond. This is one of the few transition metal-free C—S bond-forming reactions.[42]

The Inoue group tackled a transformation similar to Yeung's remote methylene oxidation (see Section 2.2.6.2) utilizing a different approach. Carbocycles substituted with a 1,2-diketone **32** can be irradiated to affect a Norish–Yang cyclization to prepare a fused hydroxycyclobutanone **33**, which can be further functionalized to acylated carbocycles.[43] Excitation of an electron in the carbonyl group leads to intramolecular hydrogen-atom abstraction. Intra-molecular radical combination closes the cyclobutanone ring.

In the course of investigating the scope of the reaction, they noticed ethereal C—H bonds are preferentially cleaved in the hydrogen-atom abstraction step over alkyl C—H bonds. They investigated this further as a possible inter-molecular method of forming new C—C bonds. Indeed, using benzophenone as the external source of the oxygen-centered radical, ethereal C—H bonds can be cleaved to carbon-centered radicals. The radical can either add to an isocyanate to append an amide **34**,[44] abstract a nitrile from tosyl cyanide to append a nitrile **35**,[45] or react with a 1-tosylacetylene to append an alkyne **36**.[46] A variety of cyclic and acyclic molecules react at a tertiary carbon atom or α to an oxygen or nitrogen atom, enabling one-step construction of quaternary carbon atoms. In competition experiments, the order of preference for hydrogen atom abstraction is the carbon α to nitrogen, then α to oxygen. Beyond that, non-substituted methine hydrogen atoms are abstracted in preference to methylene hydrogen atoms.

**35** 73%

**36** 92%

## 2.3   RADICAL-MEDIATED C(sp²)−H FUNCTIONALIZATION

### 2.3.1   ALDEHYDE FUNCTIONALIZATION

Expanding on their work involving alkynylation of ethers and hydrocarbons (see Section 2.2.7.1), The Fuchs group recognized the C−H BDE of an aldehyde (88 kcal mol⁻¹) could be well suited for their methodology.[39] Treating aldehydes with acetylenic triflones **25** in the presence of radical initiators, the aldehyde C−H bond in **37** could be homolytically cleaved to afford the acyl radical. Unfortunately, in assessing the scope of this reaction, several attempts were thwarted by decarbonylation from the acyl radical. In those cases, running the reaction under high pressure carbon monoxide atmosphere suppressed this tendency. Addition of the acyl radical to the acetylenic triflones provided propargylic ketones **38** in good yields. Formates, with their stronger C−H bond (BDE 94 kcal mol⁻¹) failed under these reaction conditions. Aromatic aldehydes also failed, leaving aliphatic aldehydes as the only successful substrates tested by the Fuchs group.

**37**    **38** 90%

Hypervalent iodine can affect the amidation or azidation of aldehydes under metal-free conditions. The Chen[47] and Bols[34] groups showed that ligand exchange of acetate for azide on phenyliodine(III) diacetate can be homolytically cleaved to an iodine-centered radical and azide radical. Hydrogen atom abstraction from the aldehyde yields an acyl radical which can abstract azide to afford the acyl azide and iodobenzene. Interestingly, the two groups isolate different products from identical starting materials. The Bols group isolated carbamoyl azides **39** and **40** from a Curtius rearrangement of the intermediate acyl azide. The Chen group reports isolating the acyl azides **41** and **42**.

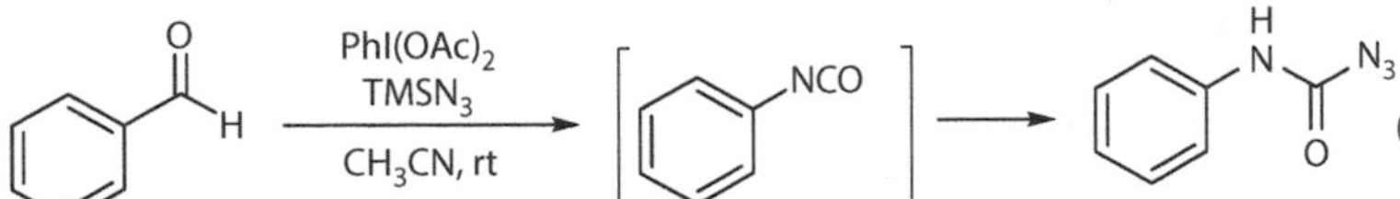

**39** 81%    (ref 47)

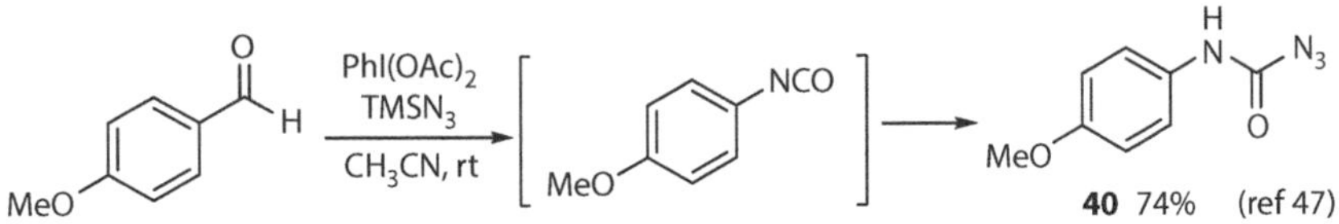

**40** 74%    (ref 47)

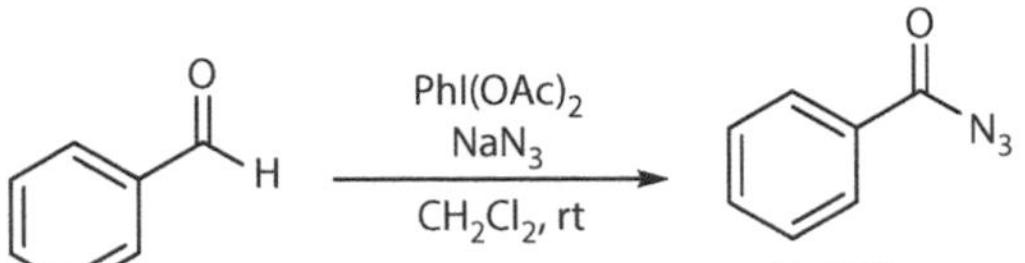

**41** 87%    (ref 34)

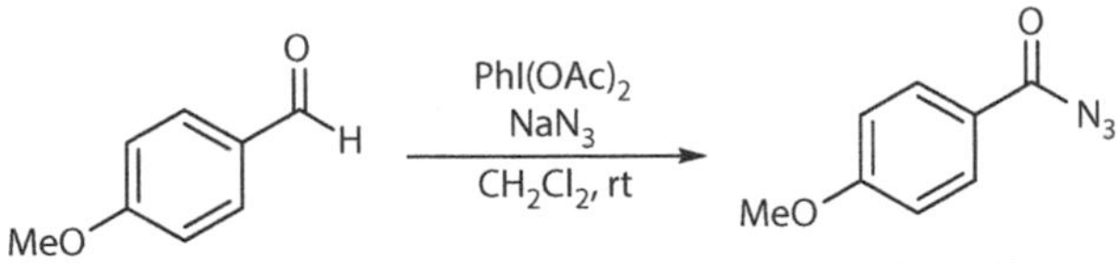

**42** 92%

Wan and coworkers used *t*-butyl hydroperoxide and Bu$_4$NI in formamides to synthesize amides. Oxygen-centered radicals (either *t*-butoxyl or *t*-butylperoxyl radical) abstract a hydrogen atom from an aldehyde and from dimethylformamide (DMF). The acyl radical from DMF decomposes to carbon monoxide and a nitrogen-centered radical. Radical combination provides the amides **43** and **44**.[48] Although, in the absence of a convenient method of generating a nitrogen-centered radical, other groups propose an ionic mechanism for the formation of the C—N bond.[49,50]

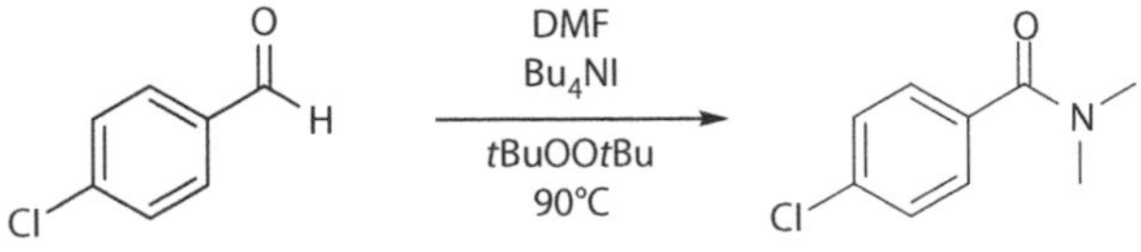

**43** 94%

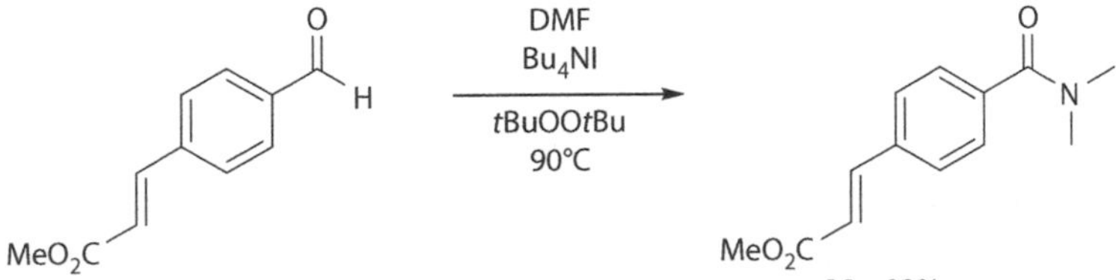

**44** 93%

### 2.3.2   ARYL C—H BOND FUNCTIONALIZATION

#### 2.3.2.1   The Minisci Reaction

The Friedel–Crafts reaction requires electron-rich aromatic rings for the reaction to occur. Friedel–Crafts reaction, with electron-poor rings under ionic conditions, is difficult to achieve. Minisci showed that by generating a nucleophilic carbon-centered radical, HAS with electron-poor heteroaromatic rings affects the same net transformation.[51] Activation of the ring through protonation is necessary.

The scope of the reaction is quite broad, and a number of functional groups are tolerated in the reaction. Radicals generated at the carbonyl of an amide or α to an alcohol, ether (**45**) or amine can be utilized in the Minisci reaction.[52] Minisci, later, showed homolysis of sugar-derived alkyl iodides **46** under thermal conditions to give carbon-centered radicals. These radicals undergo HAS to quinoline derivatives **47** and other heteroaromatics in good yields.[53] Cowden showed that decarboxylation of commercially available α-amino acids **48** can form a carbon-centered radical for use in the Minisci reaction.[54]

#### 2.3.2.2   Potassium *t*-Butoxide Mediated

Biaryls are ubiquitous substructures in biologically active molecules. The direct synthesis of biaryls is well known, and a focus on environmentally friendly methods makes such couplings more attractive. Hypervalent iodine reagents can be a greener alternative to transition metal catalysis for the coupling of unactivated aryl rings.

A 2008 paper from the Itami group noted the direct coupling of an aryl iodide to an unactivated aryl ring in the absence of transition metals typically employed for this purpose.[55] The Itami group was attempting biaryl formation through the use of iridium catalysis with potassium $t$-butoxide as base, but control reactions found the reaction occurred solely in the presence of potassium $t$-butoxide. Optimized conditions of an aryl iodide with a variety of electron deficient heteroaromatic rings and potassium $t$-butoxide for 5 min under microwave conditions provided a number of successfully coupled biaryls.

Aware of the possibility for trace metal contamination, the reaction was extensively studied by Itami and others in an attempt to rule in a radical-mediated pathway and rule out advantageous metal catalysis. The use of freshly sublimed base did not diminish the yield. Analysis by ICP-AES in the original Itami correspondence showed no transition metals above the detection limit of the instrument (60–300 ppb, depending on the element analyzed),[55] and while ICP-MS data from the Shi group did show 10–1000 ppb Pd, Cu, and Fe contaminants, intentionally adding various metal salts in much larger concentrations led to no improvement or even diminished efficiency.[56,57] Radical scavengers inhibited the reaction.[55] EPR experiments showed no signal with combinations of some (but not all) of the reagents; only when an aryl halide, potassium $t$-butoxide and 1,10-phenanthroline were mixed did the resulting EPR signal show evidence for the presence of radicals during the course of the reaction.[58] This reaction, like the benzylic oxidation of $C(sp^3)$–H bonds (see Section 1.6), has generated significant controversy over the appropriateness of the "C—H activation" label. Studer and Curran have published an essay on this very topic featuring this very reaction.[2] They posit the HAS method proposed for this reaction does not qualify as C—H activation.

Several groups have probed the conditions necessary to affect this transformation, and while reaction can occur solely in the presence of potassium $t$-butoxide under forcing conditions,[55,59] amine additives (dimethyl ethylene diamine (DMEDA),[60] 1,10-phenanthroline or derivatives,[56,61] phenylhydrazine,[62] proline,[63] quinoline-1-amino-2-carboxylic acid[64]), $N$-heterocyclic carbene,[57] or alcohol additives[65] often allow the same transformations to be achieved under milder reaction conditions. Metal-organic frameworks also promote reaction.[66] Amine additives can be eliminated for intra-molecular reactions with an amine tether[67,68] or by photostimulation.[69] The use of ionic liquids as recyclable solvent has the potential of improving the environmental profile of this reaction.[70] Furthermore, potassium $t$-butoxide is privileged among bases at inducing biaryl coupling. Other counter-cations are ineffective, as are other potassium bases.[55]

The mechanism is believed to begin by SET from the base to the aryl iodide to form a radical anion. Inter-molecular reaction can occur through a HAS, the radical variant of the more commonly known EAS reaction. For intermolecular reaction, addition of the radical to the coupling partner results in a resonance-stabilized radical. Deprotonation with base forms a biaryl radical anion which propagates the mechanism by donating an electron to another aryl halide.[2] For intra-molecular reaction, the aryl radical likely adds to the ipso position of the tether. Subsequent ring expansion and re-aromatization yields the product.[68]

An amide linker allows for intra-molecular coupling to provide biaryl lactam substructures, present in many biologically active molecules.[71] This methodology has been applied to the total synthesis of a number of phenanthridinone derivatives isolated from *Amaryllidaceae*. Intra-molecular coupling of precursor **49**, with the two aryl rings tethered together by an amide linker, provided oxoassoanine **50** in moderate yield. Others in this series were prepared in a similar manner.[67,72]

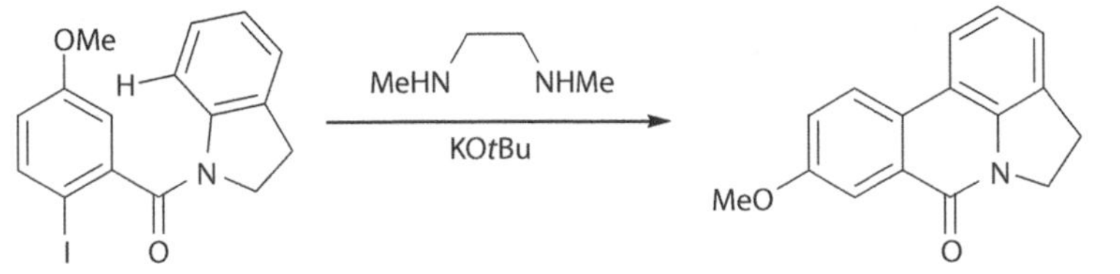

**49**                          Oxoassoanine    **50**    62%

Recent studies on organic "super electron donors" (SEDs) show that some neutral, highly electron-rich pi bonds **52** are capable of donating an electron from their neutral ground state.[73] Aryl and alkyl carbanions can be created from halides through SET from the SEDs. Application of SED precursors **51** to the direct arylation coupling partners in such examples as described above successfully formed coupled products. This suggests a reason the potassium *t*-butoxide reaction is more active with amine additives. To study this, investigation of products from a "blank" reaction between potassium *t*-butoxide and phenanthroline **53** or pyridine, followed by quench with electron-accepting $I_2$, yielded dimerized phenanthroline **54** and pyridine with a similar structure as the SEDs the Murphy group had been synthesizing. The authors contend these dimers could be the initial source of the single electron for these potassium *t*-butoxide-mediated reactions.[73]

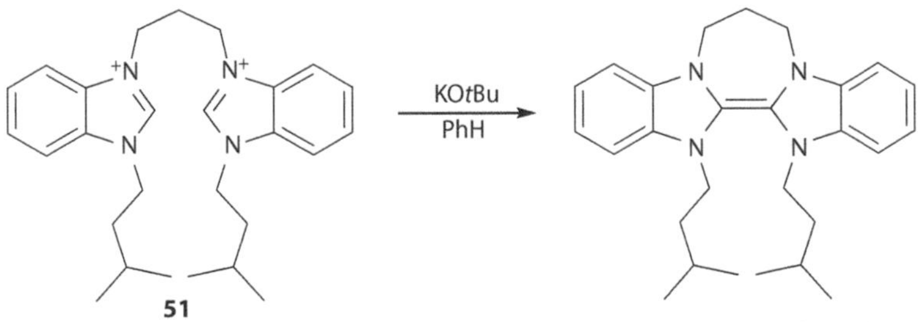

**51**

"Super electron donor"
**52**

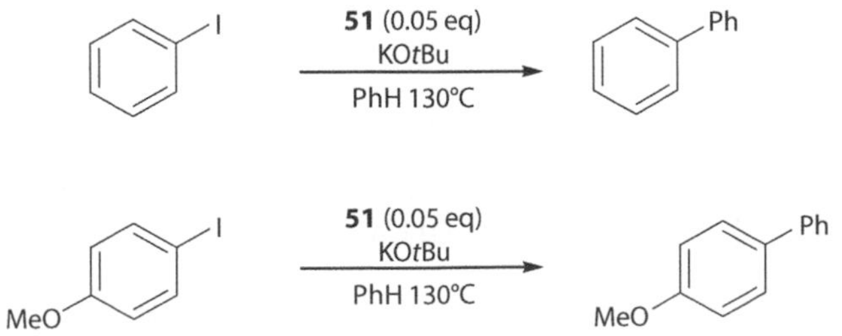

### 2.3.2.3 Hypervalent Iodine Mediated

Pioneering work by Kita demonstrated that SET from electron-rich aromatic rings to phenyliodine(III) bis(trifluoroacetate) generates a radical cation which can be attacked by nucleophilic species with concomitant transfer of a second electron to the iodine species. Rearomatization by deprotonation yields the aromatic substitution product.[74] While this particular methodology does not fit the model for this review since the bond-forming event is ionic in nature, extensions and modifications do proceed through a radical-mediated C–H activation reaction and will be discussed below.

Zhu and coworkers developed the cyclization of *N*-arylamidines **55** and **56** mediated by phenyliodine(III) diacetate.[75] Ligand substitution of the amidine for the acetate on the iodine center generates an *N*-iodoimido intermediate. Homolysis gives an *N*-centered radical which can undergo HAS. Formation of the N–C bond provides a resonance-stabilized radical. The iodine species accepts a second electron to form a cationic intermediate similar to EAS. Final rearomatization by deprotonation provides the benzimidazoles **57** and **58**.

The Yokoyama group found that irradiating 3-phenyl-1-propanol and its derivatives **59** and **60** with phenyliodine(III) diacetate and $I_2$ formed a hypoiodite. Homolytic O—I bond cleavage results in an oxygen-centered radical which can undergo HAS to synthesize a chromane. Once formed, however, the now-activated electron-rich aromatic ring is primed for a classical EAS under the reaction conditions. AcOI acts as the electrophile for this tandem reaction to place an iodine substituent para to the new oxygen substituent in products **61** and **62**.[76,77]

A trifluoromethyl group can be appended to an aryl ring mediated by hypervalent iodine(III). The Shibata group generated a radical cation through SET from an electron rich aryl ring to phenyliodine(III) bis(trifluoroacetate).[78] Combination with a trifluoromethyl radical generated from fragmentation of trifluoromethanesulfonate adds a trifluoromethyl group. Electron-poor rings perform poorly in this reaction.

Diaryl-$\lambda^3$-iodanes (often referred to as diaryliodonium salts) such as [Ph$_2$I]OTf or derivatives can couple the phenyl group with another aromatic ring to form a biaryl.[79,80] Electron rich and electron poor aromatic rings, as well as heteroaromatic rings, are tolerated.

Some disagreement exists as to the precise mechanism for this reaction. The groups cited above propose a mechanism involving aryl radicals, likely through HAS. EPR studies and failed reactions in the presence of TEMPO as a radical scavenger support that theory. Others propose a mechanism where iodine acts like a transition metal center.[81,82] Ligand displacement by the external aryl ring, followed by something analogous to reductive elimination yields the biaryl product.

### 2.3.2.4   Bioisostere Formation

Bioisosteres are substituents with similar physical properties to common functional groups, but, for medicinal purposes, often do not share the same toxicity or metabolic liability in a potential drug candidate. The Baran group has developed a method for facile introduction of a number of fluorinated bioisosteres onto aromatic rings using radical chemistry. Original work showed that treatment of aromatic or heteroaromatic systems with sodium trifuloromethylsulfinate and t-butyl hydroperoxide directly appended the trifluoromethyl group to the aromatic ring.[83] Unfortunately, some sodium sulfinate salts required metal salt (silver or zinc) additives in stoichiometric amounts.[84]

Further development of the reaction led to zinc bis(sufinate) salts as stable, scalable reagents for this reaction, many of which are now commercially available.[85,86] Treatment of the zinc salt with t-butyl hydroperoxide generates a trifluoromethyl radical. Heteroaromatic systems can undergo HAS with this electrophilic radical at innately nucleophilic sites in the aromatic ring, though choice of solvent can impact the regioselectivity of this reaction.

R = CF₃, CF₂H, CH₂F, CH₂CF₃

R = $CF_3$, $CF_2H$, $CH_2F$, $CH_2CF_3$
89%, 73%, 80%, 51%

R = $CF_3$, $CF_2H$, $CH_2F$, $CH_2CF_3$
35%, 66%, 18%, 73%

A number of bioisosteres can be introduced using this methodology. Tri-, di- and monofluoromethyl groups (methyl, thiol/hydroxyl, and hydroxymethyl bioisosteres), and trifluoroethyl groups (ethyl bioisostere) can be appended with zinc sulfonate salts.[85] Introduction of a difluoroethyl group (methoxy bioisostere) can be accomplished with the sodium sulfinate and zinc chloride.[84] Isopropyl groups (steric bulk and lipophilicity promoter), and triethylene glycol groups (solubilizing unit) can be introduced as well.[85] The methodology is versatile enough that more exotic difluoroalkyl groups can be appended to aromatic rings, as long as the corresponding sulfinate can be prepared. Substrate scope is quite extensive, as a variety of biologically relevant heteroaromatic systems with a variety of other functional groups already incorporated are tolerated, though with diminished yields in some cases. The reaction is conducted in water as cosolvent and can be run open to air.

### 2.3.2.5 Witkop Photocyclization

Indole rings with a pendant chloroacetamide can be irradiated to produce medium-ring lactams in a reaction known as the Witkop photocyclization.[87] Electron transfer from the aromatic ring 63 and homolytic bond cleavage of the C—Cl bond provide a cyclic radical cation and a carbon-centered radical 64. Intra-molecular radical combination and rearomatization delivers the lactam 65. Interestingly, the indole can be targeted at the C4 position, its least reactive site.[88] Appending the chloroacetamide to C3 forms the 3,4-bridged indole, while appending the chloroacetamide to C2 forms the 2,3-annulated indole.

The Witkop photocyclization has been utilized in several natural product total syntheses. The Pagenkopf group formed the central ring of (±)-quebrachamine 66 with the Witkop photocyclization as the penultimate step in the synthesis.[89] Final amide reduction revealed the natural product. The lactam and furan rings of decursivine 67 can be formed simultaneously in a cascade reaction beginning with Witkop cyclization. Both the Mascal and Jia groups realized the synthesis of decursivine in this manner.[90,91] The Jia group explored and expanded this cascade reaction into its own methodology.[86] The Feldmen group tethered the C3- and C4-position of the indole core of dragmacidin E 68 through a Witkop cyclization reaction. Further manipulation completed the synthesis of dragmacidin E.[92]

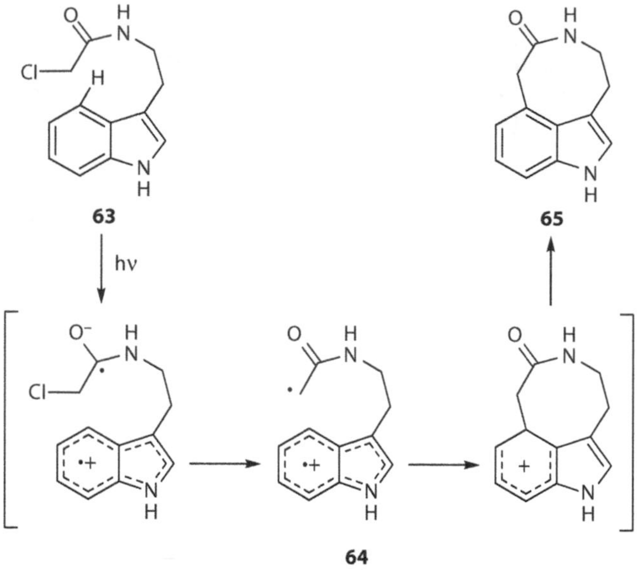
63
hν
64
65

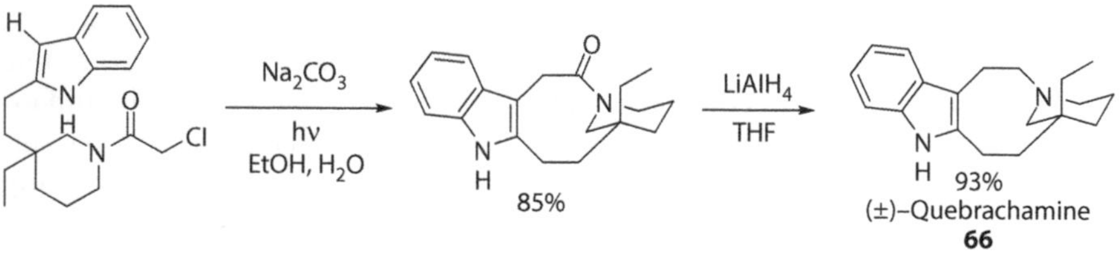
Na2CO3
hν
EtOH, H2O
85%
LiAlH4
THF
93%
(±)-Quebrachamine
66

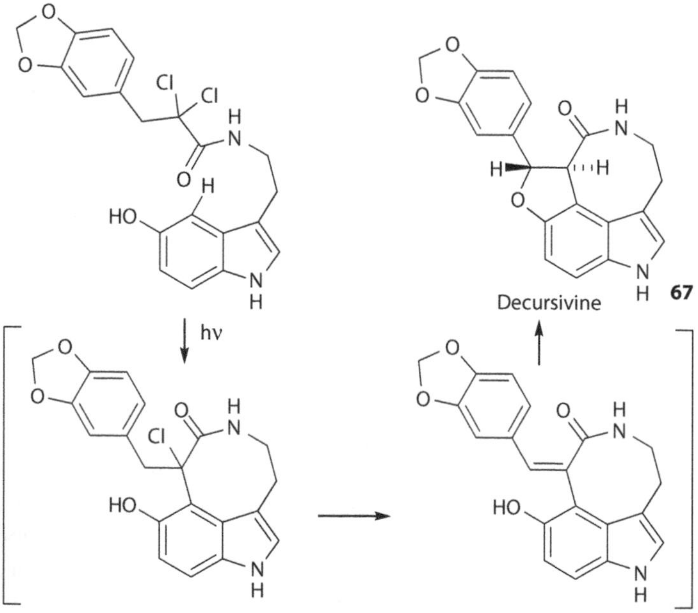
hν
Decursivine
67

## 2.3.2.6 Other Aryl C—H Bond Functionalizations

The Baran group set upon developing a methylation of aromatic rings that tolerated otherwise-reactive sites. One of the challenges they sought to overcome was the similar polarity of the starting materials and methylated products, impeding easy separation. Expanding on their bioisostere work (see Section 2.3.2.4), and drawing inspiration from $S$-adenosylmethionine, zinc bis(phenylsulfonylmethanesulfinate) was found to be a robust, scalable reagent which addressed their concerns.[93]

While not a direct source of methyl radical, the electrophilic carbon-centered radical generated when the sulfinate is treated with $t$-butyl hydroperoxide can react with nucleophilic sites on most aromatic and heteroaromatic rings. Free N—H and O—H bonds are tolerated, though electron-poor aromatic rings do not perform well in this reaction. The intermediate (phenylsulfonyl)methyl-substituted aromatic ring can be easily separated from the starting material. The intermediate can be proto-desulfonylated in one of three interchangeable ways, allowing for various incompatibilities elsewhere in the molecule to be taken into account. The intermediate can also be converted into a $CD_3$ group, $CF_2H$ group, xanthate, or alkene.

Amination of aromatic rings historically involves nitration followed by reduction. For some aromatic rings, such as pyrroles, this reaction is historically problematic. The Baran group has developed a one-pot radical imidation of aromatic rings followed by deprotection of the succinimide to furnish aromatic amines directly.[10] A succinimide derivative **69** was developed to easily decompose into a nitrogen-centered radical. HAS follows. Heating the crude product mixture deprotects the imide **70** and releases the aromatic amine **71**. Interestingly for this chapter, the reaction is catalyzed by ferrocene. The iron serves as an electron shuttle to facilitate the decomposition of the starting imide to the nitrogen-centered radical, and accepts an electron from the final intermediate to re-aromatize the system. Since the metal plays no role in the key bond-forming event or in the cleavage of the C—H bond, the reaction is included here.

A variety of electron-rich and electron-poor aromatic and heteroaromatic rings are tolerated in the reaction, as are pendant esters, halogens, and silyl protecting groups. Regioselectivity follows EAS preferences, where electron donating groups direct the nitrogen to the *ortho-* and *para-*positions. A similar reaction has been demonstrated by the Cho group utilizing hypervalent iodine, though the authors propose an ionic mechanism for their key-bond forming event.[30]

An aryl radical generated from an unsubstituted quinazoline can perform radical addition into an alkyne. The resulting carbon-centered radical can propagate the chain by abstracting a hydrogen atom from another quinazoline.[94]

## 2.3.3 ALKENYL C—H BOND FUNCTIONALIZATION

The potassium *t*-butoxide methodology for preparing biaryls from aromatic radicals (see Section 2.3.2.2) has proven quite versatile when a suitable aromatic ring serves as the coupling partner. When the coupling partner is a styrene derivative, a metal-free Heck-type reaction occurs.[95,96] Addition of the aryl radical generated as described above into the β-carbon of the styrene generates a benzylic carbon-centered radical. One possible mechanistic pathway invokes deprotonation of this intermediate to a radical anion which can transfer an electron to another aryl iodide to propagate the chain reaction, though the authors also propose two other mechanistic possibilities for product formation. The reaction has been proven inter- and intra-molecularly.

## 2.4   RADICAL MEDIATED C–H/C–H CROSS COUPLING

### 2.4.1   C(sp³)–C(sp³) Coupling

Many C(sp³)–C(sp³) coupling reactions not transition metal-mediated involve the
two electron oxidation of ethers or amines to oxocarbenium ions or iminium ions.
These are then attacked in an ionic mechanism by *in situ*-formed enolates or other
carbanion nucleophiles to generate product.[97–99] As such, they will not be covered in
this review.

### 2.4.2   C(sp³)–C(sp²) Coupling

The Antonchick group demonstrated the oxidative coupling of nitrogen heterocycles
with alkanes.[100] Combination of phenyliodine(III) bis(trifluoroacetate) with TMSN₃
is a known method of generating azide radicals. The azide radical abstracts a hydrogen atom from an alkane to form a carbon-centered radical. The carbon-centered

radical engages in a HAS reaction to yield the substituted heterocycle. A variety of nitrogen heterocycles with a variety of alkyl or electron-withdrawing groups are tolerated. The authors note that completion of the alkylation event results in a product C—H bond with a lower BDE than starting material, yet over-oxidation products were not observed. Stronger C—H bonds were preferentially functionalized over weaker C—H bonds.

Proposing a similar mechanism with *t*-butyl hydroperoxide as the oxidant, the Wang group showed alcohols and ethers could be coupled to benzimidazoles, benzoxazoles **72**, and benzthiazoles **73** in good yield.[101] Hydrogen-atom abstraction α to the oxygen atom provides a carbon-centered radical, which, similar to above, can under HAS.

Qu and coworkers extended the scope of the reaction to include purine bases **74** and **75** using di-*t*-butyl peroxide as the oxidant.[102] Here, electron-rich systems fared better than electron-poor systems. The free N—H bond remains intact during the course of the reaction, and alkylation preferentially occurred at the C2 position. Interestingly, when copper(I) iodide is added to the reaction, if the chemoselectivity switches are present, the N6 nitrogen atom is preferentially alkylated. The reaction is run neat and open to air.

Reflecting on their pioneering work in formation of aryl radicals from aryl iodides mediated by potassium *t*-butoxide,[55] the Itami group noticed cyclohexane-incorporated byproducts when the reaction was run in the presence of cyclohexane. They postulated a radical transfer from the aryl radical to form the alkyl carbon-centered radical. This hypothesis caused them to wonder if a more direct generation of alkyl radicals could affect a direct C—H/C—H coupling reaction.

Simply switching to conventional radical initiation techniques (di-*t*-butyl peroxide) led to only trace product formation, suggesting the mere formation of the alkyl radical is not sufficient for productive reaction to occur. Previous work in aryl iodide systems suggests "nitrogen activation" by coordination of nitrogen to a potassium ion or proton helps boost the receptivity of the heteroaromatic ring.[103] To replicate those conditions without added potassium salts, the Itami group used pyridine N-oxide **76** as starting material.[104] With this modification, the coupling reaction proceeded smoothly with conventional radical initiators, and a variety of cycloalkyl groups could be added to the electron-poor sites of the pyridine N-oxide, often with peralkylation.

## 2.4.3 C(sp²)—C(sp²) COUPLING

The methodology developed by Antonchick to abstract a hydrogen atom from alkanes (see Section 2.4.2) also abstracts a hydrogen atom from an aldehyde.[105] The azide radical generated from combination of phenyliodine(III) bis(trifluoroacetate) and TMSN₃ abstracts a hydrogen atom from the aldehyde to form an acyl radical. The acyl radical adds to historically electrophilic sites in protonated nitrogen heterocycles, typically *ortho* to the nitrogen atom if not already blocked, to form a radical cation. SET to the iodine species re-aromatizes the newly acylated heterocycle **77** and **78**.

## 2.5 CONCLUSIONS

Not covered in this review are radical oxidations of alcohols,[49] ethers,[106,107] thioethers[108] and amines[109] to carbonyls, oxocarbenium ions, thiocarbenium ions, and iminium ions where *in situ* nucleophilic attack through an ionic mechanism either precedes or follows radical oxidation. The radical process is only responsible for the oxidation of the starting functional group. Since the radical process does not involve the key bond formation, it is excluded here.

Ionic, nucleophilic attack of heteroatoms,[110,111] enolates,[112] or other aromatic rings[113] on aryl radical cations (which then happen to give up another electron to an electron acceptor) are excluded because the actual bond forming event, while it does involve radicals, does not proceed through single-electron movement. Rather, the bond forming event is ionic in nature, and the "leaving group" of sorts is an electron, given up to the radical mediator. Additionally, under certain conditions, oxidative de-aromatization/cyclization occurs rather than C—H substitution.

Two electron oxidation of benzylic carbons[30,114] or allylic carbons[115] to benzylic or allylic carbocations followed by $S_N1$-like trapping of the carbocation does not meet this review's criteria for a radical-mediated C—H functionalization reaction. Interestingly, the allylic oxidation paper by the Wan group[115] suggests concurrent hydrogen-atom abstraction from a carboxylic acid and an allylic carbon to form an oxygen-centered radical and an allylic radical. Even though other papers by this group using identical reaction conditions suggest a second oxidation of carbon to form carbocations in other systems,[106] this paper suggests major product formation by radical combination of these two radicals. Given the facile oxidation of other systems to carbocations under identical reaction conditions, the combination of two presumably low-concentration radicals as the major product forming step seems suspicious.

Oxidation of carbon atoms α to a carbonyl, which proceeds through an enol intermediate, is not included. The C—H bond is cleaved in an acid–base process prior to any radical pathway; thus, while the radical is essential to the key bond forming event, it is not essential in the C—H functionalization.[116] Similarly, radical oxidation of methyl imines to resonance-stabilized carbocations, trapped by nucleophiles such as azides, is not covered.[117]

## REFERENCES

1. Wencel-Delord, J.; Dröge, T.; Liu, F.; Glorius, F. *Chem. Soc. Rev.* **2011**, *40*, 4740–4761.
2. Studer, A.; Curran, D. P. *Angew. Chem. Int. Ed.* **2011**, *50*, 5018–5022.
3. Takai, K.; Tagashira, M.; Kuroda, T.; Oshima, K.; Utimoto, K.; Nozaki, H. *J. Am. Chem. Soc.* **1986**, *108*, 6048–6050.
4. Arvela, R. K.; Leadbeater, N. E.; Sangi, M. S.; Williams, V. A.; Granados, P.; Singer, R. D. *J. Org. Chem.* **2005**, *70*, 161–168, DOI: 10.1021/jo048531j.
5. Bedford, R. B.; Nakamura, M.; Gower, N. J.; Haddow, M. F.; Hall, M. A.; Huwe, M.; Hashimoto, T.; Okopie, R. A. *Tetrahedron Lett.* **2009**, *50*, 6110–6111, DOI: 10.1016/j. tetlet.2009.08.02.
6. Lauterbach, T.; Livendahl, M.; Rosellón, A.; Espinet, P.; Echavarren, A. M. *Org. Lett.* **2010**, *12*, 3006–3009, DOI: 10.1021/ol101012n.
7. Wang, X.; Zhang, B.; Wang, D. Z. *J. Am. Chem. Soc.* **2011**, *133*, 5160–5160, DOI: 10.1021/ja904224y.
8. Leadbeater, N. E. *Nat. Chem.* **2010**, *2*, 1007–1009, DOI: 10.1038/nchem.913.
9. Arancon, R. A. D.; Lin, C. S. K.; Vargas, C.; Luque, R. *Org. Biomol. Chem.* **2013**, *12*, 10–35, DOI: 10.1039/C3OB41768A.
10. Foo, K.; Sella, E.; Thomé, I.; Eastgate, M. D.; Baran, P. S. *J. Am. Chem. Soc.* **2014**, *136*, 5279–5282, DOI: 10.1021/ja501879c.

11. Thaler, W. A. *Methods Free Radic. Chem* **1969**, *2*, 121.

12. Nechvatal, A. *Adv. Free Radic.* **1972**, *4*, 175.

13. Deno, N. C.; Pohl, D. G. *J. Org. Chem.* **1975**, *40*, 380–381, DOI: 10.1021/jo00891a027.

14. Russell, G. A. In *Free Radicals*; Kochi, J. K., Ed.; Wiley: New York, **1973**; Vol. 2, p. 289.

15. Russell, G. A. In *Free Radicals*; Kochi, J. K., Ed.; Wiley: New York, **1973**; Vol. 2, p. 311.

16. Thaler, W. *J. Am. Chem. Soc.* **1963**, *85*, 2607–2613, DOI: 10.1021/ja00900a019.

17. Majetich, G.; Wheless, K. *Tetrahedron* **1995**, *51*, 7095–7129, DOI: 10.1016/0040-4020(95)00406-X.

18. Ban, Y.; Kimura, M.; Oishi, T. *Chem. Pharm. Bull. (Tokyo)* **1976**, *24*, 1490–1496, DOI: 10.1248/cpb.24.1490.

19. Francisco, C. G.; Herrera, A. J.; Suárez, E. *J. Org. Chem.* **2003**, *68*, 1012–1017, DOI: 10.1021/jo026314 h.

20. Muraki, T.; Togo, H.; Yokoyama, M. *J. Chem. Soc. [Perkin 1]* **1999**, *1999*, 1713–1716, DOI: 10.1039/A900791A.

21. Katohgi, M.; Togo, H.; Yamaguchi, K.; Yokoyama, M. *Tetrahedron* **1999**, *55*, 14885–14900, DOI: 10.1016/S0040-4020(99)00974-6.

22. Chen, K.; Richter, J. M.; Baran, P. S. *J. Am. Chem. Soc.* **2008**, *130*, 7247–7249, DOI: 10.1021/ja802491q.

23. Barton, D. H. R.; Beaton, J. M.; Geller, L. E.; Pechet, M. M. *J. Am. Chem. Soc.* **1960**, *82*, 2640–2641, DOI: 10.1021/ja01495a061.

24. Petrović, G.; Čeković, Ž. *Tetrahedron Lett.* **1997**, *38*, 627–630, DOI: 10.1016/S0040-4039(96)02357-X.

25. Dohi, T.; Kita, Y. *Chem. Commun.* **2009**, *2009*, 2073–2085, DOI: 10.1039/B821747E.

26. Newhouse, T.; Baran, P. S. *Angew. Chem. Int. Ed.* **2011**, *50*, 3362–3374, DOI: 10.1002/anie.201006368.

27. Dohi, T.; Kita, Y. *Chem Cat Chem* **2014**, *6*, 76–78, DOI: 10.1002/cctc.201300666.

28. Zhdankin, V. V.; Krasutsky, A. P.; Kuehl, C. J.; Simonsen, A. J.; Woodward, J. K.; Mismash, B.; Bolz, J. T. *J. Am. Chem. Soc.* **1996**, *118*, 5192–5197, DOI: 10.1021/ja954119x.

29. Moteki, S. A.; Usui, A.; Zhang, T.; Solorio Alvarado, C. R.; Maruoka, K. *Angew. Chem. Int. Ed.* **2013**, *52*, 8657–8660, DOI: 10.1002/anie.201304359.

30. Kim, H. J.; Kim, J.; Cho, S. H.; Chang, S. *J. Am. Chem. Soc.* **2011**, *133*, 16382–16385, DOI: 10.1021/ja207296y.

31. Fan, R.; Li, W.; Pu, D.; Zhang, L. *Org. Lett.* **2009**, *11*, 1425–1428, DOI: 10.1021/ol900090f.

32. Lamar, A. A.; Nicholas, K. M. *J. Org. Chem.* **2010**, *75*, 7644–7650, DOI: 10.1021/jo1015213.

33. Amaoka, Y.; Kamijo, S.; Hoshikawa, T.; Inoue, M. *J. Org. Chem.* **2012**, *77*, 9959–9969, DOI: 10.1021/jo301840e.

34. Pedersen, C. M.; Marinescu, L. G.; Bols, M. *Org. Biomol. Chem.* **2005**, *3*, 816–822, DOI: 10.1039/B500037H.

35. Dohi, T.; Takenaga, N.; Goto, A.; Fujioka, H.; Kita, Y. *J. Org. Chem.* **2008**, *73*, 7365–7368, DOI: 10.1021/jo8012435.

36. Ochiai, M.; Ito, T.; Takahashi, H.; Nakanishi, A.; Toyonari, M.; Sueda, T.; Goto, S.; Shiro, M. *J. Am. Chem. Soc.* **1996**, *118*, 7716–7730, DOI: 10.1021/ja9610287.

37. Zhao, Y.; Yeung, Y.-Y. *Org. Lett.* **2010**, *12*, 2128–2131, DOI: 10.1021/ol100603q.

38. Zhao, Y.; Yim, W.-L.; Tan, C. K.; Yeung, Y.-Y. *Org. Lett.* **2011**, *13*, 4308–4311, DOI: 10.1021/ol2016466.

39. Gong, J.; Fuchs, P. L. *Tetrahedron Lett.* **1997**, *38*, 787–790, DOI: 10.1016/S0040-4039(96)02413-6.

40. Xiang, J. S.; Fuchs, P. L. *Tetrahedron Lett.* **1996**, *37*, 5269–5272, DOI: 10.1016/0040-4039(96)01117-3.

41. Zhang, J.; Li, P.; Wang, L. *Org. Biomol. Chem.* **2014**, *12*, 2969–2978, DOI: 10.1039/C4OB00002A.
42. Tang, R.-Y.; Xie, Y.-X.; Xie, Y.-L.; Xiang, J.-N.; Li, J.-H. *Chem. Commun.* **2011**, *47*, 12867–12869, DOI: 10.1039/C1CC15397H.
43. Kamijo, S.; Hoshikawa, T.; Inoue, M. *Tetrahedron Lett.* **2010**, *51*, 872–874, DOI: 10.1016/j.tetlet.2009.1027.
44. Kamijo, S.; Hoshikawa, T.; Inoue, M. *Tetrahedron Lett.* **2011**, *52*, 2885–2888, DOI: 10.1016/j.tetlet.2011.03.128.
45. Kamijo, S.; Hoshikawa, T.; Inoue, M. *Org. Lett.* **2011**, *13*, 5928–5931, DOI: 10.1021/ol202659e.
46. Hoshikawa, T.; Kamijo, S.; Inoue, M. *Org. Biomol. Chem.* **2012**, *11*, 164–169, DOI: 10.1039/C2OB26785C.
47. Chen, D.-J.; Chen, Z.-C. *Tetrahedron Lett.* **2000**, *41*, 7361–7363, DOI: 10.1016/S0040-4039(00)00990-4.
48. Liu, Z.; Zhang, J.; Chen, S.; Shi, E.; Xu, Y.; Wan, X. *Angew. Chem. Int. Ed.* **2012**, *51*, 3231–3235, DOI: 10.1002/anie.201108763.
49. Tan, B.; Toda, N.; Barbas, C. F. *Angew. Chem. Int. Ed.* **2012**, *51*, 12538–12541, DOI: 10.1002/anie.201205921.
50. Prasad, V.; Kale, R. R.; Mishra, B. B.; Kumar, D.; Tiwari, V. K. *Org. Lett.* **2012**, *14*, 2936–2939, DOI: 10.1021/ol3012315.
51. Minisci, F.; Galli, R.; Cecere, M.; Malatesta, V.; Caronna, T. *Tetrahedron Lett.* **1968**, *9*, 5609–5612, DOI: 10.1016/S0040-4039(00)70732-5.
52. Minisci, F.; Giordano, C.; Vismara, E.; Levi, S.; Tortelli, V. *J. Am. Chem. Soc.* **1984**, *106*, 7146–7150, DOI: 10.1021/ja00335a048.
53. Vismara, E.; Donna, A.; Minisci, F.; Naggi, A.; Pastori, N.; Torri, G. *J. Org. Chem.* **1993**, *58*, 959–963, DOI: 10.1021/jo00056a034.
54. Cowden, C. J. *Org. Lett.* **2003**, *5*, 4497–4499, DOI: 10.1021/ol035814+.
55. Yanagisawa, S.; Ueda, K.; Taniguchi, T.; Itami, K. *Org. Lett.* **2008**, *10*, 4673–4676, DOI: 10.1021/ol8019764.
56. Sun, C.-L.; Li, H.; Yu, D.-G.; Yu, M.; Zhou, X.; Lu, X.-Y.; Huang, K.; Zheng, S.-F.; Li, B.-J.; Shi, Z.-J. *Nat. Chem.* **2010**, *2*, 1044–1049, DOI: 10.1038/nchem.86
57. Chen, W.-C.; Hsu, Y.-C.; Shih, W.-C.; Lee, C.-Y.; Chuang, W.-H.; Tsai, Y.-F.; Chen, P. P.-Y.; Ong, T.-G. *Chem. Commun.* **2012**, *48*, 6702–6704, DOI: 10.1039/C2CC32519E.
58. Zhang, H.; Shi, R.; Ding, A.; Lu, L.; Chen, B.; Lei, A. *Angew. Chem. Int. Ed.* **2012**, *51*, 12542–12545, DOI: 10.1002/anie.201206518.
59. Cuthbertson, J.; Gray, V. J.; Wilden, J. D. *Chem. Commun.* **2014**, *50*, 2575–2578, DOI: 10.1039/C3CC49019 J.
60. Liu, W.; Cao, H.; Zhang, H.; Zhang, H.; Chung, K. H.; He, C.; Wang, H.; Kwong, F. Y.; Lei, A. *J. Am. Chem. Soc.* **2010**, *132*, 16737–16740, DOI: 10.1021/ja103050x.
61. Shirakawa, E.; Itoh, K.; Higashino, T.; Hayashi, T. *J. Am. Chem. Soc.* **2010**, *132*, 15537–15539, DOI: 10.1021/ja108082.
62. Dewanji, A.; Murarka, S.; Curran, D. P.; Studer, A. *Org. Lett.* **2013**, *15*, 6102–6105, DOI: 10.1021/ol402995e.
63. Tanimoro, K.; Ueno, M.; Takeda, K.; Kirihata, M.; Tanimori, S. *J. Org. Chem.* **2012**, *77*, 7844–7849, DOI: 10.1021/jo3008594.
64. Qiu, Y.; Liu, Y.; Yang, K.; Hong, W.; Li, Z.; Wang, Z.; Yao, Z.; Jiang, S. *Org. Lett.* **2011**, *13*, 3556–3559, DOI: 10.1021/ol2009208.
65. Liu, W.; Tian, F.; Wang, X.; Yu, H.; Bi, Y. *Chem. Commun.* **2013**, *49*, 2983–2985, DOI: 10.1039/C3CC40695D.
66. Liu, H.; Yin, B.; Gao, Z.; Li, Y.; Jiang, H. *Chem. Commun.* **2012**, *48*, 2033–2035, DOI: 10.1039/C2CC16790E.

67. De, S.; Mishra, S.; Kakde, B. N.; Dey, D.; Bisai, A. *J. Org. Chem.* **2013**, *78*, 7823–7844, DOI: 10.1021/jo400890 k.
68. Roman, D. S.; Takahashi, Y.; Charette, A. B. *Org. Lett.* **2011**, *13*, 3242–3245, DOI: 10.1021/ol201160s.
69. Budén, M. E.; Guastavino, J. F.; Rossi, R. A. *Org. Lett.* **2013**, *15*, 1174–1177, DOI: 10.1021/ol3034687.
70. Vakuliuk, O.; Gryko, D. T. *Eur. J. Org. Chem.* **2011**, *2011*, 2854–2859, DOI: 10.1002/ejoc.201100004.
71. Bhakuni, B. S.; Kumar, A.; Balkrishna, S. J.; Sheikh, J. A.; Konar, S.; Kumar, S. *Org. Lett.* **2012**, *14*, 2838–2841, DOI: 10.1021/ol301077y.
72. De, S.; Ghosh, S.; Bhunia, S.; Sheikh, J. A.; Bisai, A. *Org. Lett.* **2012**, *14*, 4466–4469, DOI: 10.1021/ol3019677.
73. Murphy, J. A. *J. Org. Chem.* **2014**, *79*, 3731–3746, DOI: 10.1021/jo500071u.
74. Kita, Y.; Tohma, H.; Hatanaka, K.; Takada, T.; Fujita, S.; Mitoh, S.; Sakurai, H.; Oka, S. *J. Am. Chem. Soc.* **1994**, *116*, 3684–3691, DOI: 10.1021/ja00088a003.
75. Huang, J.; He, Y.; Wang, Y.; Zhu, Q. *Chem. – Eur. J.* **2012**, *18*, 13964–13967, DOI: 10.1002/chem.201202271.
76. Muraki, T.; Togo, H.; Yokoyama, M. *Tetrahedron Lett.* **1996**, *37*, 2441–2444, DOI: 10.1016/0040-4039(96)00313-9.
77. Togo, H.; Muraki, T.; Hoshina, Y.; Yamaguchi, K.; Yokoyama, M. *J. Chem. Soc. [Perkin 1]* **1997**, *1997*, 787–794, DOI: 10.1039/A603446B.
78. Yang, Y.-D.; Iwamoto, K.; Tokunaga, E.; Shibata, N. *Chem. Commun.* **2013**, *49*, 5510–5512, DOI: 10.1039/C3CC41667D.
79. Wen, J.; Zhang, R.-Y.; Chen, S.-Y.; Zhang, J.; Yu, X.-Q. *J. Org. Chem.* **2012**, *77*, 766–771, DOI: 10.1021/jo202150t.
80. Castro, S.; Fernández, J. J.; Vicente, R.; Fañanás, F. J.; Rodríguez, F. *Chem. Commun.* **2012**, *48*, 9089–9091, DOI: 10.1039/C2CC34592G.
81. Merritt, E. A.; Olofsson, B. *Angew. Chem. Int. Ed.* **2009**, *48*, 9052–9070, DOI: 10.1002/anie.200904689.
82. Eastman, K.; Baran, P. S. *Tetrahedron* **2009**, *65*, 3149–3154, DOI: 10.1016/j.tet.2008.09.028.
83. Ji, Y.; Brueckl, T.; Baxter, R. D.; Fujiwara, Y.; Seiple, I. B.; Su, S.; Blackmond, D. G.; Baran, P. S. *Proc. Natl. Acad. Sci.* **2011**, *108*, 14411–14415, DOI: 10.1073/pnas.1109059108.
84. Zhou, Q.; Ruffoni, A.; Gianatassio, R.; Fujiwara, Y.; Sella, E.; Shabat, D.; Baran, P. S. *Angew. Chem. Int. Ed.* **2013**, *52*, 3949–3952, DOI: 10.1002/anie.201300763.
85. Fujiwara, Y.; Dixon, J. A.; O'Hara, F.; Funder, E. D.; Dixon, D. D.; Rodriguez, R. A.; Baxter, R. D. et al. *Nature* **2012**, *492*, 95–99, DOI: 10.1038/nature11680.
86. Fujiwara, Y.; Dixon, J. A.; Rodriguez, R. A.; Baxter, R. D.; Dixon, D. D.; Collins, M. R.; Blackmond, D. G.; Baran, P. S. *J. Am. Chem. Soc.* **2012**, *134*, 1494–1497, DOI: 10.1021/ja211422 g.
87. Yonemitsu, O.; Cerutti, P.; Witkop, B. *J. Am. Chem. Soc.* **1966**, *88*, 3941–3945, DOI: 10.1021/ja00969a010.
88. Gritsch, P. J.; Leitner, C.; Pfaffenbach, M.; Gaich, T. *Angew. Chem. Int. Ed.* **2014**, *53*, 1208–1217, DOI: 10.1002/anie.201307391.
89. Bajtos, B.; Pagenkopf, B. L. *Eur. J. Org. Chem.* **2009**, *2009*, 1072–1077, DOI: 10.1002/ejoc.200801154.
90. Hu, W.; Qin, H.; Cui, Y.; Jia, Y. *Chem. – Eur. J.* **2013**, *19*, 3139–3147, DOI: 10.1002/chem.201204137.
91. Mascal, M.; Modes, K. V.; Durmus, A. *Angew. Chem. Int. Ed.* **2011**, *50*, 4445–4446, DOI: 10.1002/anie.201006423.
92. Feldman, K. S.; Ngernmeesri, P. *Org. Lett.* **2011**, *13*, 5704–5707, DOI: 10.1021/ol202535f.

93. Gui, J.; Zhou, Q.; Pan, C.-M.; Yabe, Y.; Burns, A. C.; Collins, M. R.; Ornelas, M. A.; Ishihara, Y.; Baran, P. S. *J. Am. Chem. Soc.* **2014**, *136*, 4853–4856, DOI: 10.1021/ja5007838.

94. Zhao, D.; Shen, Q.; Zhou, Y.-R.; Li, J.-X. *Org. Biomol. Chem.* **2013**, *11*, 5908–5912, DOI: 10.1039/C3OB41083H.

95. Rueping, M.; Leiendecker, M.; Das, A.; Poisson, T.; Bui, L. *Chem. Commun.* **2011**, *47*, 10629–10631, DOI: 10.1039/C1CC14297F.

96. Shirakawa, E.; Zhang, X.; Hayashi, T. *Angew. Chem. Int. Ed.* **2011**, *50*, 4671–4674, DOI: 10.1002/anie.201008220.

97. Nobuta, T.; Tada, N.; Fujiya, A.; Kariya, A.; Miura, T.; Itoh, A. *Org. Lett.* **2013**, *15*, 574–577, DOI: 10.1021/ol303389t.

98. Liu, Q.; Li, Y.-N.; Zhang, H.-H.; Chen, B.; Tung, C.-H.; Wu, L.-Z. *Chem. – Eur. J.* **2012**, *18*, 620–627, DOI: 10.1002/chem.201102299.

99. Alagiri, K.; Devadig, P.; Prabhu, K. R. *Chem. – Eur. J.* **2012**, *18*, 5160–5164, DOI: 10.1002/chem.201200100.

100. Antonchick, A. P.; Burgmann, L. *Angew. Chem. Int. Ed.* **2013**, *52*, 3267–3271, DOI: 10.1002/anie.201209584.

101. He, T.; Yu, L.; Zhang, L.; Wang, L.; Wang, M. *Org. Lett.* **2011**, *13*, 5016–5019, DOI: 10.1021/ol201779n.

102. Xia, R.; Niu, H.-Y.; Qu, G.-R.; Guo, H.-M. *Org. Lett.* **2012**, *14*, 5546–5549, DOI: 10.1021/ol302640e.

103. Minisci, F.; Vismara, E.; Fontana, F. *Heterocycles* **1989**, *28*, 489, DOI: 10.3987/REV-88-SR1.

104. Deng, G.; Ueda, K.; Yanagisawa, S.; Itami, K.; Li, C.-J. *Chem. – Eur. J.* **2009**, *15*, 333–337, DOI: 10.1002/chem.200801893.

105. Matcha, K.; Antonchick, A. P. *Angew. Chem. Int. Ed.* **2013**, *52*, 2082–2086, DOI: 10.1002/anie.201208851.

106. Chen, L.; Shi, E.; Liu, Z.; Chen, S.; Wei, W.; Li, H.; Xu, K.; Wan, X. *Chem. – Eur. J.* **2011**, *17*, 4085–4089, DOI: 10.1002/chem.20110019

107. Zhang, Y.; Li, C.-J. *J. Am. Chem. Soc.* **2006**, *128*, 4242–4243, DOI: 10.1021/ja060050p.

108. Li, Z.; Li, H.; Guo, X.; Cao, L.; Yu, R.; Li, H.; Pan, S. *Org. Lett.* **2008**, *10*, 803–805, DOI: 10.1021/ol702934 k.

109. Ushakov, D. B.; Gilmore, K.; Kopetzki, D.; McQuade, D. T.; Seeberger, P. H. *Angew. Chem. Int. Ed.* **2014**, *53*, 557–561, DOI: 10.1002/anie.201307778.

110. Kantak, A. A.; Potavathri, S.; Barham, R. A.; Romano, K. M.; DeBoef, B. *J. Am. Chem. Soc.* **2011**, *133*, 19960–19965, DOI: 10.1021/ja2087085.

111. Hamamoto, H.; Hata, K.; Nambu, H.; Shiozaki, Y.; Tohma, H.; Kita, Y. *Tetrahedron Lett.* **2004**, *45*, 2293–2295, DOI: 10.1016/j.tetlet.2004.01.104.

112. Turner, T. C.; Shibayama, K.; Boger, D. L. *Org. Lett.* **2013**, *15*, 1100–1103, DOI: 10.1021/ol400135n.

113. Dohi, T.; Ito, M.; Itani, I.; Yamaoka, N.; Morimoto, K.; Fujioka, H.; Kita, Y. *Org. Lett.* **2011**, *13*, 6208–6211, DOI: 10.1021/ol202632 h.

114. Huang, J.; Li, L.-T.; Li, H.-Y.; Husan, E.; Wang, P.; Wang, B. *Chem. Commun.* **2012**, *48*, 10204–10206, DOI: 10.1039/C2CC35450 K.

115. Shi, E.; Shao, Y.; Chen, S.; Hu, H.; Liu, Z.; Zhang, J.; Wan, X. *Org. Lett.* **2012**, *14*, 3384–3387, DOI: 10.1021/ol3013606.

116. Tada, N.; Cui, L.; Ishigami, T.; Ban, K.; Miura, T.; Uno, B.; Itoh, A. *Green Chem.* **2012**, *14*, 3007–3009, DOI: 10.1039/C2GC36238D.

117. Chen, F.; Huang, X.; Cui, Y.; Jiao, N. *Chem. – Eur. J.* **2013**, *19*, 11199–11202, DOI: 10.1002/chem.201301933.

# 3 Pd-Catalyzed C—H Functionalization

*Jesse D. Carrick*

## CONTENTS

## 3.1 INTRODUCTION

This chapter highlights a number of innovative applications of C—H functionalization of various substrates using palladium catalysis. As the field continues to experience diverse and expansive growth, methods continue to be developed through frontier areas toward the realization of the carbon–hydrogen bond serving as a unique functional group for the preparation of advanced synthons. Significant efforts in this area have been undertaken to provide expeditious access to functionalized synthons from simple starting materials. Recent reviews have highlighted important advances in the areas of computational,[1] oxidation,[2] site-selectivity,[3] cross-coupling,[4] synthetic work,[5] medicinal chemistry,[6] and the construction of heterocycles[7] utilizing Pd-mediated C—H functionalization.

## 3.2 ARYLATION

Formation of C—C bonds between arenes and aromatic substrates with catalytic Pd(II)[8] species has provided entry into diverse structures.[9] Methods have been developed to prepare materials with high levels of *ortho-* and *para-* regioselectivity depending on reaction conditions employed. Daugulis has described *ortho*-regioselective arylation using benzamide directing groups with functionalized iodonium salts,[10] aryl chlorides,[11] as well as aryl iodides.[12] Dong has accomplished high levels of *ortho*-regioselectivity via direct arylation of benzamides,[13] *N*-aryloxazolidinones,[14] and *O*-phenylcarbamates[15] (Scheme 3.1) with the corresponding unactivated arene

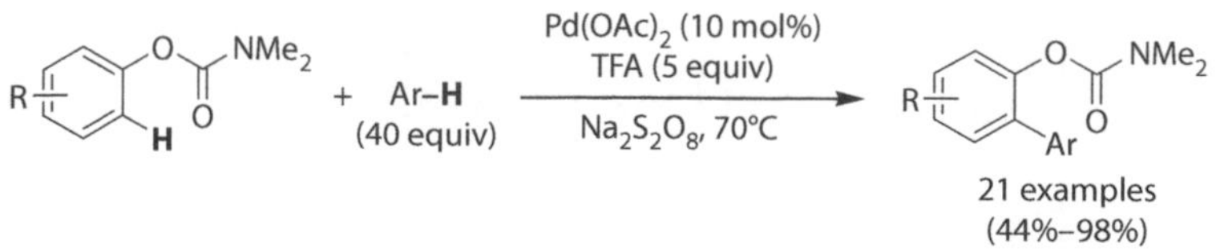

SCHEME 3.1   Directed C—H arylation.

using sodium persulfate as an oxidant. Approaches to *para*-selective arylation have
been published by Yu.[16]

The use of a Lewis-basic donating group to facilitate metal coordination prior
to the C—H functionalization event is not required in all cases. Direct arylation of
arenes with aryl iodides[17] and bromides[18] in the absence of a directing group has been
described by Fagnou (Scheme 3.2).

The synthesis of relevant biaryls has also been reported with aryltrifluoroborates
directed by the carboxylic acid functional group[19] in addition to oxidative dimerization.[20] Sanford has described the preparation of biaryls via two discrete C—H
functionalizations through the intermediacy of a Pd(IV) *bis*-cyclopalladated moiety
which reductively eliminates to afford the new C—C bond (Scheme 3.3).[21] An analogous procedure affording the homo-coupled arene from the unactivated starting
material has also been reported.[22]

A significant advance in C—H functionalization has been realized in the area
of cross-coupling. The ability to couple diverse structures without the use of
activated or pre-metallated intermediates represents an important achievement.
Oxidative arylation of non-activated arenes has been reported (Scheme 3.4).[23] An
extremely innovative outcome of this study is the ability to influence regioselectivity of the arylation predicated on the employment of X-type ligand. In the case of
carboxylate ligands concomitantly added with less than one equivalent of quinone
regioisomer **1** is preferentially formed. If carbonate moieties are utilized, regioisomer **2** is favored.

Arylation via C—H functionalization is not limited to electron-rich arenes as
demonstrated by Gaunt using trifluoroborate salts and an electron-deficient aryl
imine donor (Scheme 3.5).[24] The method displays broad substrate scope and provides
a mild alternative to procedures requiring elevated temperatures.

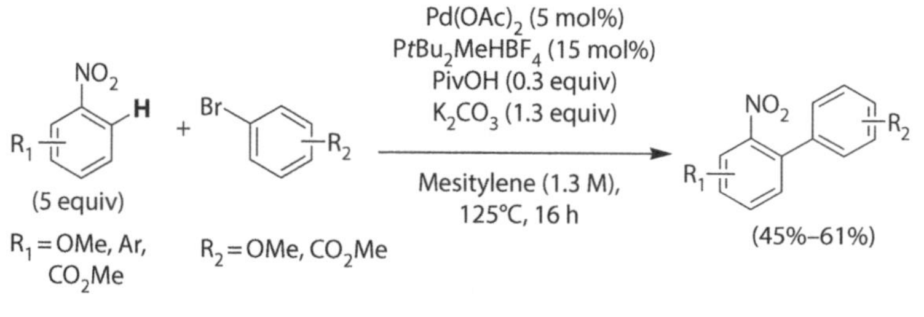

SCHEME 3.2   Direct arylation of substituted arenes.

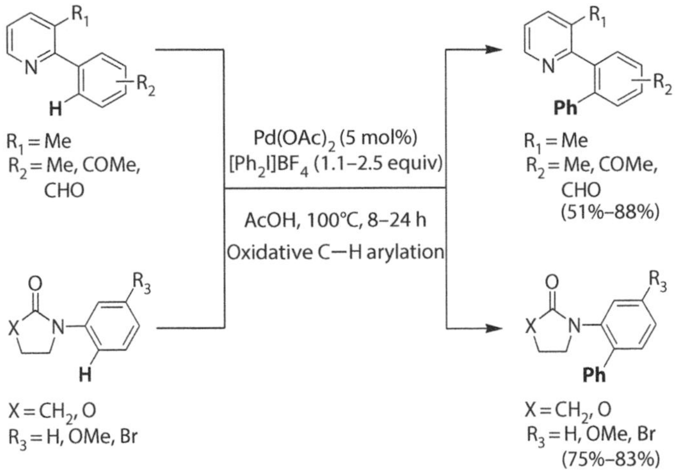

**SCHEME 3.3**　Directed oxidative C—H arylation of arenes.

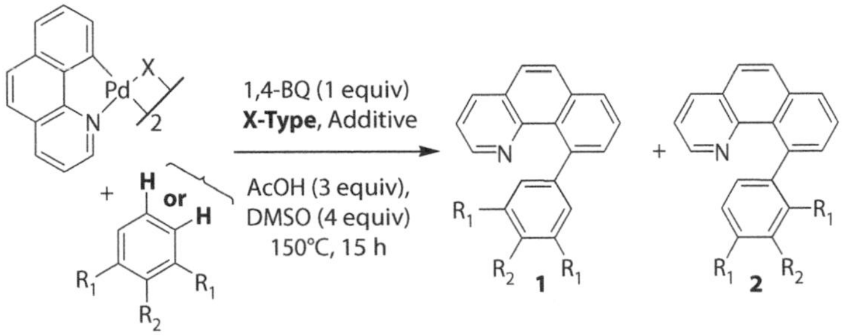

| Entry | Arene | X-Type | Additive | A:B | Yield$^{GC}$ |
|-------|-------|--------|----------|-----|--------------|
| 1 | 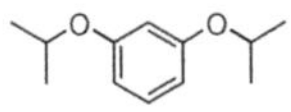 | OAc | AcOH | 15:1 | 94 |
| 2 | 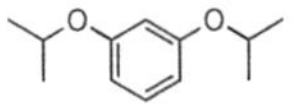 | CO$_3$ | | 1:6 | 85 |
| 3 | 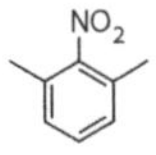 | OAc | | 64:1 | 100 |
| 4 | 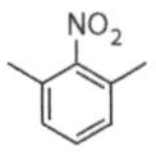 | CO$_3$ | | 1:5 | 69 |
| 5 | 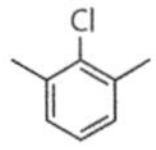 | OAc | | 60:1 | 95 |
| 6 | 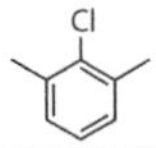 | CO$_3$ | | 1:2 | 67 |

**SCHEME 3.4**　Oxidative arylation of arenes.

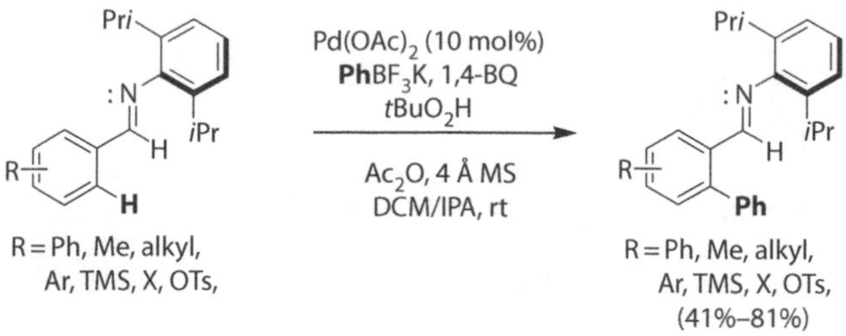

**SCHEME 3.5**  Arylation of electron-deficient aldimines.

Template-directed arylation on arenes with high levels of *meta*-regioselectivity has been described by Yu.[25] Ligand-accelerated processes involving weak coordination through substituted homobenzoic acids have been leveraged for C—C bond formation with arylboron reagents.[26] Yu has also described the importance of matching an appropriate ligand to the desired substrate when using homogeneous C—H functionalization with Pd(II) catalysis (Scheme 3.6).[27] Examination of arylboron reagents during the course of this study revealed that phenylpinacolborinate performed comparably to the corresponding phenyltrifluoroborate salt for the *o*-trifluoromethyl substrate. Both of the aforementioned reagents were superior to the boronic acid, tetralone, and MDMA derivatives. In the absence of an indicated ligand-poor conversion, all coupling experiments were observed.

Phenylation of heterocycles in the absence of pre-functionalized reagents has been described by Fagnou (Scheme 3.7).[28] Oxidative coupling of *N*-acylated indoles with benzene proceeded in moderate to good yield with high levels of regioselectivity for the 2-position while minimizing *bis*-arylation or the alternative regioisomer.

The prevalence of heterocycles in natural products, materials, pharmaceuticals, and agrochemicals is widely known. Frequent utility of metal-mediated coupling reactions, via the Kumada, Negishi, and Suzuki protocols,[29] is often used to diversify heterocyclic scaffolds. C—H functionalization of non-activated heterocycles provides a strategic alternative which instantly broadens the scope of scaffolds available for the formation of C-Aryl bonds through Pd catalysis.[30] Recent advances using directing groups and Pd° catalysis have been disseminated. Direct arylation of heterocyclic structures with Pd(II) has been extensively investigated.[31] Arylation of *1H-*[32] and *2H*-indazoles,[33] indoles,[34] pyridines,[35] pyrroles,[36] furans,[37] and thiazoles[38] has been reported using Pd(II)-mediated C—H functionalization. Approaches to

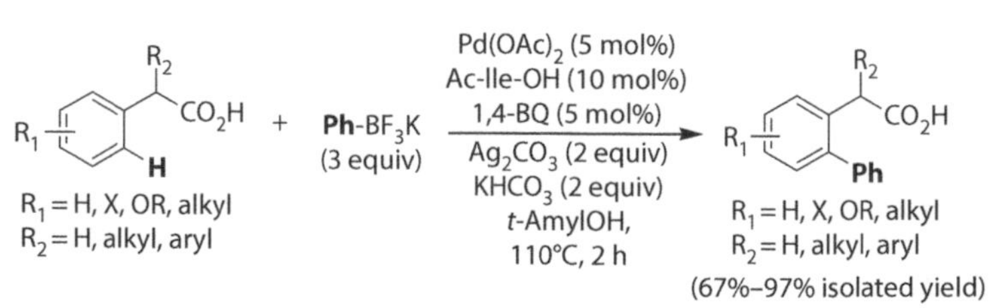

**SCHEME 3.6**  Ligand-accelerated *ortho*-regioselective arylation.

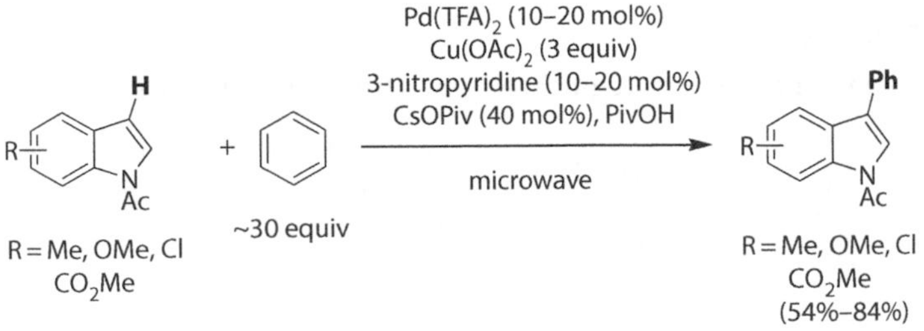

**SCHEME 3.7**  Arylation of unactivated indoles.

cross-coupling using aryl bromides[39] and aryl chlorides[40] have also been accomplished. Intramolecular arylations have been leveraged in the construction of biaryl scaffolds using direct methods[41] and tandem processes.[42] Approaches to heterocycle synthesis via intra-molecular C—H functionalization through arylation have also been studied.[43]

Approaches to C—C bond formation through decarboxylative C—H activation of oxazoles and thiazoles for inter-molecular coupling of unactivated azole moieties[44] have been demonstrated by Greaney (Scheme 3.8). Treatment of carboxylic acid-functionalized heteroarenes with metal carbonates using Pd catalysis promotes rapid decarboxylation resulting in a new Pd–C bond. Carbopalladation at C-2 of the oxazole coupling partner results in an intermediate which reductively eliminates to afford the coupled heteroarene. The scope of the transformation is accommodating to a variety of functionalities. Decarboxylative cross-coupling of thiazoles with aryl halides exploiting $PdCl_2$ catalysis has also been reported.[45]

Extensive investigation of the direct arylation of thiophenes via C—H functionalization has established the method's compatibility with numerous functional groups.[46] Inherent substrate bias exists for metalation of the α-position of the heterocycle. Recently, methods have been described, which lead to regioselective arylation of the less reactive β-position[47] in addition to the ability to modulate regiodivergency with the implementation of electron-deficient ligands (β-selectivity) or electron-rich ligands (α-selectivity) with readily available aryl iodides and boronic acids using $PdCl_2$ catalysis (Table 3.1).[48] Sequential processes have also found widespread utility.[49]

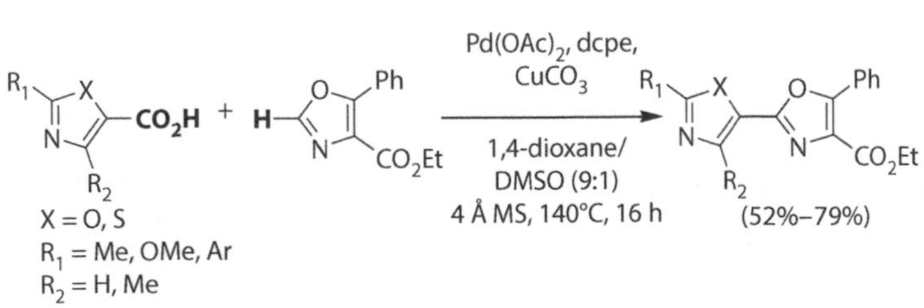

**SCHEME 3.8**  Pd(II)-catalyzed decarboxylative heteroarene arylation.

**TABLE 3.1**

**Ligand Influenced Regiodivergent Thiophene Arylation**

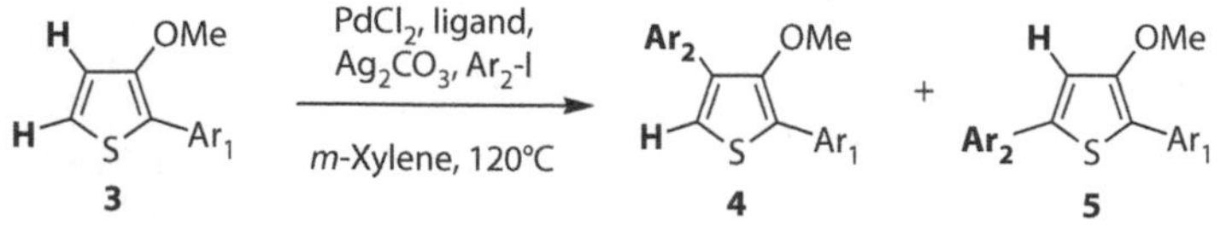

| Entry | $Ar_2$ | Ligand | Yield | (4:5) |
|-------|--------|--------|-------|-------|
| 1 | $C_6H_6$ | $P[OCH(CF_3)_2]_3$ | 80 | (97:3) |
| 2 | $C_6H_5$ | 2,2'-bipyridyl | 89 | (1:99) |
| 3 | $o\text{-MeC}_6H_4$ | $P[OCH(CF_3)_2]_3$ | 88 | (98:2) |
| 4 | $o\text{-MeC}_6H_4$ | 2,2'-bipyridyl | 98 | (2:98) |
| 5 | $p\text{-MeC}_6H_4$ | $P[OCH(CF_3)_2]_3$ | 83 | (96:4) |
| 6 | $p\text{-MeC}_6H_4$ | 2,2'-bipyridyl | 91 | (1:99) |
| 7 | $p\text{-OMeC}_6H_4$ | $P[OCH(CF_3)_2]_3$ | 75 | (98:2) |
| 8 | $p\text{-OMeC}_6H_4$ | 2,2'-bipyridyl | 83 | (1:99) |
| 9 | $p\text{-AcC}_6H_4$ | $P[OCH(CF_3)_2]_3$ | 71 | (95:5) |
| 10 | $p\text{-AcC}_6H_4$ | 2,2'-bipyridyl | 80 | (1:99) |

C—H functionalization of the *N*-oxide derivative can lead to enhanced reactivity.[50] These reagents can provide strategic advantages over analogous organometallic species which often inherently display sensitivity to moisture or air.[51] Differential biarylation of pyridine-*N*-oxides through sequential processes using the corresponding triflate has also been demonstrated.[52] Functionalization of the benzylic position of substituted picoline *N*-oxides has also been highlighted (Scheme 3.9).[53]

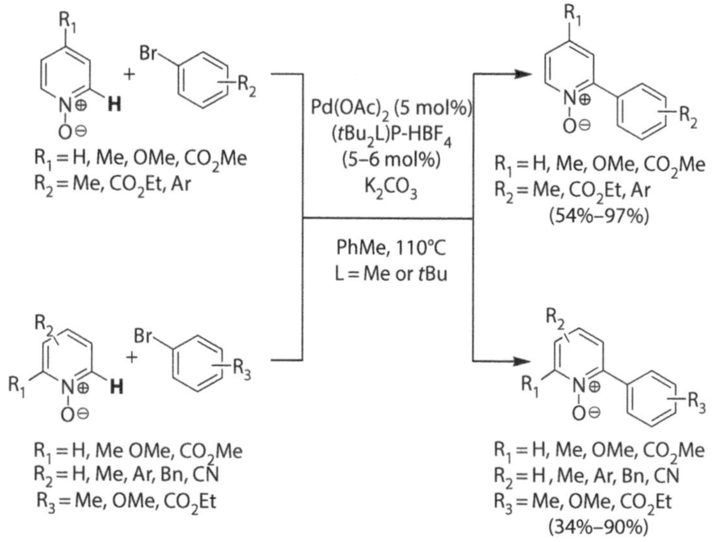

**SCHEME 3.9**   Arylation of heterocyclic *N*-oxides catalyzed by Pd(OAc)$_2$.

## 3.3 OLEFINATION

The utility of the Heck reaction in its many forms has been extensively studied for the construction of useful synthons and complex molecules. Advances in C—H functionalization have enabled the synthesis of Heck products directly from unactivated arenes via Pd(II) catalysis.[54] Olefination of electron-rich and -deficient arenes has been reported.[55] Approaches to olefination of arenes have included directed olefination with haloolefins.[56]

A variety of innovative methods have been developed to address issues of regioselective olefination of arenes. Two common approaches to arene olefination are frequently employed via ligand modulation[57] or the use of a resident directing group[58] on the arene scaffold. Alternative directing groups through the use of silanols have also been reported for Pd(II)-mediated C—H olefination.[59] Regiodivergent olefination outcomes on aryl-ether scaffolds can be accomplished with the choice of directing group.[60] Weak coordination through the requisite carboxylate salt and appropriate ligand can promote the *ortho*-olefinated product.[61] If *meta*-selective olefination is desired, the development of a nitrile-directing template has enabled high-levels of regioselectivity; typically greater than 7:1 favoring the *meta*-olefinated product (Scheme 3.10).[62]

Olefination via Pd(II) C—H functionalization is not limited to carbon-based arenes. Olefination of the C-2 position of pyridines can be accomplished through the *N*-oxide.[63] Intra-molecular-directed alkenylation of pyridines leading to

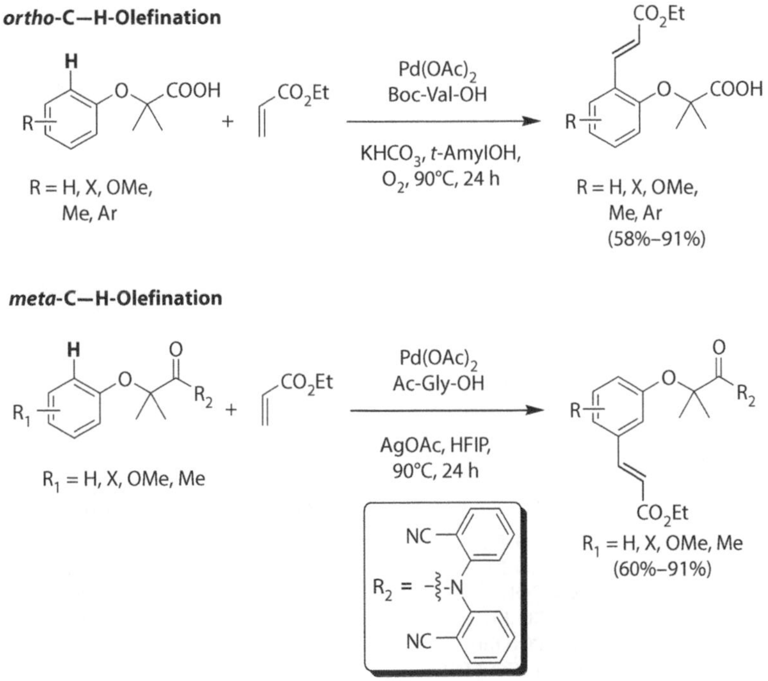

**SCHEME 3.10** Regiodivergent directed C—H olefination of arenes.

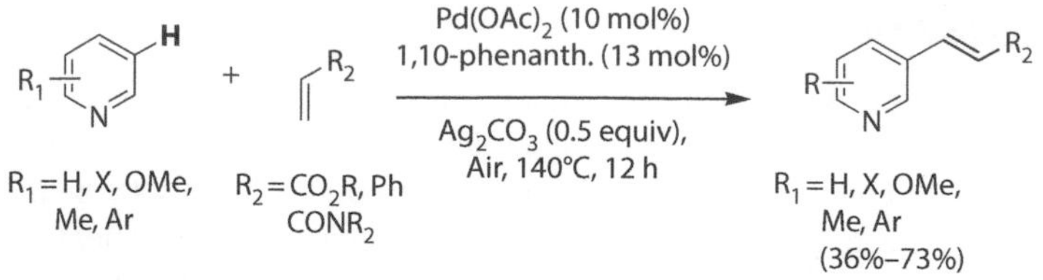

**SCHEME 3.11**  Ligand-promoted C-3 olefination of pyridine.

regioselective olefination at C-3 has been reported.[64] Regioselective olefination at the C-3 position of pyridines without the use of a pendant directing group was accomplished by Yu[65] (Scheme 3.11). A wide variety of electron-donating and -withdrawing substituents were tolerated with this method including alkoxy and halo- which can be further elaborated through functional group interconversion.

Olefination of heterocycles is not limited to pyridines. Gaunt has reported the ability to regioselectively olefinate pyrroles predicated on the *N*-protecting group.[66] When a Boc group is employed, olefination is regioselective for the C-2 position. When a sterically demanding TIPS group is utilized, the C-2 position is effectively sterically encumbering, necessitating olefination regioselectively for the C-3 position (Table 3.2).

The use of the intra-molecular Heck reaction in concert with a subsequent direct arylation catalyzed by Pd(II) has been reported by Fagnou.[67] Iterative Pd(II) mediated C—H olefination leading to diolefinated arenes has been reported.[68] Tandem processes synergistically leveraging C—H olefination and carbonylation have been employed toward the construction of β-quaternary carbons[69] (Scheme 3.12). Treatment of electron-deficient amides with Pd(OAc)$_2$ initiates a β-regioselective *sp$^3$* C—H activation, which, upon treatment with ethyl acrylate, affords the olefination

**TABLE 3.2**

**Regiodivergent Olefination of Pyrrole**

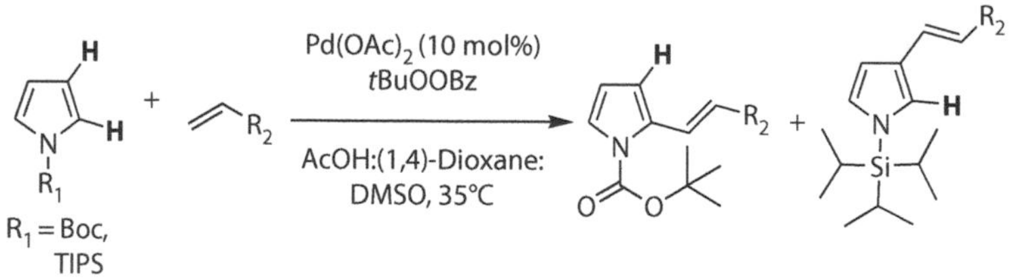

| Entry | R$_2$ | Yield (C-2) | Yield (C-3) |
|-------|-------|-------------|-------------|
| 1 | R$_2$ = CO$_2$Bn | 73 | 78 |
| 2 | R$_2$ = CO$_2$Bu | 75 | 81 |
| 3 | R$_2$ = COEt | 69 | 69 |
| 4 | R$_2$ = SO$_2$Me | 69 | 71 |
| 5 | R$_2$ = PO(OEt)$_2$ | 60 | 70 |
| 6 | R$_2$ = 4-(CO$_2$Me)Ph | 60 | 63 |
| 7 | R$_2$ = CN | 63 | 60 |

R2
R1
H
O
NHArF
+
CO2Et
R1,R2 = alkyl
Pd(OAc)2 (10 mol%)
Ligand (40 mol%)
TEMPO (10 mol%)
AgOAc (2 equiv)
K2HPO4 (1.1 equiv)
NaHCO3(2 equiv)
O2, C6F5CF3, 120°C
NArF
(49%–87%)
N
Ligand

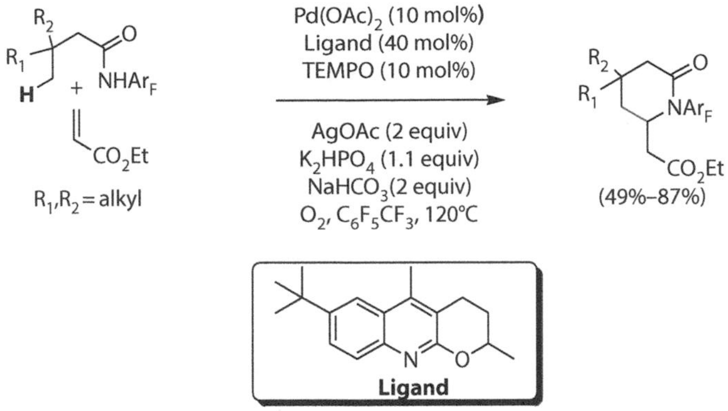

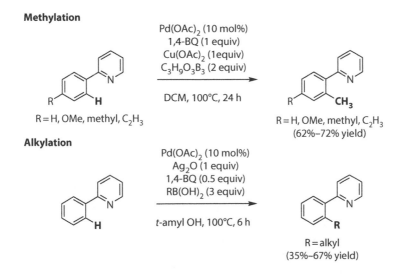

**SCHEME 3.12** Tandem C—H olefination/conjugate addition.

product. Spontaneous intra-molecular conjugate addition affords the listed products in moderate to high yield. The preparation of Z-enamides has also been reported.[70]

## 3.4 FUNCTIONAL GROUP INTER-CONVERSION OF ARENES

Numerous methods have been developed toward the construction of advanced arenes employing Pd(II) mediated C—H functionalization.[71] Installation of a benzylic methyl group via arene methylation[72] was first described by Yu. Treatment of the functionalized pyridine substituted arenes with methyl boroxine provided access to methylated arenes as well as methylated pyridines. The method was extended to application of generic alkylboronic acids for a comprehensive treatise of C—H alkylation catalyzed by Pd(II) (Scheme 3.13).

Methylation
Pd(OAc)2 (10 mol%)
1,4-BQ (1 equiv)
Cu(OAc)2 (1equiv)
C3H9O3B3 (2 equiv)
DCM, 100°C, 24 h
R = H, OMe, methyl, C2H3
R = H, OMe, methyl, C2H3
(62%–72% yield)
CH3
Alkylation
Pd(OAc)2 (10 mol%)
Ag2O (1 equiv)
1,4-BQ (0.5 equiv)
RB(OH)2 (3 equiv)
t-amyl OH, 100°C, 6 h
R = alkyl
(35%–67% yield)

**SCHEME 3.13** Methylation and alkylation of arenes.

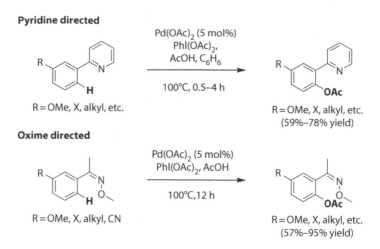

**SCHEME 3.14**   Directed acetoxylation of arenes.

Pyridine[73] and oxime-directed acetoxylation[74] has been reported by Sanford (Scheme 3.14). Treatment of the arenes below with Pd(OAc)$_2$ and *bis*-acetoxyiodobenzene (BAIB) afforded the desired acetoxylated arenes in good yields with high levels of regioselectivity, greater than 20:1 in most cases, for the more sterically accessible *ortho*-position. The transformation displayed wide functional group tolerance with respect to the *meta*-substituent on the arene. The utilization of oxone as the terminal oxidant displayed comparable reactivity to BAIB and provides a safe, greener alternative. Tandem acetoxylation processes have also been described.[75]

Functionalization via cyanation[76] and deuteration[77] of arenes provides an additional avenue of derivatization. Given the ubiquitous presence of fluorine in medicinally relevant compounds—the commercial anti-depressants, Prozac, Zoloft, and Paxil, to name a few—methods to introduce fluorine into synthons have attracted significant attention. Early efforts focused on the regioselective *ortho-* fluorination of arenes with an electrophilic fluorine source (Scheme 3.15).[78] Functional group inter-conversion of the triflamide directing group is straightforward and provides access to the amine, azide, nitrile, aldehyde, or malonic ester functionalities. An alternative to the triflamide procedure has been reported leveraging arylbenzamides as the directing group.[79]

Intra-molecular, *ortho*-regioselctive trifluoromethylation[80] examples employing pyridine and electron-deficient benzamides as the directing groups with electrophilic fluorine sources have been disseminated (Scheme 3.16). A wide variety of functionality, including activating and deactivating substituents, is compatible, affording the desired trifluoromethylated arenes in good yields and high levels of regioselectivity.

Borylation of the benzylic position[81] of arenes via C—H functionalization has been achieved. Yu has also reported the ability construct C-B bonds via directed C—H functionalization of arylbenzamides catalyzed by Pd(II).[82] Formation of carbon–carbon bonds between arenes and alkyltrifluoroborates,[83] alkylstannanes,[84] and alkyl halides[85] has been reported. The incorporation of amino acid derived

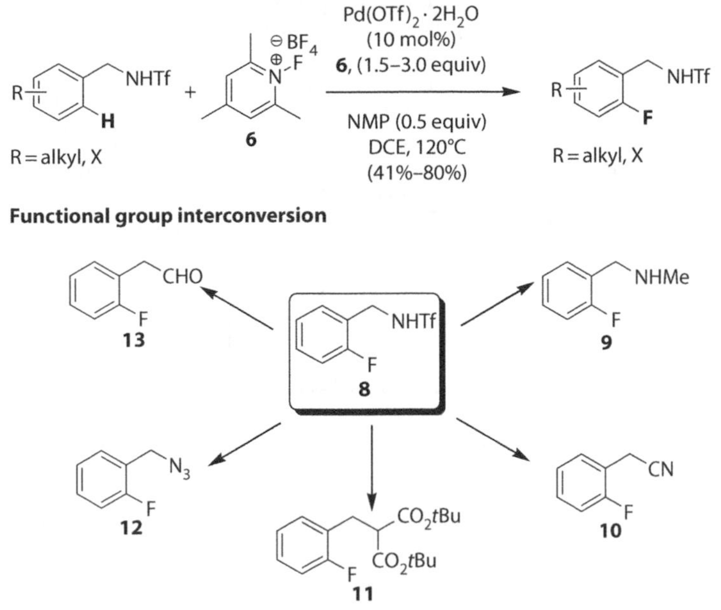

**SCHEME 3.15** Directed fluorination of benzyltriflamides.

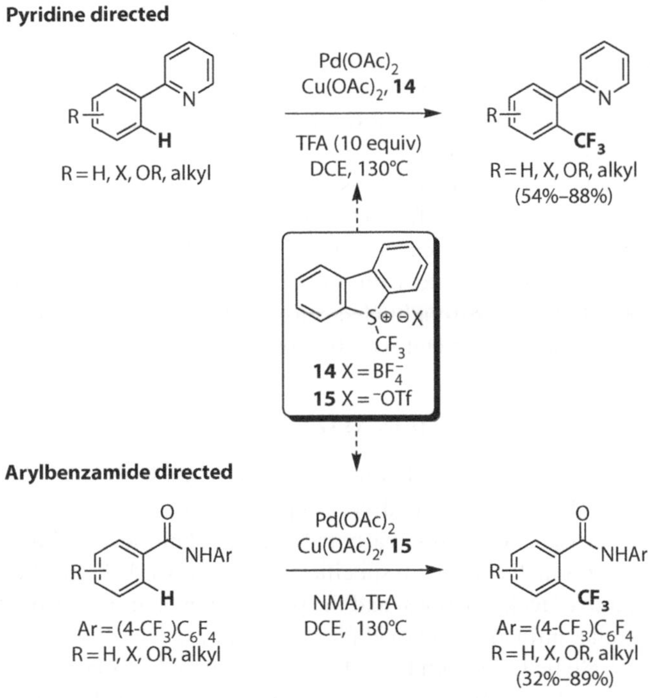

**SCHEME 3.16** Directed trifluoromethylation of arenes.

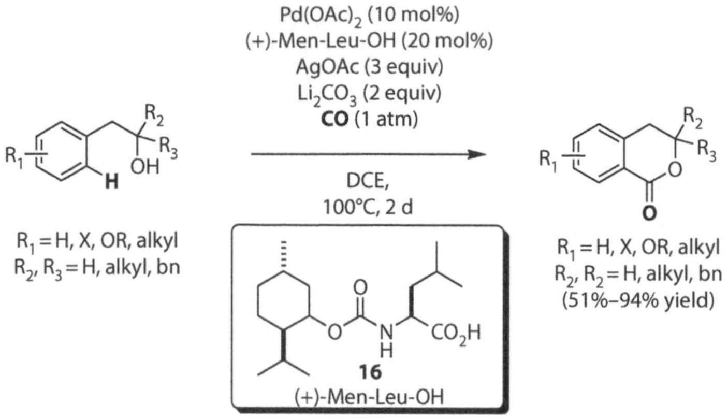

**SCHEME 3.17**  Preparation of 1-isochromanones via C—H carbonylation.

ligands has led to dramatic accelerations in conversion of starting materials.[86] Intra-molecular amination toward the construction of *N*-containing heterocycles has been described for the synthesis of indoles[87] catalyzed by Pd(dba)$_2$ and indolines[88] cata-lyzed by Pd(OAc)$_2$. Recently, inter-molecular methods for the preparation of amines using Pd(II) catalysis have been achieved using *N*-arylpthalimides[89] and alkyl amines[90] under directed conditions. Advanced functional group inter-conversions of arenes via iodidination,[91] phosphorylation,[92] and sulfonamide formation have been disclosed.[93]

The synthesis of 1-*iso*chromanones has been accomplished using C—H carbonyl-ation[94] mediated by Pd(II) in the presence of one atmosphere of carbon monoxide via a hydroxyl-directed C—H activation followed by intra-molecular hydroxyl-trapping of the resulting intermediate (Scheme 3.17).

The preparation of 1,2-dibenzoic acids[95] catalyzed by Pd(II) can be accomplished via directed C—H carbonylation followed by the *in situ* hydrolysis of an intermedi-ary cyclic anhydride. Electron-rich substituted benzoic acids performed better over-all with respect to yield and rate with the *ortho*-directed C—H functionalization occurring more efficiently. Strongly electron-withdrawing groups on the aromatic ring render such substrates difficult with this protocol.

## 3.5  ALLYLIC C—H FUNCTIONALIZATION

The continuous exploration of C—H functionalization methods has led to new reac-tivity manifolds. The utility of the allyl group has provided entry into a variety of functionalizations via electrophilic carbopalladation. This section highlights some recent developments in the area with specific attention toward allylic alkylation, ace-toxylation, oxidative Heck reactions, and the preparation of amino alcohols and diols via intra-molecular amination and oxidation procedures.

Activation of allylic C—H bonds for further functionalization via alkylation has been achieved (Scheme 3.18).[96] Treatment of functionalized allyl benzenes with *bis*-sulfinylethane Pd(II) acetate as a catalyst facilitates an allylic C—H activation leading

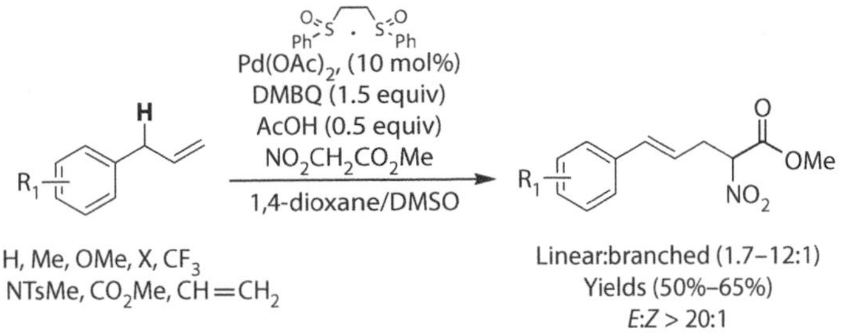

**SCHEME 3.18** Inter-molecular Pd-catalyzed allylic C—H alkylation.

to the formation of a dimeric Pd-π-allyl complex. When treated with nitromethylac-etate the formation of the alkylation product shown in good linear to branched ratios with excellent E:Z stereoselectivity is afforded. The substrate scope is broad and allyl benzenes, with a variety of electron-withdrawing and donating functionality at various regioisomeric positions, are compatible. An interesting observation during the course of the study revealed that aryl halides and triflates do not react under the listed conditions as would be expected in a Pd(0)-mediated process. This method has been recently applied to aliphatic substrates utilizing benzoyl nitromethane with similar outcomes.[97]

When the use of nucleophiles other than water in the presence of terminal alkenes under Pd(II) catalysis Wacker-type products frequently predominate. Use of White's electrophilic *bis*-sulfinyl Pd(II) acetate catalyst with terminal alkenes in the presence of acetic acid enabled the preparation of terminal acetoxylated ole-fins via allylic C—H oxidation and subsequent regioselective nucleophilic trap-ping of a Pd π-allyl complex (Table 3.3).[98] The substrates screened afforded the desired end products with high levels of E-selectivity and in moderate yields with excellent linear:branched ratios. Recent updates to this method include the use of N-tosylcarbamates as nucleophiles.[99]

C—H functionalization of N-toyslated carbamates via intra-molecular allylic ami-nation has been demonstrated for the construction of 1,2-*syn*-[100] and 1,3-*syn*-amino alcohols[101] (Scheme 3.19). Allylic C—H activation via the familiar Pd-π-allyl com-plex followed by regioselective intra-molecular nucleophilic trapping with the teth-ered electron-deficient carbamate and subsequent β-hydride elimination afforded the desired 1,2 or 1,3-amino alcohols. The overall scope of both transformations accommodates a variety of functionality including alkyl, carbonyl derivative, inter-nal olefins, and other relevant functional groups. The measured crude diaster-eoselectivities for both reactions were modest, but the minor diastereomer for both cases was typically purified by column chromatography affording excellent 1,3-*syn* diastereoselectivities and in workable yields. Further synthetic derivatization of the prepared materials via known methods can provide direct access toward functional-ized α-amino acids. The prevalence of the β-lactam pharmacophore in medicinal chemistry is well known. A recent extension of this method has been applied toward the functionalization of β-lactams leveraging Lewis acid co-catalysis.[102]

**TABLE 3.3**

**Regioselective Acetoxylation via Allylic C—H Activation**

**Typical Wacker/Wacker-type reactions**

**White**

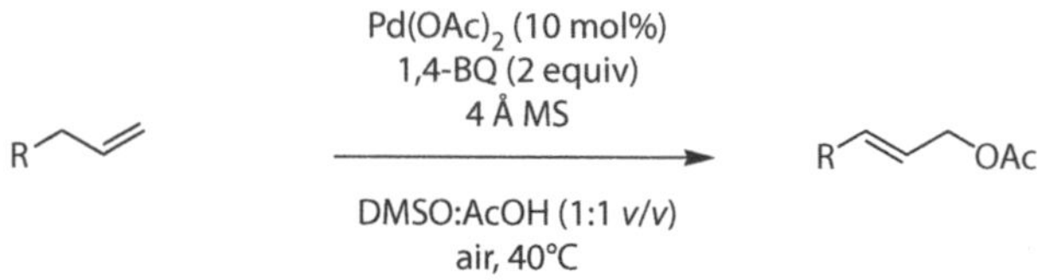

| Entry | Product | L:B | E:Z | Yield (%) |
|-------|---------|-----|-----|-----------|
| 1 | 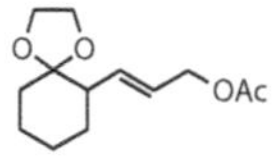 | 99:1 | 20:1 | 50 |
| 2 | 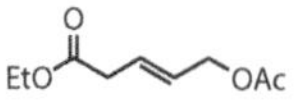 | 20:1 | 13:1 | 54 |
| 3 | 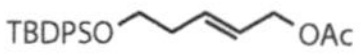 | 31:1 | 11:1 | 50 |
| 4 | 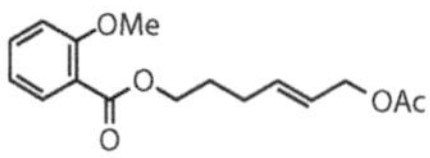 | 17:1 | 13:1 | 56 |
| 5 | 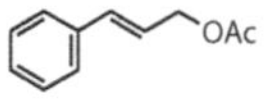 | 99:1 | 13:1 | 65 |

Oxidative Heck reactions via Pd(II) C—H functionalization of terminal alkenes with pinacol boranes have been described for the preparation of styrenes and derivatives through electrophilic Pd(II) catalysis (Scheme 3.20).[103] Treatment of a functionalized allylic precursor with the Pd(II) catalysts listed facilitated an allylic C—H activation. Subsequent transmetallation of the aryl boronic acid and reductive elimination afforded the desired olefin with excellent stereoselectivity. The scope of the transformation allows for a variety of activating and deactivating substituents on the aryl boronic acid as well as numerous functional groups on the starting alkene. A tandem allylic C—H oxidation/vinylic arylation protocol has also been reported.[104]

A number of biologically relevant natural products, (+)-discodermolide, amphidinolide C, and macrolactin A, possess a 1,3-conjugated diene motif making this functionality of direct relevance. White has demonstrated the utility of the

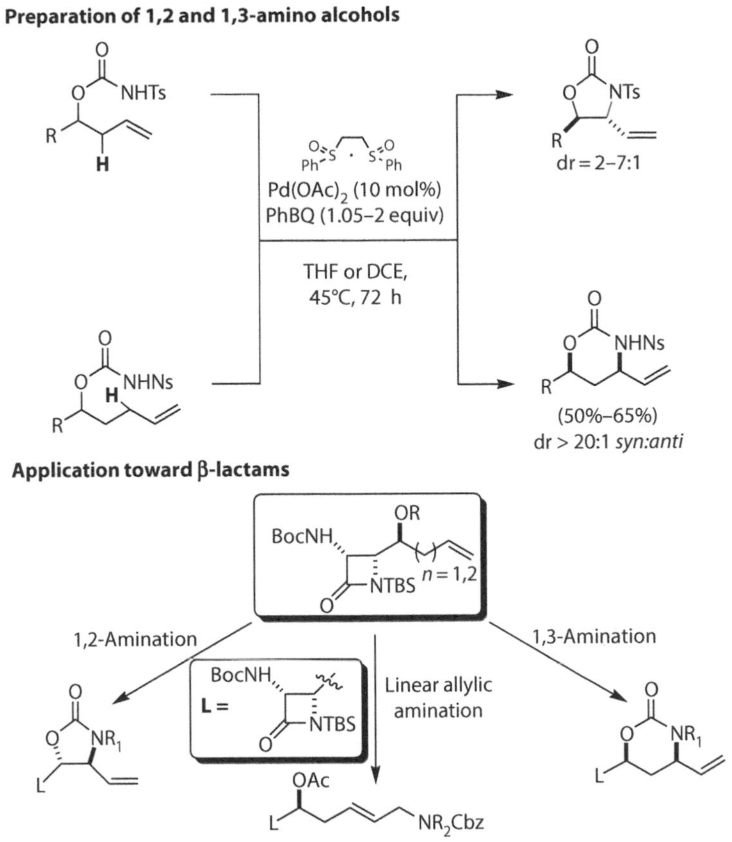

**Preparation of 1,2 and 1,3-amino alcohols**

**Application toward β-lactams**

**SCHEME 3.19** Diastereoselective C—H amination and relevant applications.

bis-phenylsulfinyl palladium(II) acetate catalyst toward the preparation of dienes with high levels of stereoselectivity for the $E$ isomer in good to excellent yield through terminal $sp^2$ C—H activation and cross-coupling with pinacol boranes via an oxidative Heck protocol (Table 3.4).[105]

Preparation of multi-oxygenated species via tethered intra-molecular allylic C—H functionalization has provided substrate-controlled catalytic entry into differentiated 1,2-syn diols with moderate to good levels of diastereoselectivity and excellent

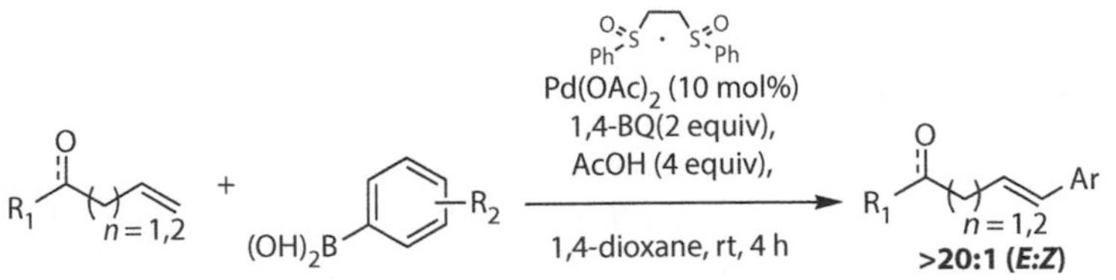

**SCHEME 3.20** Oxidative Heck arylation of terminal alkenes.

## TABLE 3.4
## Vinylation of Terminal Alkenes via an Oxidative Heck Reaction

| Entry | Product | E:Z | Yield (%) |
|-------|---------|-----|-----------|
| 1 | | 20:1 | 62 |
| 2 | | 20:1 | 50 |
| 3 | | 20:1 | 53 |
| 4 | | 20:1 | 66 |
| 5 | | 10:1 | 72 |
| 6 | | 6:1 | 51 |

regioselectivity (Table 3.5).[106] Such synthetic motifs were previously accessed by alkoxyallylboration[107] or similar methods. Various substituents including alkyl, aryl, and tetrasubstituted carbons are accommodated under these conditions. Applications toward the preparation of pyrans through the Ireland–Claisen rearrangement[108] also highlight the method's synthetic versatility.

Functional group inter-conversion of the end products can be accomplished through a straightforward sequence (Scheme 3.21). Treatment of cyclic substrate **17** under basic conditions affords the requisite hydroxy acid. Protection of the secondary alcohol as the silyl ether followed alkylation of the acid affords the methyl ester. Samarium mediated reduction and subsequent protection affords the orthogonally protected diol **19**.

This approach has also been leveraged toward the construction of differentiated imidazolidinone and aminooxazoline structures.[109] Tandem approaches toward allylic C–H amination and sequential arylation have been demonstrated to access functionalized oxazolidinones toward homophenylalanine derivatives directly from the linear alkenyl carbamate (Scheme 3.22).[110]

**TABLE 3.5**

**Preparation of *syn*-1,2-diols via C—H Functionalization**

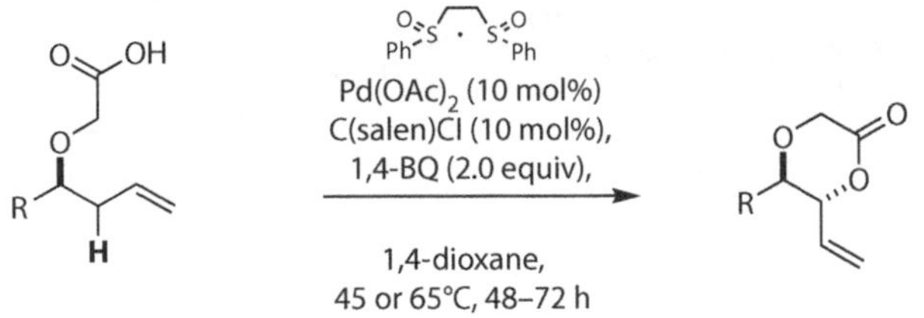

| Entry | Product | dr (*anti:syn*) | Yield (%) |
|-------|---------|-----------------|-----------|
| 1 | 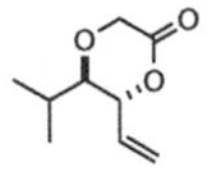 | 9:1 | 83 |
| 2 | 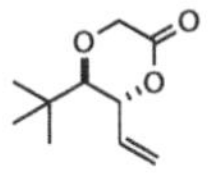 | 11:1 | 22 |
| 3 | 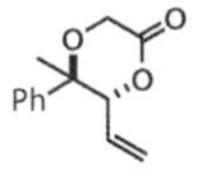 | 3:1 | 57 |
| 4 | 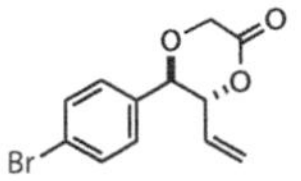 | 3:1 | 57 |
| 5 | 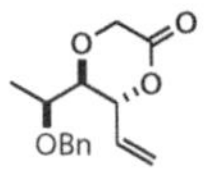 | 8:1 | 62 |
| 6 | 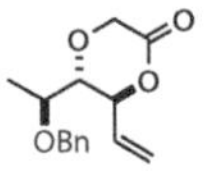 | 4:1 | 62 |

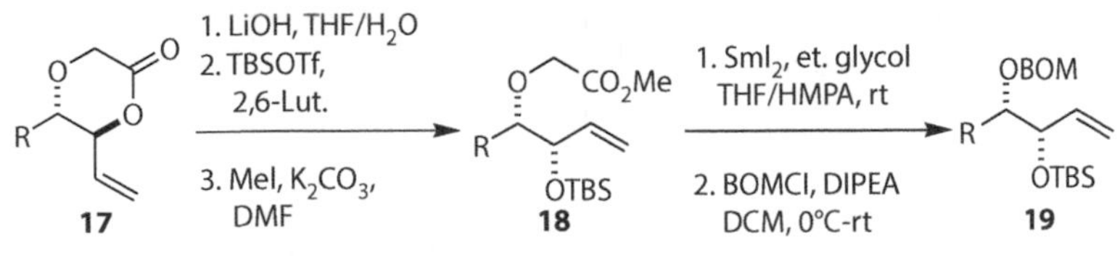

**SCHEME 3.21**  Functional group inter-conversions of C—H functionalized diols.

## 3.6 C—H FUNCTIONALIZATION OF *sp³*-HYBRIDIZED CARBON ATOMS

A frontier of research in the area of Pd-mediated C—H activation lies with chemo-, regio-, and stereoselectively activating the *sp³* hybridized carbon for subsequent functionalization. Methods have been developed to acetoxylate β-methyl groups relative to

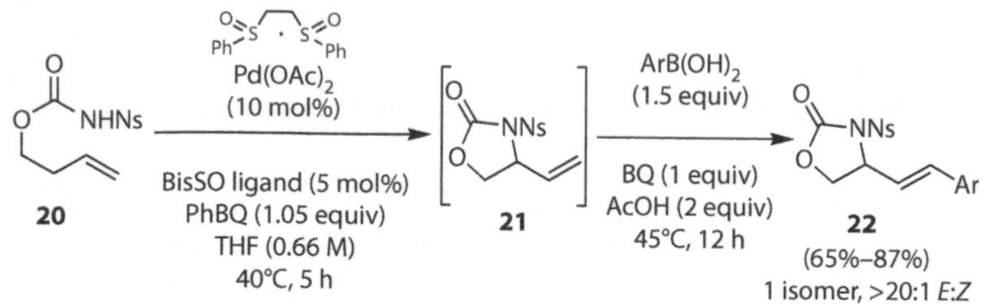

**SCHEME 3.22** Sequential allylic C—H amination/vinylic arylation.

an oxime directing group (Table 3.6).[111] Treatment of the listed oximes with Pd(OAc)$_2$ and BAIB afforded the desired *ortho*-acetoxylated arenes in good yields with high levels of regioselectivity. The scope of the method describes alternative substrates to the arenes presented including aliphatic and pthalimide examples. Functional group inter-conversion of the resulting acetoxylated products provides synthetic versatility.

**TABLE 3.6**

**O-Acetyl Oxime Directed C—H Acetoxylation**

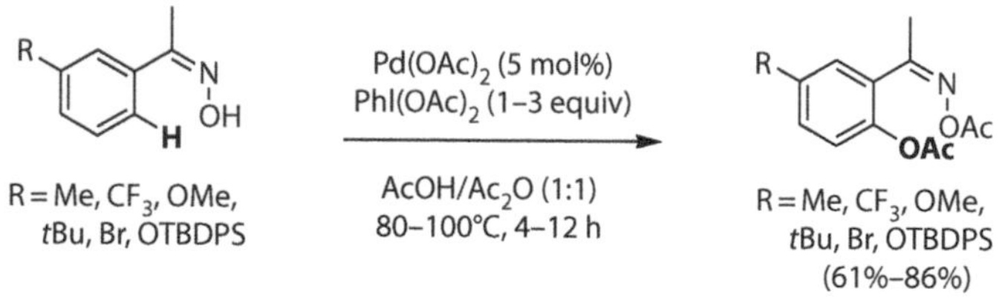

**Functional group interconversions**

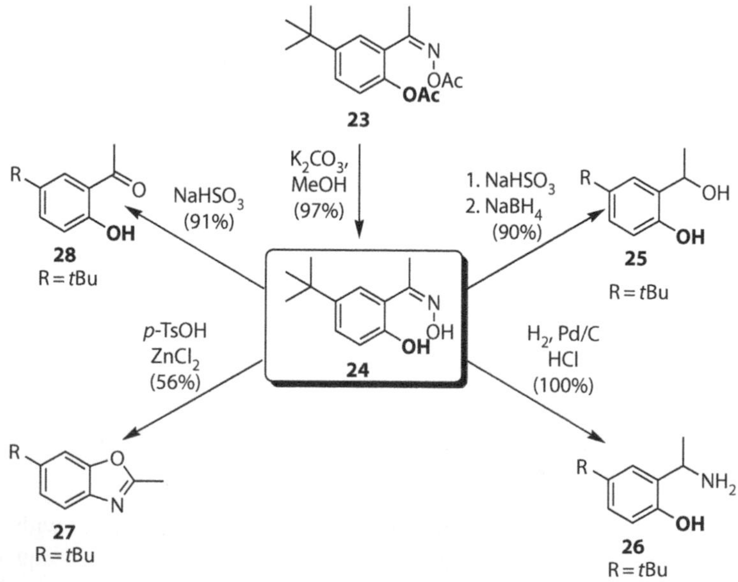

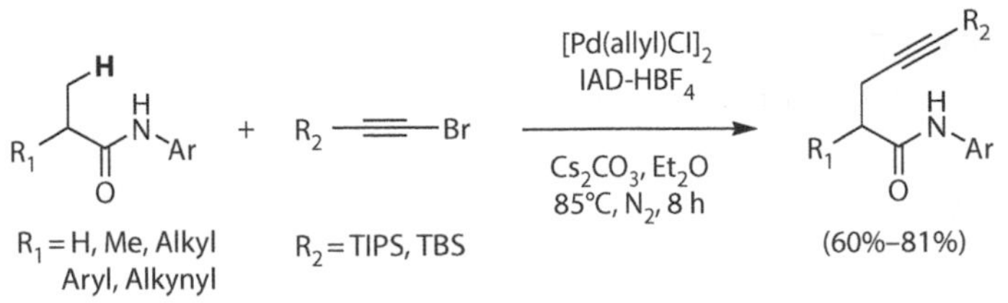

**SCHEME 3.23** Amide-directed intra-molecular alkynylation of $sp^3$ C—H bonds.

Access to ketones, alcohols, amines, and oxazoles can be achieved in a straightforward manner. Acetoxylation of unactivated $sp^3$ C—H bonds has also been demonstrated by Sanford with the use of an $O$-methyl oxime directing group.[112]

Yu has used aliphatic amides in conjunction with $Pd^0/N$-heterocyclic carbene complexes to activate $sp^3$ hybridized carbons of $\beta$-amides for functionalization as the corresponding alkyne without the use of a co-oxidant (Scheme 3.23).[113] High levels of chemoselectivity and mono alkynylation were observed providing access to synthetically useful alkynyl amides.

Arylation of unactivated $sp^3$-hybridized carbons using Pd-catalysis has been an area of intense focus.[114] Inter-molecular arylation methods have focused on the use of a resident directing group to influence C—H functionalization. Amides have been widely utilized toward this objective for the preparation of $sp^3$-aryl bonds.[115] Daugulis has demonstrated the ability to directly arylate picolinamides (Table 3.7).

Yu has leveraged aryl amides for directed arylation. Treatment of the listed amide with $Pd(TFA)_2$ facilitates a regioselective $\beta$-$sp^3$-C—H activation leading to the formation of arylated products (Scheme 3.24). The substrate scope for the listed transformation allows for a variety of substitutents on the $\alpha$- and $\beta$-carbons allowing for synthetic versatility. Efforts describing sulfonamide directing-groups have also been reported.[116]

The use of amide-directed $sp^3$-arylation has been successfully utilized in the asymmetric preparation of unnatural $\alpha$-amino acids (Scheme 3.25).[117] Treatment of electron-deficient phthalimide protected $\alpha$-amino amides with $Pd(TFA)_2$ and aryl iodides led the formation of novel phenylalanine derivatives in high yield and excellent chemoselectivity. The method displays robust tolerance for a multitude of substituted aromatics with varying electronic properties. Suppression of *bis*-arylation was excellent under the reported conditions. This method was also extended toward the preparation homoarylated phenylalanines affording novel aryl differentiated moieties with excellent diastereoselectivities.

Pd-mediated $\beta$-arylation of esters (Scheme 3.26),[118] $\alpha$-aminoesters,[119] and silylketene acetals[120] has been reported. The scope of the listed method was broad with respect to functionality on the aryl halide with the exception of weak alkyl donating substituents, which did not perform well. An extension to an asymmetric variant using a chiral davephos affords asymmetric products. Enantioselectivities were modest with 3:1 favoring the $R$-enantiomer being the best for the substrates screened. Asymmetric arylation of fused cyclopentanes via intra-molecular C—H functionalization using Pd(0) has been accomplished.[121]

**TABLE 3.7**
**Directed Arylation of Picolinamides**

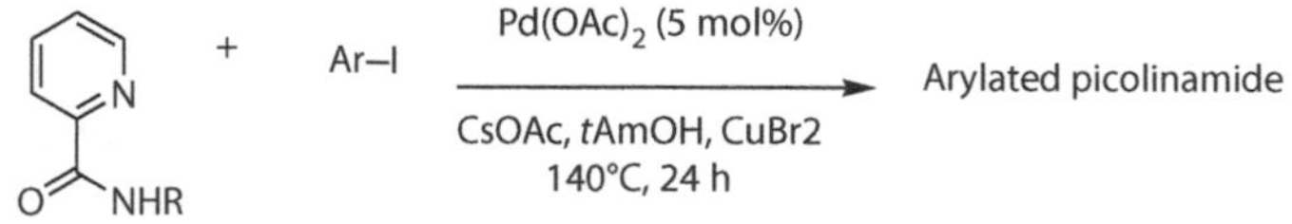

| Entry | Picolinamide | Aryl Iodide | Arylated Picolinamide | Yield (%) |
|-------|-------------|-------------|----------------------|-----------|
| 1 | 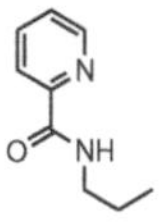 | 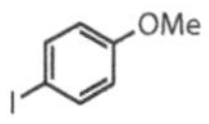 | 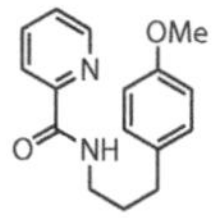 | 56 |
| 2 | 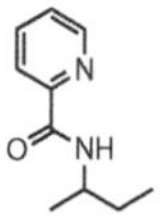 | 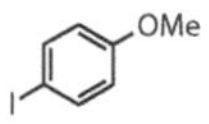 | 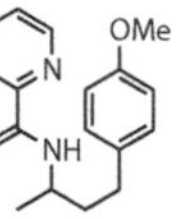 | 75 |
| 3 | 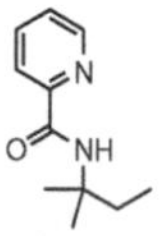 | 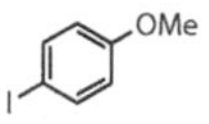 | 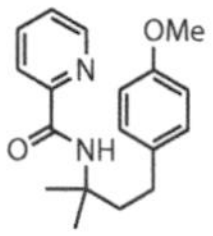 | 91 |
| 4 | 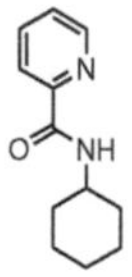 | 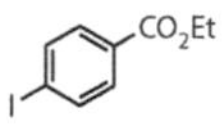 | 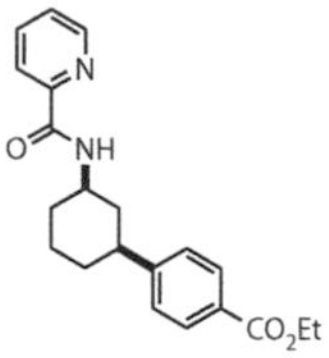 | 86 |

Approaches to the preparation of functionalized succinimides and their ring-opened analogs via intra-molecular carbonylation of fluoroamide-directed $sp^3$-hybridized C—C bonds represent an important step forward for Pd(II) catalysis and can provide direct access toward the preparation of differentiated 1,4-dicarbonyl compounds; a common structural motif in complex molecule synthesis (Scheme 3.27).[122]

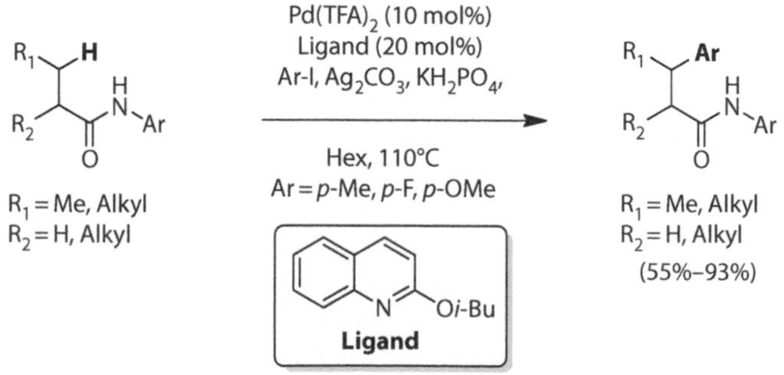

**SCHEME 3.24**   Directed arylation of aryl amides.

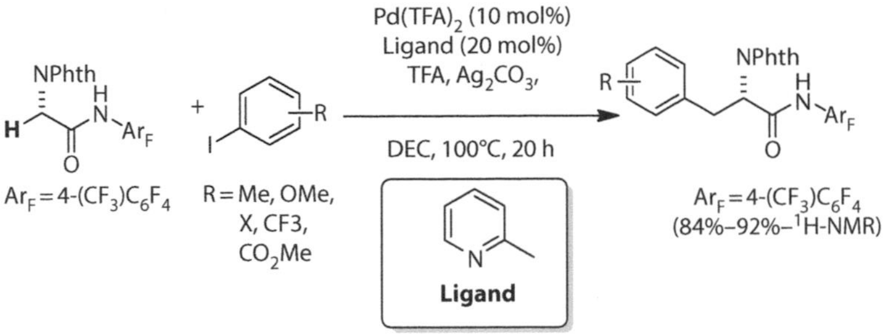

**SCHEME 3.25** Preparation of novel chiral α-amino acids.

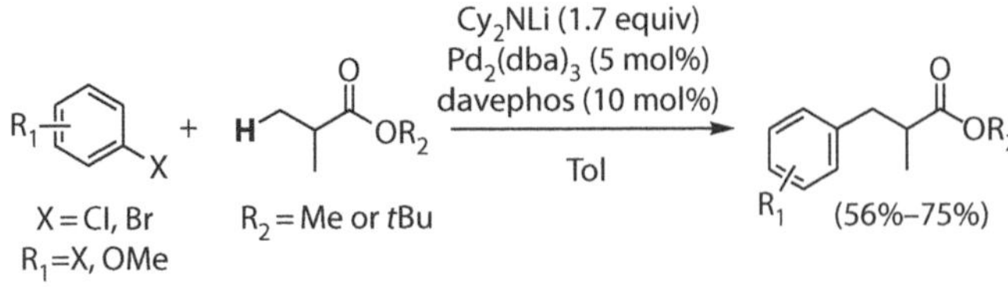

**SCHEME 3.26** Arylation of enolates with $Pd_2(dba)_3$.

The substrate scope of this method accommodated a variety of substituents α to the amide affording end products in good to excellent yields.

Intra-molecular cross-coupling experiments with monodentate directing groups on unactivated $sp^3$-hybridized carbons have been demonstrated using ethyl pyridines and aryl iodides[123] as well as hydroxamic acids coupled with boronic acids.[124]

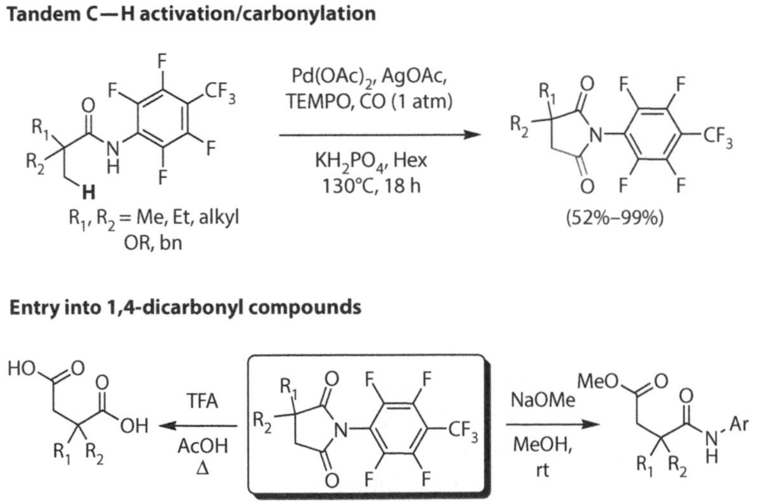

**Tandem C—H activation/carbonylation**

**Entry into 1,4-dicarbonyl compounds**

**SCHEME 3.27** Tandem C—H activation/carbonylation of amides.

**TABLE 3.8**

**Nucleophilic Fluorination with AgF**

Pd(OAc)$_2$ (10 mol%)
PhI(OPiv)$_2$, (2 equiv)
AgF (5 equiv)

MgSO$_4$ (2 equiv)
DCM, 60°C, 16 h

| Entry | R$_1$ | R$_2$ | Yield (%) |
|-------|-------|-------|-----------|
| 1 | H | H | 49 |
| 2 | NO$_2$ | H | 41 |
| 3 | CN | H | 70 |
| 4 | CO$_2$CH$_3$ | H | 59 |
| 5 | F | H | 30 |
| 6 | Br | H | 44 |
| 7 | H | Br | 42 |
| 8 | I | H | 55 |
| 9 | Ph | H | 39 |
| 10 | Me | H | 39 |

Fluorination of C—H bonds of 8-methylquinoline was described by Sanford using an oxidative pathway with an electrophilic fluoride source as well as using nucleophilic AgF in concert with an oxidant[125] via oxidative addition of a hypervalent PhI(OPiv)$_2$ to Pd(II), followed by a ligand substitution with the nucleophilic fluoride source, and subsequent reductive elimination from Pd(IV) to form the new C—F bond (Table 3.8). An alternative mechanistic proposal includes the oxidation of the cyclometal-lated Pd(II) with PhIF$_2$ formed *in situ* via ligand substitution of F⁻ on the prerequisite PhI(OPiv)$_2$.[126] This innovative approach provides a solution to the use of extremely cost-prohibitive sources of "F⁺."

Olefination of unactivated $sp^3$ bonds via C—H activation was demonstrated by Baudoin via oxidative addition to aryl bromides with Pd(0) followed by successive β-elimination and concomitant reductive elimination to afford functionalized ole-finic arenes.[127] Intra-molecular olefination of unactivated $sp^3$ C—H bonds through the corresponding cyclohexenylbromide[128] provided entry into the hexahydroindole core of the aeruginosin family of natural products (Table 3.9).

Further methods have been realized for olefination/cyclization toward nitrogen-containing heterocycles. Yu employed electron-deficient amides for C—H activa-tion using Pd(OAc)$_2$ followed by inter-molecular olefination with acrylates. The resulting α/β-unsaturated esters underwent spontaneous *in situ* 5-*exo*-trig cycli-zation to afford the cyclized Michael adduct.[129] All products were isolated as the BF$_4^-$ salt. Sanford expanded the substrate scope of this transformation and dem-onstrated that the olefination product or cyclic adduct could be afforded with pru-dent application of specific reaction conditions (Scheme 3.28).[130] Further synthetic manipulation of the cyclized product through the use of excess DBU affords the acyclic olefinic free base.

**TABLE 3.9**

**Preparation of Indanes via Intra-Molecular $sp^3$ C—H Olefination**

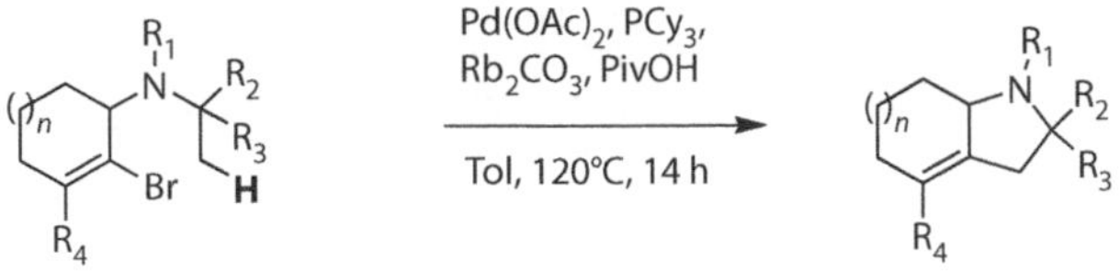

| Entry | Bromoalkene | Product | Yield (%) |
|-------|-------------|---------|-----------|
| 1 | | | 73 |
| | | (dr = 95:5) | |
| 2 | 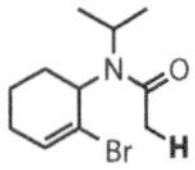 | 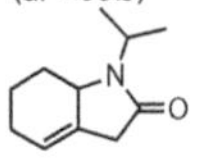 | 83 |
| 3 | 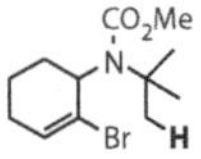 | 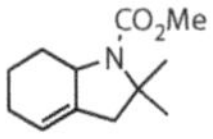 | 86 |
| 4 | 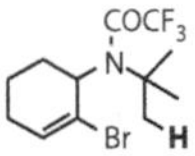 | 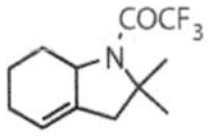 | 78 |
| 5 | 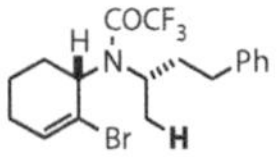 | 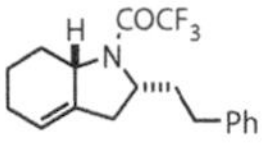 | 71 |
| 6 | 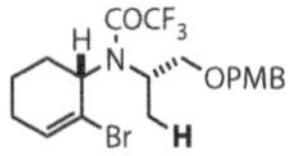 | 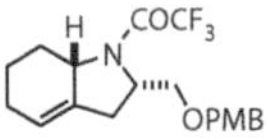 | 84 |

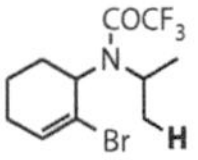

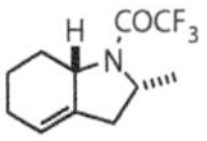

C—H activation of $sp^3$ bonds is not limited to arylation, olefination, or oxidative methods. Recently, Baudoin has leveraged Pd(0)-mediated C—H functionalization toward the construction of benzocyclobutenes through oxidative addition to aryl halides followed by intra-molecular carbopalladation leading to a five-membered palladacycle which reductively eliminates to afford the listed benzocyclobutene[131]

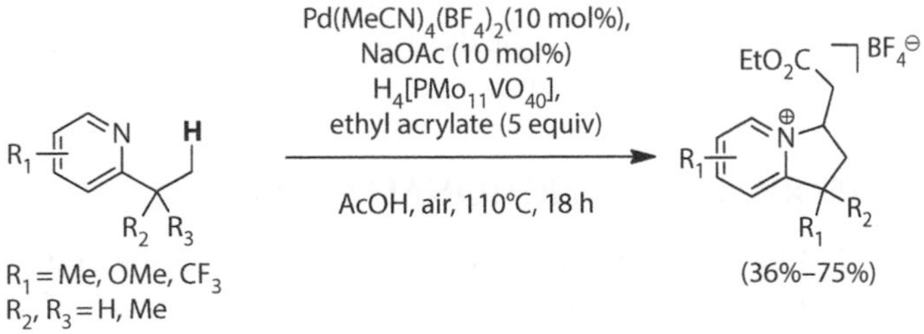

**SCHEME 3.28** Tandem C—H olefination/conjugate addition.

**TABLE 3.10**

**Preparation of Functionalized Benzocyclobutenes via sp³ C—H Functionalization/Intra-Molecular Coupling**

$R_1, R_2, R_3 =$ H, Me, OMe
CN, $CO_2R$

| Entry | Bromoarene | Product | Yield (%) |
|---|---|---|---|
| 1 | | | 78 |
| 2 | | | 85 |
| 3 | | | 77 |
| 4 | | | 87 |
| 5 | | | 62 |

(Table 3.10). Construction of indanes, indanones, dihydrobenzofurans, and azain-dolines from aryl chloride precursors via intra-molecular $sp^3$-arylation has also been communicated.[132]

This method in the content of microwave-assisted cycloaddition reactions with relevant dienophiles via electrocyclic ring-opening which facilitates generation of a reactive diene **31** for an inter-molecular Diels–Alder [4 + 2] cycloaddition (Scheme 3.29) has also been demonstrated.[133] The overall scope of this method is broad and amenable to a variety of functional group substitutions on the aromatic as well as the cyclobutane moiety. The reported yields range from modest to excellent and the overall diastereoselectivity is impressive.

## 3.7 ASYMMETRIC C—H FUNCTIONALIZATION

Progressive C—H functionalization methods have focused on the ability to tune a specific optically pure ligand to a discrete substrate for the purposes of asymmetric catalysis.[134] Yu has demonstrated the ability to use chiral mono-*N*-protected amino

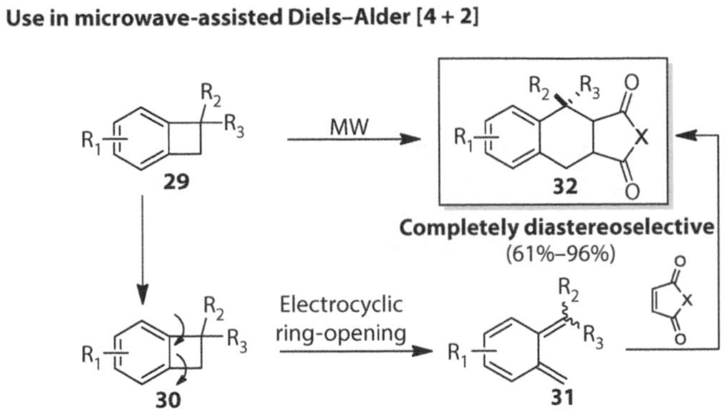

**Use in microwave-assisted Diels–Alder [4 + 2]**

**SCHEME 3.29** Microwave-assisted inter-molecular Diels–Alder reaction of benzocyclobutenes.

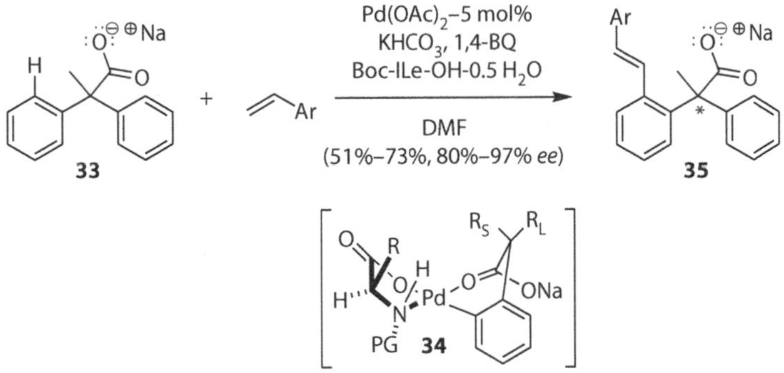

**SCHEME 3.30** Enantioselective C—H olefination of diphenylacetic acids.

acid ligands in an efficient desymmetrization of prochiral diphenylacetic acids (Scheme 3.30).[135]

Extension to the use of acrylates afforded the desired olefination products as a mixture of the desired alkene and intra-molecular alkoxy-Michael addition products in high *ee*. Analogous methodology has been extended to the asymmetric intra-molecular lactonization of diphenylacetic acids toward benzofuranones[136] as well as the enantioselective iodination of diaryl pyridines[137] and methylamines.[138] Asymmetric C—H activation of cyclopropanes has also been reported.[139]

## 3.8 APPLICATIONS IN COMPLEX MOLECULE SYNTHESIS

C—H bond functionalization has not only revolutionized the construction of synthons, but also completely transformed retrosynthetic analysis of complex molecules leading to improvements in efficiency, minimization of oxidation-state transformations, and access to privileged scaffolds at a rapid rate.[140] This section highlights

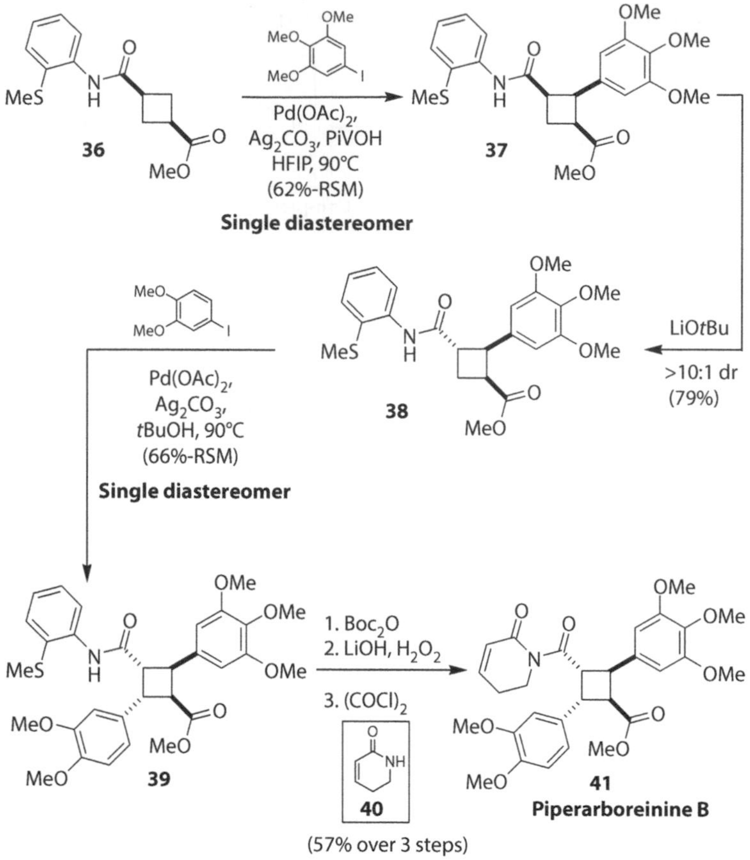

**SCHEME 3.31** Baran's total synthesis of piperarboreine B.

several recent reports of complex molecule synthesis in which one or more critical C—C bond forming steps utilized a Pd-catalyzed C—H functionalization.

Baran successfully employed a regioselective substrate-controlled sequential arylation strategy via Pd-mediated C—H functionalization toward the construction of the piperarborenine cyclobutane containing family of alkaloids[141] (Scheme 3.31). At the time of publication this was the first example of *sp³* hybridized arylation of the cyclobutane ring. *Cis*-disubstituted cyclobutane **36** underwent a regio- and completely diastereoselective *sp³* C—H arylation with the listed aryl iodide to afford advanced intermediate **37**. Epimerization of the resident amide followed by a successive *sp³* C—H arylation afforded tetra-substituted cyclobutane **39**.

Protection of the amide-*N* with Boc₂O followed by oxidative hydrolysis and lactamization afforded piperarboreinine B (**41**) in excellent yield over three steps from **39**.

The diversity of C—H functionalization methods currently available allows for rapid analog preparation of natural products. Baran's total synthesis (+)-hongoquercin A[142] leveraged a ligand-accelerated regioselective methylation of functionalized

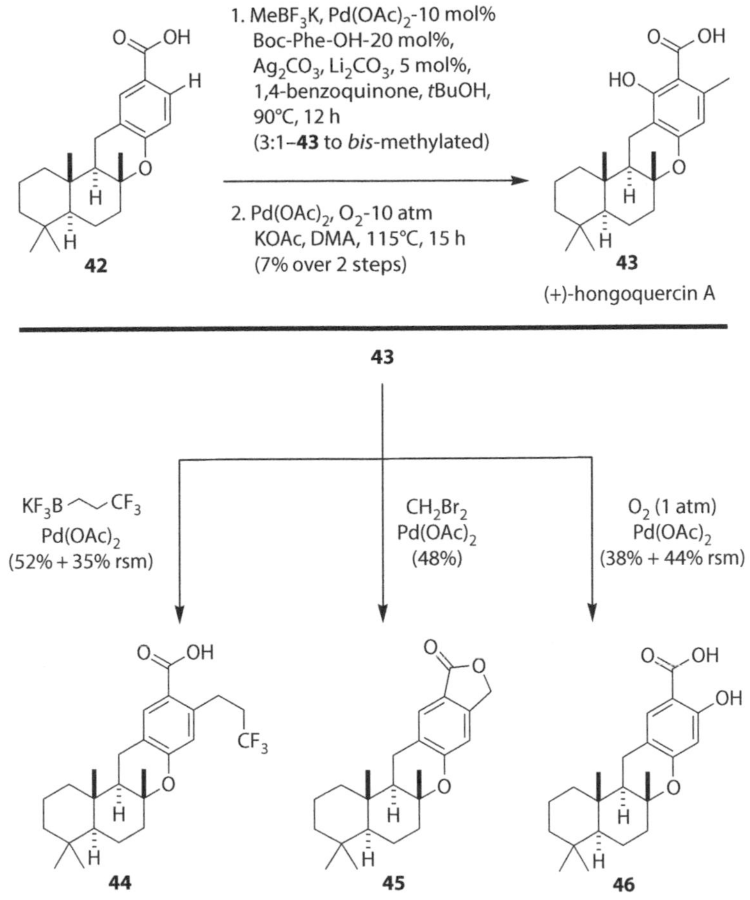

**SCHEME 3.32** Baran's hongoquercin A synthesis and analog preparation.

arene **42** followed by subsequent functional group inter-conversions toward the preparation of the desired target (**43**) (Scheme 3.32). Advanced intermediate **42** afforded a scaffold for analog preparation via Pd-catalyzed C−H functionalization. Intramolecular lactonization, as well as regioselective oxidation, alkylation, amination, and arylation were all successful.

The construction of the serine-threonine phosphatase inhibiting *bis*-indole alkaloid dragmacidin D[143] highlighted the convergent efficiency of which natural products can be constructed using strategic C−H functionalization (Scheme 3.33). Arylation of tosylindole **47** with unactivated 3-silyloxythiophene **48** using Pd(II) catalysis in concert with an electron-deficient ligand regioselectively afforded the desired coupling product **49**.

Functional group inter-conversion followed by C−H/C−H coupling of pyrazinone *N*-oxide with indole **51** afforded advanced intermediate **52** which was converted to the desired pyrazinone for a subsequent C−H/C−H coupling with indole **53**. Completion of the natural product synthesis leveraged three direct C−H

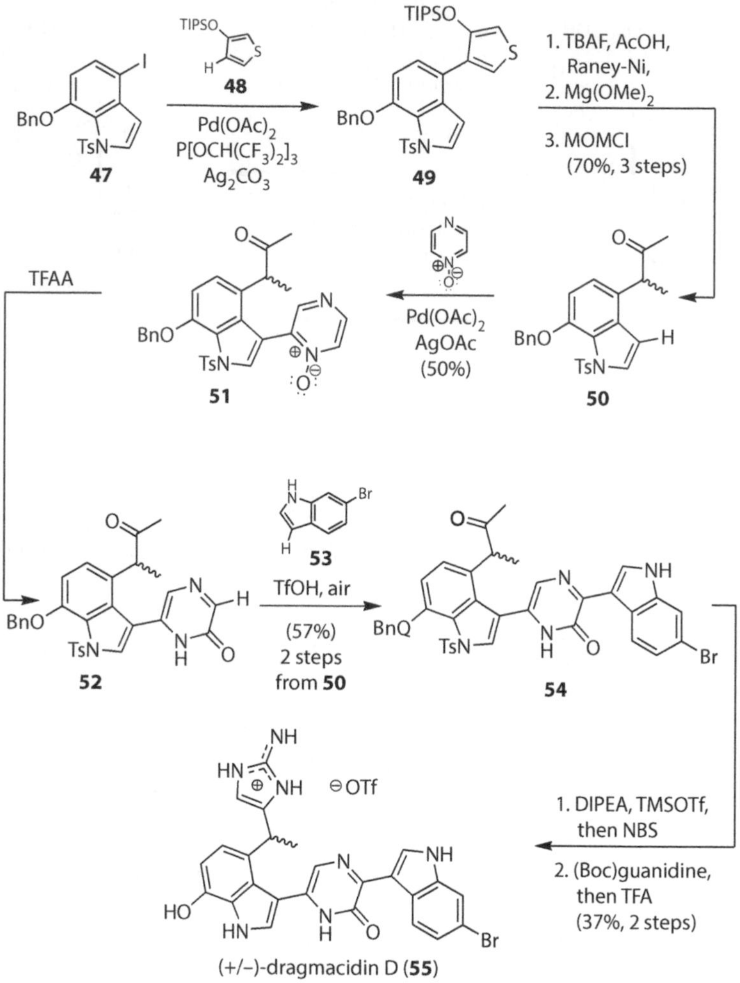

**SCHEME 3.33**   Itami's dragmacidin D total synthesis.

functionalizations toward the completion of (+/−)-dragmacidin D (**55**) in 37% yield over two steps from **54**.

Late-stage ligand-accelerated inter-molecular olefination via Pd-mediated C—H activation was utilized toward the convergent construction of the biologically relevant aromatic natural product (+)-lithospermic acid (**58**) (Scheme 3.34)[144]. Subsequent functionalization completed the natural product in efficient 12 steps for the longest linear sequence in 11% overall yield from commercially available *o*-eugenol.

The synthesis of macrolide natural products has captured the attention and inspired significant research effort over many decades. Compounds in this class frequently display potent biological activity. Often, completion of the target requires a late stage macrocyclization event between an activated carboxylic acid derivative and

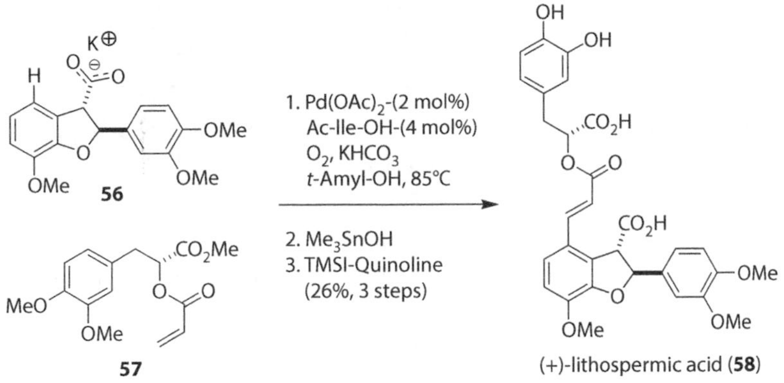

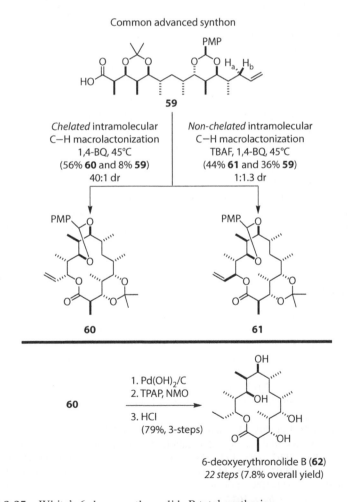

**SCHEME 3.34** Yu's (+)-lithospermic acid total synthesis.

**SCHEME 3.35** White's 6-deoxyerythronolide B total synthesis.

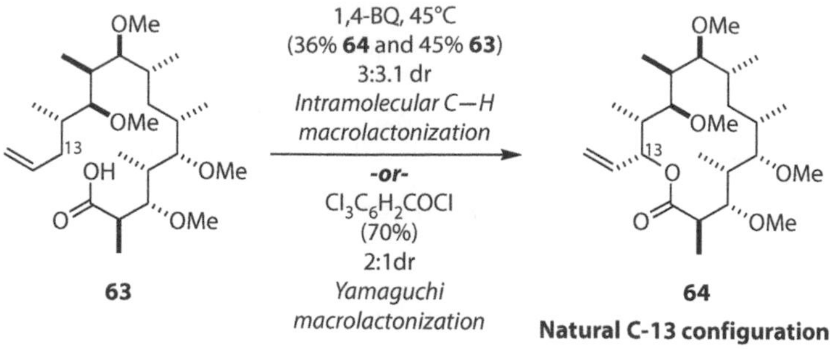

**SCHEME 3.36**   Macrolactonization of erythromycin analogs.

subsequent intra-molecular trap with a resident alcohol via the familiar Yamaguchi, Keck, Corey–Nicolaou, and related procedures.[145] In 2006 White demonstrated that treatment of linear alkenoic acids with palladium *bis*-phenylsulfinylethane diacetate could afford macrolactones ring sizes of 14 to 19 by oxidative cyclization via allylic C—H activation.[146] An innovative approach to macrocyclization was showcased by White in the total synthesis of 6-deoxyerythronolide B (Scheme 35).[147] A late-stage chelate assisted allylic C—H oxidation of synthon **59** was employed to complete macrolactone in 64% overall yield based on recovered starting material with a substrate-controlled diastereoselectivity of 40:1 for the listed isomer.

Subsequent efforts by White have proven that resident conformational rigidity of the acyclic polypropionate precursor **59** was not required to complete macrolactonization as previously postulated (Scheme 3.36).[148] The presence of conformationally biasing elements does, however, play a significant role in obtaining high-levels of diastereocontrol in the cyclization event. This important experimental discovery challenged over 30 years of previous work on the total synthesis of erythromycin which postulated that resident rigidity of the acyclic precursor was intimately associated with the fidelity of the ring closure.

## REFERENCES

1. Balcells, D.; Clot, E.; Eisenstein, O. *Chem. Rev.* **2010**, *110*, 749–823.
2. Newhouse, T.; Baran, P. S. *Angew. Chem. Int. Ed.* **2011**, *50*, 3362–3374.
3. Neufeldt, S. R.; Sanford, M. S. *Acc. Chem. Res.* **2012**, *45*, 936–946.
4. Chen, X.; Engle, K. M.; Wang, D.-H.; Yu, J.-Q. *Angew. Chem. Int. Ed.* **2009**, *48*, 5094–5115.
5. White, M. C. *Synlett.* **2012**, *23*, 2746–2748.
6. Dai, H.-X.; Stepan, A. F.; Plummer, M. S.; Zhang, Y.-H.; Yu, J.-Q. *J. Am. Chem. Soc.* **2011**, *133*, 7222–7228.
7. Thansandote, P.; Lautens, M. *Chem. Eur. J.* **2009**, *15*, 5874–5883.
8. For recent reviews on Pd catalyzed arylation see: a) Alberico, D.; Scott, M. E.; Lautens, M. *Chem. Rev.* **2007**, *107*, 174–238. b) Campeau, L.-C.; Stuart, D. R.; Fagnou, K. *Aldrichim. Acta.* **2007**, *40*, 35–44. c) Daugulis, O.; Do, H.-Q.; Shabashov, D. *Acc. Chem. Res.* **2009**, *42*, 1074–1086. For mechanistic studies see: Gorelsky, S. I.; Lapointe, D.; Fagnou, K. *J. Org. Chem.* **2012**, *77*, 658–668.

9. For early work on Pd$^o$ arylation see: a) Hennings, D. D.; Iwasa, S.; Rawal, V. H. *J. Org. Chem.* **1997**, *62*, 2–3. b) Satoh, T.; Kawamura, Y.; Miura, M.; Nomura, M. *Angew. Chem. Int. Ed.* **1997**, *109*, 1740–1742. c) Satoh, T.; Kawamura, Y.; Miura, M.; Nomura, M. *Angew. Chem. Int. Ed.* **1997**, *109*, 1820–1822.

10. For mechanistic work see: Deprez, N. R.; Sanford, M. S. *J. Am. Chem. Soc.* **2009**, *131*, 11234–11241. Daugulis, O.; Zaitsev, V. G. *Angew. Chem. Int. Ed.* **2005**, *44*, 4046–4048. For use in tandem processes see: Shabashov, D.; Daugulis, O. *J. Org. Chem.* **2007**, *72*, 7720–7725.

11. Chiong, H. A.; Pham, Q.-N.; Daugulis, O. *J. Am. Chem. Soc.* **2007**, *129*, 9879–9884.

12. a) Shabashov, D.; Daugulis, O. *Org. Lett.* **2006**, *8*, 4947–4949. b) Lazareva, A.; Daugulis, O. *Org. Lett.* **2006**, *8*, 5211–5213. c) Shabashov, D.; Molina-Maldonado, J. R.; Daugulis, O. *J. Org. Chem.* **2008**, *73*, 7818–7821.

13. Yeung, C. S.; Zhao, X.; Borduas, N.; Dong, V. M. *Chem. Sci.* **2010**, *1*, 331–336.

14. Yeung, C. S.; Dong, V. M. *Synlett.* **2011**, *7*, 974–978.

15. Zhao, X.; Yeung, C. S.; Dong, V. M. *J. Am. Chem. Soc.* **2010**, *132*, 5837–5844.

16. Wang, X.; Leow, D.; Yu, J.-Q. *J. Am. Chem. Soc.* **2011**, *133*, 13864–13867.

17. René, O.; Fagnou, K. *Org. Lett.* **2010**, *12*, 2116–2119.

18. a) Lafrance, M.; Rowley, C. N.; Wook, T. K.; Fagnou, K. *J. Am. Chem. Soc.* **2006**, *128*, 8754–8756. b) Lafrance, M.; Fagnou, K. *J. Am. Chem. Soc.* **2006**, *128*, 16496–16497. c) Caron, L.; Campeau, L.-C.; Fagnou, K. *Org. Lett.* **2008**, *10*, 4533–4536. For intramolecular cases see: Lafrance, M.; Lapointe, D.; Fagnou, K. *Tetrahedron.* **2008**, *64*, 6015–6020.

19. Wang, D.-H.; Mei, T.-S.; Yu, J.-Q. *J. Am. Chem. Soc.* **2008**, *130*, 17676–17677.

20. Pintori, D. G.; Greaney, M. F. *Org. Lett.* **2011**, *13*, 5713–5715.

21. a) Kalyani, D.; Deprez, N. R.; Desai, L. V.; Sanford, M. S. *J. Am. Chem. Soc.* **2005**, *127*, 7330–7331. b) Deprez, N. R.; Sanford, M. S. *J. Am. Chem. Soc.* **2009**, *131*, 11234–11241.

22. Hull, K. L.; Lanni, E. L.; Sanford, M. S. *J. Am. Chem. Soc.* **2006**, *128*, 14047–14049.

23. Hull, K. L.; Sanford, M. S. *J. Am. Chem. Soc.* **2007**, *129*, 11904–11905. For mechanistic studies see: a) Kalyani, D.; Deprez, N. R.; Desai, L. V.; Sanford, M. S. *J. Am. Chem. Soc.* **2005**, *127*, 7330–7331. b) Hull, K. L.; Sanford, M. S. *J. Am. Chem. Soc.* **2009**, *131*, 9651–9653. c) Lyons, T. W.; Hull, K. L.; Sanford, M. S. *J. Am. Chem. Soc.* **2011**, *133*, 4455–4464.

24. Tredwell, M. J.; Gulias, M.; Bremeyer, N. G.; Johansson, C. C. C.; Collins, B. S. L.; Gaunt, M. J. *Angew. Chem. Int. Ed.* **2011**, *50*, 1076–1079.

25. Wan, L.; Dastbaravardeh, N.; Li, G.; Yu, J.-Q. *J. Am. Chem. Soc.* **2013**, *135*, 18056–18059.

26. Engle, K. M.; Thuy-Boun, P. S.; Dang, M.; Yu, J.-Q. *J. Am. Chem. Soc.* **2011**, *133*, 18183–18193. Wasa, M.; Chan, K. S. L.; Yu, J.-Q. *Chem. Lett.* **2011**, *40*, 1004–1006.

27. Engle, K. M.; Yu, J.-Q. *J. Am. Chem. Soc.* **2013**, *78*, 8927–8955.

28. Stuart, D. R.; Fagnou, K. *Science.* **2007**, *316*, 1172–1175.

29. For a recent review on Pd-mediated C—C bond formation under the aforementioned protocols see: Jana, R.; Pathak, T.; Sigman, M.; *Chem. Rev.* **2011**, *111*, 1417–1492.

30. Wasa, M.; Worrell, B. T.; Yu, J.-Q. *Angew. Chem. Int. Ed.* **2010**, *49*, 1275–1277. Liégault, B.; Petrov, I.; Gorelsky, S. I.; Fagnou, K. *J. Org. Chem.* **2010**, *75*, 1047–1060.

31. For a recent review see: Mei, T.-S.; Kou, L.; Ma, S.; Engle, K. M.; Yu, J.-Q. *Synthesis.* **2012**, *44*, 1778–1791. For mechanistic studies see: Gorelsky, S. I.; Lapointe, D.; Fagnou, K. *J. Am. Chem. Soc.* **2008**, *130*, 10848–10849.

32. Ye, M.; Edmunds, A. J. F.; Morris, J. A.; Sale, D.; Zhang, Y.; Yu, J.-Q. *Chem. Sci.* **2013**, *4*, 2374–2379.

33. Hattori, K.; Yamaguchi, K.; Yamaguchi, J.; Itami, K. *Tetrahedron.* **2012**, *68*, 7605–7612.

34. Deprez, N. R.; Kalyani, D.; Krause, A.; Sanford, M. S. *J. Am. Chem. Soc.* **2006**, *128*, 4972–4973.

35. Ye, M.; Gao, G.-L.; Edmunds, A. J. F.; Worthington, P. A.; Morris, J. A.; Yu, J.-Q. *J. Am. Chem. Soc.* **2011**, *133*, 19090–19093.

36. Wagner, A. M.; Sanford, M. S. *Org. Lett.* **2011**, *13*, 288–291.

37. For C-2 arylation see: Battace, A.; Lemhadri, M.; Zair, T.; Doucet, H.; Santelli, M. *Organometallics*. **2007**, *26*, 472–474. For C-4 arylation see: Gottumukkala, A. L.; Doucet, H. *Adv. Synth. Catal.* **2008**, *350*, 2183–2188.

38. Tani, S.; Uehara, T. N.; Yamaguchi, J.; Itami, K. *Chem. Sci.* **2014**, *5*, 126–135. Roger, J.; Požgan, F.; Doucet, H. *J. Org. Chem.* **2009**, *74*, 1179–1186. For benzylation of thiazoles see: Lapointe, D.; Fagnou, K. *Org. Lett.* **2009**, *11*, 4160–4163. For thiazole C—H arylation in the context of medicinally relevant structures see: Sekizawa, H.; Amaike, K.; Itoh, Y.; Suzuki, T.; Itami, K.; Yamaguchi, J. *ACS Med. Chem. Lett.* **2014**, *5*, 582–586.

39. Liégault, B.; Lapointe, D.; Caron, L.; Vlassova, A.; Fagnou, K. *J. Org. Chem.* **2009**, *74*, 1826–1834.

40. a) Chiong, H. A.; Daugulis, O. *Org. Lett.* **2007**, *9*, 1449–1451. b) Nadres, E. T.; Lazareva, A.; Daugulis, O. *J. Org. Chem.* **2011**, *76*, 471–483. c) Derridj, F.; Roger, J.; Geneste, F.; Djebbar, S.; Doucet, H. *J. Organomet. Chem.* **2009**, *694*, 455–465. d) Roy, D.; Mom, S.; Beaupérin, M.; Doucet, H.; Hierso, J.-C. *Angew. Chem. Int. Ed.* **2010**, *49*, 6650–6654.

41. a) Campeau, L.-C.; Parisien, M.; Leblanc, M.; Fagnou, M. *J. Am. Chem. Soc.* **2004**, *126*, 9186–9187. b) Campeau, L.-C.; Fagnou, K. *Chem. Commun.* **2006**, *42*, 1253–1264. c) Nadres, E. T.; Daugulis, O. *J. Am. Chem. Soc.* **2012**, *134*, 7–10.

42. a) Mariampillai, B.; Alliot, J.; Mengzhou, L.; Lautens, M. *J. Am. Chem. Soc.* **2007**, *129*, 15372–15379. b) Blaszykowski, C.; Aktoudianakis, E.; Alberico, D.; Bressy, C.; Hulcoop, D. G.; Jafarpour, F.; Joushaghani, A.; Laleu, B.; Lautens, M. *J. Org. Chem.* **2008**, *73*, 1888–1897. c) Bryan, C. S.; Lautens, M. *Org. Lett.* **2008**, *10*, 4633–4636. d) Panteleev, J.; Geyer, K.; Aquilar-Aguilar, A.; Wang, L.; Lautens, M. *Org. Lett.* **2010**, *12*, 5092–5095. e) Nicolaus, N.; Franke, P. T.; Lautens, M. *Org. Lett.* **2011**, *13*, 4236–4239. f) Schulman, J. M.; Friedman, A. A.; Panteleev, J.; Lautens, M. *Chem. Commun.* **2012**, *48*, 55–57. g) Liégault, B.; Petrov, I.; Gorelsky, S. I.; Fagnou, K. *J. Org. Chem.* **2010**, *75*, 1047–1060. h) Lapointe, D.; Markiewicz, T.; Whipp, C. J.; Toderian, A.; Fagnou, K. *J. Org. Chem.* **2011**, *76*, 749–759.

43. Hwang, S. J.; Cho, S. H.; Chang, S. *J. Am. Chem. Soc.* **2008**, *130*, 16158–16159.

44. Zhang, F.; Greaney, M. F. *Angew. Chem. Int. Ed.* **2010**, *49*, 2768–2771.

45. Zhang, F.; Greaney, M. F. *Org. Lett.* **2010**, *12*, 4745–4747.

46. a) Roger, J.; Požgan, F.; Doucet, H. *Adv. Synth. Catal.* **2010**, *352*, 696–710. b) Derridj, F.; Roger, J.; Djebbar, S.; Doucet, H. *Org. Lett.* **2010**, *12*, 432–4323. c) Bheeter, C. B.; Bera, J. K.; Doucet, H. *J. Org. Chem.* **2011**, *76*, 6407–6413. d) Chen, L.; Roger, J.; Bruneau, C.; Dixneuf, P. H.; Doucet, H. *Chem. Commun.* **2011**, *47*, 1872–1874.

47. Yuan, K.; Doucet, H. *Chem. Sci.* **2014**, *6*, 392–396. Ueda, K.; Yanagisawa, Y.; Yamaguchi, J.; Itami, K. *Angew. Chem. Int. Ed.* **2010**, *49*, 8946–8949.

48. Yanagisawa, S.; Itami, K. *Tetrahedron.* **2011**, *67*, 4425–4430. Fagnou has demonstrated α-selective arylation of thiophenes with alternative electron-deficient phosphines. René, O.; Fagnou, K. *Adv. Synth. Catal.* **2010**, *352*, 2116–2120.

49. Yanagisawa, S.; Ueda, K.; Sekizawa, H.; Itami, K. *J. Am. Chem. Soc.* **2009**, *131*, 14622–14623.

50. For mechanistic studies see: Sun, H.-Y.; Gorelsky, S. I.; Stuart, Campeau, L.-C.; Fagnou, K. *J. Org. Chem.* **2010**, *75*, 8180–8189. Tan, Y.; Barrios-Landeros, F.; Hartwig, J. F. *J. Am. Chem. Soc.* **2012**, *134*, 3683–3686.

51. Campeau, L.-C.; Bertrand-Laperle, M.; Leclerc, J.-P.; Villemure, E.; Gorelsky, S.; Fagnou, K. *J. Am. Chem. Soc.* **2008**, *130*, 3276–3277. Campeau, L.-C.; Stuart, D. R.; Leclerc, J.-P.; Bertrand-Laperle, M.; Villemure, E.; Sun, H.-Y.; Lasserre, S.; Gulmond, N.; Lecavallier, M.; Fagnou, K. *J. Am. Chem. Soc.* **2009**, *131*, 3291–3306.

52. Schipper, D. J.; El-Salfiti, M.; Whipp, C. J.; Fagnou, K. *Tetrahedron.* **2009**, *65*, 4977–4983. Campeau, L.-C.; Fagnou, K. *Org. Synth.* **2011**, *88*, 22–32.

53. Campeau, L.-C.; Schipper, D. J.; Fagnou, K. *J. Am. Chem. Soc.* **2008**, *130*, 3266–3267.

54. For mechanistic work see: Baxter, R. D.; Sale, D.; Engle, K. M.; Yu, J.-Q.; Blackmond, D. G. *J. Am. Chem. Soc.* **2012**, *134*, 4600–4606.

55. Zhang, Y.-H.; Shi, B.-F.; Yu, J.-Q. *J. Am. Chem. Soc.* **2009**, *131*, 5072–5074.

56. Zaitsev, V. g.; Daugulis, O. *J. Am. Chem. Soc.* **2005**, *127*, 4156–4157.

57. Wang, D.-H.; Engle, K. M.; Shi, B.-F. *Science.* **2010**, *327*, 315–319. For use of pyridine ligands in C—H olefination see: Kubota, A.; Emmert, M. H.; Sanford, M. S. *Org. Lett.* **2012**, *14*, 1760–1763.

58. Lu, Y.; Wang, D.-H.; Engle, K. M.; Yu, J.-Q. *J. Am. Chem. Soc.* **2010**, *132*, 5916–5921.

59. Huang, C.; Chattopadhyay, B.; Gevorgyan, V. *J. Am. Chem. Soc.* **2011**, *133*, 12406–12409.

60. Dai, H.-X.; Li, G.; Zhang, X.-G.; Stepan, A. F.; Yu, J.-Q. *J. Am. Chem. Soc.* **2013**, *135*, 7567–7571.

61. For alternative *ortho*-regioselective conditions see: Li, G.; Leow, D.; Wan, L.; Yu, J.-Q. *Angew. Chem. Int. Ed.* **2013**, *52*, 1245–1247.

62. For computational work see: Yang, Y.-F.; Cheng, G.-J.; Peng, L.; Leow, D.; Sun, T.-Y.; Chen, P.; Zhang, X.; Yu, J-Q.; Wu, Y.-D.; Houk, K. N. *J. Am. Chem. Soc.* **2014**, *136*, 344–355. Leow, D.; Li, G.; Mei, T.; Yu, J.-Q. *Nature.* **2013**, *486*, 518–522. Truong, T.; Daugulis, O. *Angew. Chem. Int. Ed.* **2012**, *51*, 11677–11679.

63. a) Campeau, L.-C.; Rousseaux, S.; Fagnou, K. *J. Am. Chem. Soc.* **2005**, *127*, 18020–18021. b) Kanyiva, K. S.; Nakao, Y.; Hiyama, T. *Angew. Chem. Int. Ed.* **2007**, *46*, 8872–8874. c) Cho, S. H.; Hwang, S. J.; Chang, S. *J. Am. Chem. Soc.* **2008**, *130*, 9254–9256. d) Wu, J.; Cui, X.; Chen, L.; Jiang, G.; Wu, Y. *J. Am. Chem. Soc.* **2009**, *131*, 13888–13889. e) Xi, P.; Yang, F.; Qin, S.; Zhao, C.; Lan, J.; Gao, G.; Hu, C.; You, J. *J. Am. Chem. Soc.* **2010**, *132*, 1822–1824. f) Larivée, A.; Mousseau, J. J.; Charette, A. B. *J. Am. Chem. Soc.* **2008**, *130*, 52–54.

64. Gürbüz, N.; Özdemir, L.; Çetinkaya, B. *Tetrahedron. Lett.* **2005**, *46*, 2273–2277. Wasa, M.; Worrell, B. T.; Yu, J.-Q. *Angew. Chem. Int. Ed.* **2010**, *49*, 1275–1277.

65. Ye, M.; Gao, G.-L.; Yu, J.-Q. *J. Am. Chem. Soc.* **2011**, *133*, 6964–6967.

66. Beck, E. M.; Grimster, N. P.; Hatley, R.; Gaunt, M. J. *J. Am. Chem. Soc.* **2006**, *128*, 2528–2529.

67. René, O.; Lapointe, D.; Fagnou, K. *Org. Lett.* **2009**, *11*, 4560–4563. For other examples of similar tandem C—H functionalization processes see: a) Ruck, R. T.; Huffman, M. A.; Kim, M. M.; Shevlin, M.; Kandur, W. V.; Davies, I. W. *Angew. Chem. Int. Ed.* **2008**, *47*, 4711–4714. b) Satyanarayana, G.; Maichle-Mössmer, C.; Maier, M. E. *Chem. Commun.* **2009**, *12*, 1571–1573. c) Hu, Y.; Yu, C.; Ren, D.; Hu, Q.; Zhang, L.; Cheng, D. *Angew. Chem. Int. Ed.* **2009**, *48*, 5448–5451. d) Beaudoin, M.; Wolfe, J. P. *Org. Lett.* **2007**, *9*, 3073–3075.

68. Engle, K. M.; Wang, D.-H.; Yu, J.-Q. *Angew. Chem. Int. Ed.* **2010**, *49*, 6169–6173.

69. Li, S.; Chen, G.; Feng, C.-G.; Gong, W.; Yu, J.-Q. *J. Am. Chem. Soc.* **2014**, *136*, 5267–5270.

70. Lee, J. M.; Ahn, D.-S.; Jung, D. Y.; Lee, J.; Do, Y.; Kim, S. K.; Chang, S. *J. Am. Chem. Soc.* **2006**, *128*, 12954–12962.

71. For reviews on ligand-directed C—H functionalization see: Lyons, T. W.; Sanford, M. S. *Chem. Rev.* **2010**, *110*, 1147–1169. Engle, K. M.; Mei, T.-S.; Wasa, M.; Yu, J.-Q. *Acc. Chem. Res.* **2012**, *45*, 788–802.

72. Chen, X.; Goodhue, C. E.; Yu, J.-Q. *J. Am. Chem. Soc.* **2006**, *128*, 12634–12635. Giri, R.; Maugel, N.; Li, J.-J.; Wang, D.-H.; Breazzano, S. P.; Saunders, L. B.; Yu, J.-Q. *J. Am. Chem. Soc.* **2007**, *129*, 3510–3511.

73. a) Dick, A. R.; Hull, K. L.; Sanford, M. S. *J. Am. Chem. Soc.* **2004**, *126*, 2300–2301. b) Kalyani, D.; Sanford, M. S. *Org. Lett.* **2005**, *7*, 4149–4152. c) Emmert, M. H.; Cook,

A. K.; Xie, Y. J.; Sanford, M. S. *Angew. Chem. Int. Ed.* **2011**, *50*, 9409–9412. d) Cook, A. K.; Emmert, M. H.; Sanford, M. S. *Org. Lett.* **2013**, *15*, 5428–5431. For directed mechanistic investigations see: Desai, L. V.; Stowers, K. J.; Sanford, M. S. *J. Am. Chem. Soc.* **2008**, *130*, 13285–13293. Stowers, K. J.; Sanford, M. S. *Org. Lett.* **2009**, *11*, 4584–4587.

74. Desai, L. V.; Malik, H. A.; Sanford, M. S. *Org. Lett.* **2006**, *8*, 1141–1144.

75. Vickers, C. J.; Mei, T.-S.; Yu, J.-Q. *Org. Lett.* **2010**, *12*, 2511–2513.

76. Kim, J.; Chang, S. *J. Am. Chem. Soc.* **2010**, *132*, 10272–10274.

77. Ma, S.; Villa, G.; Thuy-Boun, P. S.; Homs, A.; Yu, J.-Q. *Angew. Chem. Int. Ed.* **2014**, *53*, 734–737.

78. Wang, X.; Mei, T.-S.; Yu, J.-Q. *J. Am. Chem. Soc.* **2009**, *131*, 7520–7521.

79. Chan, K. S. L.; Wasa, M.; Wang, X.; Yu, J.-Q. *Angew. Chem. Int. Ed.* **2011**, *50*, 9081–9084.

80. a) Wang, X.; Truesdale, L.; Yu, J.-Q. *J. Am. Chem. Soc.* **2010**, *132*, 3648–3649. b) Zhang, X.; Dai, H.-X.; Wasa, M.; Yu, J.-Q. *J. Am. Chem. Soc.* **2012**, *134*, 11948–11951. c) Miura, M.; Feng, C.-G.; Ma, S.; Yu, J.-Q. *Org. Lett.* **2013**, *15*, 5258–5261. d) Ye, Y.; Sanford, M. S. *Synlett.* **2012**, *23*, 2005–2013. b) For mechanistic studies see: Ball, N. D.; Gary, J. B.; Sanford, M. S. *J. Am. Chem. Soc.* **2011**, *133*, 7577–7584.

81. Mkhalid, I. A. I.; Barnard, J. H.; Marder, T. B.; Murphy, J. M.; Hartwig, J. F. *Chem. Rev.* **2010**, *110*, 890–931.

82. Dai, H.-X.; Yu, J.-Q. *J. Am. Chem. Soc.* **2012**, *134*, 134–137.

83. Neufeldt, S. R.; Seigerman, C. K.; Sanford, M. S. *Org. Lett.* **2013**, *15*, 2302–2305.

84. Chen, X.; Li, J.-J.; Hao, X.-S.; Goodhue, C. E.; Yu, J.-Q. *J. Am. Chem. Soc.* **2006**, *128*, 78–79.

85. Zhang, Y.-H.; Shi, B.-F.; Yu, J.-Q. *Angew. Chem. Int. Ed.* **2009**, *48*, 6097–6100.

86. Thuy-Boun, P. S.; Villa, G.; Dang, D.; Richardson, P.; Su, S.; Yu, J.-Q. *J. Am. Chem. Soc.* **2013**, *135*, 17508–17513.

87. Tan, Y.; Hartwig, J. F. *J. Am. Chem. Soc.* **2010**, *132*, 3676–3677.

88. Mei, T.-S.; Wang, X.; Yu, J.-Q. *J. Am. Chem. Soc.* **2009**, *131*, 10806–10807.

89. Shrestha, R.; Mukherjee, P.; Tan, Y.; Litman, Z. C.; Hartwig, J. F. *J. Am. Chem. Soc.* **2013**, *135*, 8480–8483.

90. Yoo, E. J.; Ma, S.; Mei, T.-S.; Chan, K. S. L.; Yu, J.-Q. *J. Am. Chem. Soc.* **2011**, *133*, 7652–7655.

91. Li, J.-J.; Giri, R.; Yu, J.-Q. *Tetrahedron.* **2008**, *64*, 6979–6987.

92. Feng, C.-G.; Ye, M.; Xiao, K.-J.; Li, S.; Yu, J.-Q. *J. Am. Chem. Soc.* **2013**, *135*, 9322–9325.

93. Zhao, X.; Dimitrijević, E.; Dong, V. M. *J. Am . Chem. Soc.* **2009**, *131*, 3466–3467.

94. Lu, Y.; Leow, D.; Wang, X.; Engle, K. M.; Yu, J.-Q. *Chem. Sci.* **2011**, *2*, 967–971.

95. Giri, R.; Yu, J.-Q. *J. Am. Chem. Soc.* **2008**, *130*, 14082–14083. Giri, R.; Lam, J. K.; Yu, J.-Q. *J. Am. Chem. Soc.* **2010**, *132*, 686–693.

96. Young, A. J.; White, M. C. *J. Am. Chem. Soc.* **2008**, *130*, 14090–14091.

97. Young, A. J.; White, M. C. *Angew. Chem. Int. Ed.* **2011**, *50*, 6824–6827.

98. Chen, M. S.; White, M. C. *J. Am. Chem. Soc.* **2004**, *126*, 1346–1347. Chen, M. S.; Prabagaran, N.; Labenz, N. A.; White, M. C. *J. Am. Chem. Soc.* **2005**, *127*, 6970–6971.

99. Reed, S. A.; Mazzotti, A. R.; White, M. C. *J. Am. Chem. Soc.* **2009**, *131*, 11701–11706.

100. Fraunhoffer, K. J.; White, M. C. *J. Am. Chem. Soc.* **2007**, *129*, 7274–7276.

101. Rice, G. T.; White, M. C. *J. Am. Chem. Soc.* **2009**, *131*, 11707–11711.

102. Qi, X.; Rice, G. T.; Lall, M. S.; Plummer, M. S.; White, M. C. *Tetrahedron.* **2010**, *66*, 4816–4826.

103. Delcamp, J. H.; Brucks, A. P.; White, M. C. *J. Am. Chem. Soc.* **2008**, *130*, 11270–11271.

104. Delcamp, J. H.; White, M. C. *J. Am. Chem. Soc.* **2006**, *128*, 15076–15077.

105. Delcamp, J. H.; Gormisky, P. E.; White, M. C. *J. Am. Chem. Soc.* **2013**, *135*, 8460–8463.

106. Gormisky, P. E.; White, M. C. *J. Am. Chem. Soc.* **2011**, *133*, 12584–12589. For applications towards improving the synthesis of complex molecules possessing polyoxygenated scaffolds see: Covell, D. J.; White, M. C. *Tetrahedron.* **2013**, *69*, 7771–7778.

107. For examples of alkoxyallylboration in complex molecule synthesis see: a) Smith, A. L.; Pitsinos, E.; Hwang, C.-K.; Mizuno, Y.; Saimoto, H.; Scarlato, G. R.; Suzuki, T.; Nicolaou, K. C. *J. Am. Chem. Soc.* **1993**, *115*, 7612–7624. b) Mueller-Hendrix, A. J.; Jennings, M. P. *Tet. Lett.* **2010**, *51*, 4260–4262. c) Yin, N.; Wang, G.; Qian, M.; Negishi, E. *Angew. Chem. Int. Ed.* **2006**, *45*, 2916–2920.

108. For a review on the Ireland-Claisen rearrangement see: Castro, A. M. M. *Chem. Rev.* **2004**, *104*, 2939–3002.

109. Strambeanu, I. I.; White, M. C. *J. Am. Chem. Soc.* **2013**, *135*, 12032–12037.

110. Jiang, C.; Covell, D. J.; Stepan, A. F.; Plummer, M. S.; White, M. C. *Org. Lett.* **2012**, *14*, 1386–1389.

111. Neufeldt, S. R.; Sanford, M. S. *Org. Lett.* **2010**, *12*, 532–535.

112. Desai, L V.; Hull, K. L.; Sanford, M. S. *J. Am. Chem. Soc.* **2004**, *126*, 9542–9543.

113. He, J.; Wasa, M.; Chan, K. S. L.; Yu, J.-Q. *J. Am. Chem. Soc.* **2013**, *135*, 3387–3390.

114. For recent reviews see: a) Baudoin, O. *Chem. Soc. Rev.* **2011**, *40*, 4902–4911. b) Jazzar, R.; Hitce, J.; Renaudat, A.; Sofack-Kreutzer, J.; Baudoin, O. *Chem. Eur. J.* **2010**, *16*, 2654–2672. c) Giri, R.; Shi, B.-F.; Engle, K. M.; Maugel, N.; Yu, J.-Q. *Chem. Soc. Rev.* **2009**, *38*, 3242–3272.

115. a) Zaitsev, V. G.; Shabashov, D.; Daugulis, O. *J. Am. Chem. Soc.* **2005**, *127*, 13154–13155. b) Shabashov, D.; Daugulis, O. *J. Am. Chem. Soc.* **2010**, *132*, 3965–3972. c) Nadres, E. T.; Santos, G. I. F.; Shasbashov, D.; Daugulis, O. *J. Org. Chem.* **2013**, *78*, 9689–9714. d) Wasa, M.; Engle, K. M; Yu, J.-Q. *J. Am. Chem. Soc.* **2009**, *131*, 9886–9887. e) Wasa, M.; Yu, J.-Q. *Tetrahedron.* **2010**, *66*, 4811–4815. f) Wasa, J.; Chan, K. S. L.; Zhang, X.-G.; He, J.; Miura, M.; Yu, J.-Q. *J. Am. Chem. Soc.* **2012**, *134*, 18570–18572. For mechanistic insights see: Figg, T. M.; Wasa, M.; Yu, J.-Q.; Musaev, D. J. *J. Am. Chem. Soc.* **2013**, *135*, 14206–14214.

116. Rousseaux, S.; Gorelsky, S. I.; Chung, B. K. W.; Fagnou, K. *J. Am. Chem. Soc.* **2010**, *132*, 10692–10705.

117. He, J.; Li, S.; Deng, Y.; Fu, H.; Laforteza, B. N.; Spangler, J. E.; Homs, A.; Yu, J.-Q. *Science.* **2014**, *343*, 1216–1220. For mechanistic and computational insights see: Giri, R.; Lan, Y.; Liu, P.; Houk, K. N.; Yu, J.-Q. *J. Am. Chem. Soc.* **2012**, *134*, 14118–14126.

118. Renaudat, A.; Ludivine, J.-G.; Jazzar, R.; Kefalidis, C. E.; Clot, E.; Baudoin, O. *Angew. Chem. Int. Ed.* **2010**, *49*, 7261–7265.

119. Aspin, S.; Goutierre, A.-S.; Larini, P.; Jazzar, R.; Baudoin, O. *Angew. Chem. Int. Ed.* **2012**, *51*, 10808–10811.

120. Aspin, S.; López-Suárez, L.; Larini, P.; Goutierre, A.-S.; Jazzar, R.; Baudoin, O. *Org. Lett.* **2013**, *15*, 5056–5059.

121. Martin, N.; Pierre, C.; Davi, M.; Jazzar, R.; Baudoin, O. *Chem. Eur. J.* **2012**, *18*, 4480–4484.

122. Yoo, E. J.; Wasa, M.; Yu, J.-Q. *J. Am. Chem. Soc.* **2010**, *132*, 17378–17380.

123. Shabashov, D.; Daugulis, O. *Org. Lett.* **2005**, *7*, 3657–3659.

124. Wang, D.-H.; Wasa, M.; Giri, R.; Yu, J.-Q. *J. Am. Chem. Soc.* **2008**, *130*, 7190–7191.

125. McMurtrey, K. B.; Racowski, J. M.; Sanford, M. S. *Org. Lett.* **2012**, *14*, 4094–4097. Additional mechanistic reports have been published. Racowski, J. M.; Gary, J. B.; Sanford, M. S. *Angew. Chem. Int. Ed.* **2012**, *51*, 3414–3417.

126. Sun, H.; Wang, B.; DiMagno, S. G. *Org. Lett.* **2008**, *10*, 4413–4416.

127. Hitce, J.; Retailleau, P.; Baudoin, O. *Chem. Eur. J.* **2007**, *13*, 792–799.

128. Sofack-Kreutzer, J.; Martin, N.; Renaudat, A.; Jazzar, R.; Baudoin, O. *Angew. Chem. Int. Ed.* **2012**, *51*, 10399–10402.

129. Wasa, M.; Engle, K. M.; Yu, J.-Q. *J. Am. Chem. Soc.* **2010**, *132*, 3680–3681.

130. Stowers, K. J.; Fortner, K. C.; Sanford, M. S. *J. Am. Chem. Soc.* **2011**, *133*, 6541–6544.
131. Chaumontet, M.; Piccardi, R.; Audic, N.; Hitce, J.; Peglion, J.-L.; Clot, E.; Baudoin, O. *J. Am. Chem. Soc.* **2008**, *130*, 15157–15166.
132. Rousseaux, S.; Davi, M.; Sofack-Kreutzer, J.; Pierre, C.; Kefalidis, C. E.; Clot, E.; Fagnou, K.; Baudoin, O. *J. Am. Chem. Soc.* **2010**, *132*, 10706–10716.
133. Chamontet, M.; Retailleau, P.; Baudoin, O. *J. Org. Chem.* **2009**, *74*, 1774–1776.
134. For a recent review of stereoselective C−H functionalization see: Giri, R.; Shi, B.-F.; Engle, K. M.; Maugel, N.; Yu, J.-Q. *Chem. Soc. Rev.* **2009**, *38*, 3242–3272. For mechanistic investigations in enantioselective C−H activation catalyzed by chiral Pd(II) complexes see: Musaev, D. G.; Kaledin, A.; Shi, B.-F.; Yu, J.-Q. *J. Am. Chem. Soc.* **2012**, *134*, 1690–1698.
135. Shi, B.-F.; Zhang, Y.-H.; Lam, J. K.; Wang, D.-H.; Yu, J.-Q. *J. Am. Chem. Soc.* **2010**, *132*, 460–461.
136. Cheng, X.-F.; Li, Y.; Su, Y.-M.; Yin, F.; Wang, J.-Y.; Sheng, J.; Vora, H. U.; Wang, X.-S.; Yu, J.-Q. *J. Am. Chem. Soc.* **2013**, *135*, 1236–1239.
137. Shi, B.-F.; Maugel, N.; Zhang, Y.-H.; Yu, J.-Q. *Angew. Chem. Int. Ed.* **2008**, *47*, 4882–4886.
138. Chu, L.; Wang, X.-C.; Moore, C. E.; Rheingold, A. L.; Yu, J.-Q. *J. Am. Chem. Soc.* **2013**, *135*, 16344–16347.
139. Wasa, M.; Engle, K. M.; Lin, D. W.; Yoo, E. J.; Yu, J.-Q. *J. Am. Chem. Soc.* **2011**, *133*, 19598–19601.
140. For recent reviews on C−H functionalization in natural product synthesis see: McMurray, L.; O'Hara, F.; Gaunt, M. J. *Chem. Soc. Rev.* **2011**, *40*, 1885–1898. Yamaguchi, J.; Yamaguchi, A. D.; Itami, K. *Angew. Chem. Int. Ed.* **2012**, *51*, 8960–9009.
141. Gutekunst, W. R.; Baran, P. S. *J. Am Chem. Soc.* **2011**, *133*, 19076–19079.
142. Rosen, B. R.; Simke, L. R.; Thuy-Boun, P. S.; Dixon, D. D.; Yu, J.-Q.; Baran, P. S. *Angew. Chem. Int. Ed.* **2013**, *52*, 7317–7320.
143. Mandal, D.; Yamaguchi, A. D.; Yamaguchi, J.; Itami, K. *J. Am. Chem. Soc.* **2011**, *133*, 19660–19663. For another total synthesis of dragmacidin d see: Garg, N. K.; Sarpong, R.; Stoltz, B. M. *J. Am. Chem. Soc.* **2002**, *124*, 13179–13182.
144. Wang, D.-H.; Yu, J.-Q. *J. Am. Chem. Soc.* **2011**, *133*, 5767–5769. For another synthesis of the target using C−H functionalization see: O'Malley, S. J.; Tan, K. L.; Watzke, A.; Bergman, R. G.; Ellman, J. A. *J. Am. Chem. Soc.* **2005**, *127*, 13496–13497. Colby, D. A.; Bergman, R. G.; Ellman, J. A. *Chem. Rev.* **2010**, *110*, 624–655.
145. For a review on macrolactonization see: Parenty, A.; Moreaux, X.; Nicks, G.; Campagne, J. *Med. Chem. Rev.* **2013**, PR1–PR40.
146. Fraunhoffer, K. J.; Prabagaran, N.; Sirois, L. E.; White, M. C. *J. Am. Chem. Soc.* **2006**, *128*, 9032–9033.
147. Stang, E. M.; White, M. C. *Nat. Chem.* **2009**, *1*, 547–551.
148. Stang, E. M.; White, M. C. *Angew. Chem. Int. Ed.* **2011**, *50*, 2094–2097.

# 4 Rhodium-Catalyzed C—H Activation

*Micheal Fultz*

## CONTENTS

## 4.1 INTRODUCTION

Rhodium is a black powder or silver-colored solid that has been used extensively in C—H bond activation. Transition-metal-catalyzed C—H and C—C bond activation has emerged as a powerful strategy to form organic building blocks of complex structures in a step- and atom-economical fashion. Catalytic C—H bond activation has emerged as a process to produce structurally diverse organic molecules due to the minimization of stoichiometric metallic waste.[1]

Rhodium complexes provide some of the most attractive catalysts for carbon manipulation with high reactivity, regioselectivity, scope, and functional group tolerance. In particular, rhodium complexes have displayed potential for the synthesis of various heterocyclic and carbocyclic compounds through the C—H bond activation reactions.[2] Rhodium complexes have been shown to catalyze sp$^2$ C—H bond insertion into several pi bonds including alkenes, alkynes, aldehydes, and imines.[3]

With all of the advantages rhodium has in organic synthesis there is a significant drawback. Homogenous rhodium complexes remain an industrial challenge because they are expensive, cannot be easily recycled, and can be difficult to separate from a product mixture which is a particular problem in the pharmaceutical industry.[4]

## 4.2 ELECTROPHILIC AROMATIC SUBSTITUTION

Electrophilic aromatic substitution has a long history in aromatic chemistry. Using transition metals to catalyze the reaction has made significant advances to the field, notably regioselectivity, mechanisms and reagents needed.

In traditional electrophilic aromatic substitutions the reaction is governed by the electronic properties of the substituents on the aromatic ring. In transition-metal-catalyzed reactions, the regiochemistry can be controlled by numerous inter-molecular interactions. Sometimes the regiocontrol has been achieved by substituents on the arene, which bind to the catalyst and direct the reaction to either the ortho-[5] or meta-[6] positions. In other examples the regioselectivity results from the steric interactions of the substituents ortho to a reacting C—H bond.[7]

### 4.2.1 AROMATIC SILYLATION

Aromatic silylation has the potential to be a valuable tool toward the synthesis of arylsilane monomers for silicone polymers and intermediates needed for other complex molecules.[8] Organosilanes are more stable than the organoboranes to many of the conditions of typical organic transformations, and the steric properties of the substituents can modulate the regioselectivitity of the reaction. Traditional arylsilane formation has been done with hydrosilanes and only at high temperatures or with photochemical reactions with a large excess of the arene as compared to the hydrosilane. This is a significant challenge since the arene is typically more valuable than the hydrosilane. Another consideration is the substitution of the silicon. Hartwig has developed a rhodium catalyst that works at significantly lower temperatures with a simple hydrogen acceptor and a readily available arylsilanes. This reaction provides a product with high regiocontrol and yields.[8]

In this example, the regioselectivity of the silylation results from the high level of steric control by substituents ortho and meta to the reacting C—H bond. Under these conditions, the silylaltion of various 1,3-disubstituted arenes **2a–2f** occurred at the mutually meta position with greater than the 95:5 selectivity.

|  |  |
|---|---|
| **1a.** (R = Me) | **2a.** (R = Me) 83% |
| **1b.** (R = OMe) | **2b.** (R = OMe) 82% |
| **1c.** (R = Cl) | **2c.** (R = Cl) 86% |
| **1d.** (R = OSiMe$_2$tBu) | **2d.** (R = OSiMe$_2$tBu) 67% |
| **1e.** (R = NMe$_2$) | **2e.** (R = NMe$_2$) 65% |
| **1f.** (R = C(O)NEt$_2$) | **2f.** (R = C(O)NEt$_2$) 96% |

Reaction conditions: [Rh(coe)$_2$OH]$_2$ (1 mol%), HSiMe(OTMS)$_2$, cyclohexene, THF, 45°C 12–36 h, product bearing SiMe(OTMS)$_2$

Interestingly, when silylation is compared to the borylation, the regioselectivity is both higher and inverted in selectivity. Aryl silylation displays exceptional selectivity for the silyl protected amine (**3a**) for the para position of the nitrogen. One major advantage of this rhodium-catalyzed reaction is that over-silylation rarely seems to be a significant by-product. Rarely did the over-silylation occur at over 10% recovered yields. This increased selectivity is noted with the bicyclic ring **3**. The presence of the free amine is particularly noteworthy since this is a challenge in the traditional electrophilic aromatic substitution pathway.

**3a.** R = TIPS
**3b.** R = Me

**4a.** R = TIPS
M = [Si] 53% (>99:1 a:b)
M = [B] 53% 49:45:6 a:b:c)
**4b.** R = Me
M = [Si] 95% (83:17 a:b)
M = [B] 79% (46:47:7 a:b:c)

## 4.2.2 AROMATIC ACYLATION

Rhodium-catalyzed direct C—H bond functionalization has emerged as a tool for the construction of carbon–carbon or carbon–heteroatom bonds in the regioselective amidation of isocyanates. Using acyl azides and taking advantage of their propensity to proceed through the Curtius Rearrangement,[9] Shin and coworkers[10] developed a pathway to form aromatic amides in both high yields and regioselectivity. One of the major challenges with using isocynates (example **6**) is due to the difficulty in controlling the dual regioselectivity of the electrophile leading to a mixture of both C—C and C—N amidated products. This dual nature can be seen in the production of 2 different regioisomers of the aromatic amide **8** and **9**.

The catalyst system [RhCp*Cl₂]₂ with AgSbF₆, which is widely used in C—H functionalization, resulting in both C—C and C—N amidations, provided no regioselectivity. Examining other silver additives or other solvents had little effect on the selectivity of the reaction. Only with the addition of 100 mole percent of acetonitrile was significant selectivity achieved. If excess acetonitrile was used, the reaction yields were cut in half. The scope of the reaction was compatible with sensitive functional groups such as ketones, esters, aldehydes, and halides. The presence of electron withdrawing or donating groups on either of the arenes had minimal effect on the yields of selectivity of the reaction. The selectivity of the addition comes from a five-membered transition metallacycle starting from the pyridine nitrogen to activate the *ortho* C—H bond.

## 4.2.3  AROMATIC CYANATION

Aryl nitriles are structural motifs that frequently occur as the core structures of many pharmaceuticals and agrochemicals. In addition, the widespread synthetic utility of the nitrile is highlighted by serving as a versatile building-block leading to the formation of numerous functional groups such as aldehydes, amides, and amines.[11] To incorporate the nitrile into the arene, many toxic metal cyanide sources M—CN (M = K,[12] Na,[13] Zn[14]) have been used. Non-metallic cyano group sources have many advantages; most notably, they avoid producing stoichiometric metal waste and hazardous HCN gas.

Rhodium-catalyzed cyanation can be highly efficient, selective, and practical. Pyrimidines and their derivatives have attracted attention as motifs for use in material and medicinal chemistry. The development of a readily available regioselective pathway for the functionalization of arylpyrimidines **10a** is a significant accomplishment. Previous efforts using standard palladium-catalyzed conditions, Xu and coworkers[15] were able to achieve only 50% yield using palladium with 2-phenylpyrimidine **10a** after 24 h at elevated temperatures. Using a rhodium catalysis system, Xu improved the yield of the same product **12a** to 84% by stabilizing the five-membered ring in the transition state **11**. This protocol was also successfully applied to the vinyl C—H bond cyanation of vinyl proton of **10b**, which represents the first example of metal-catalyzed direct C—H cyanation reaction of alkenes with isocyanide providing **12b**.

10a = aromatic ring                          11                          12a = 84%
10b = alkene                                                            12b = 64%

## 4.3   INDOLE AND ISOINDOLE FORMATION

Substituted indoles exist widely in pharmaceuticals, agrochemicals, natural products, and intermediates in synthetic organic chemistry. As a result, considerable attention has been on the development of new synthetic methods to provide indoles. Among them, transition-metal-catalyzed direct coupling of commercially available arylamines and alkynes via C—H activation is of particular interest because of its sustainable and environmentally friendly features. One concern is the need for stoichiometric oxidants, such as copper and silver salts, which are typically required to complete the oxidative cycle. One promising method to overcome this problem is to use oxygen as the sole oxidant providing water as the only byproduct.[16] Jiao and coworkers[17] have accomplished this using a palladium catalyst with arylamines; however, only electron deficient alkynes could be used.

Huang reported the successful development of an Rh (I) to Rh (III) catalytic system. Using this knowledge of rhodium oxidation, the team developed a cyclization process using simple aryl amines and multi-substituted alkynes to provide substituted indoles in moderate yields. As seen in the yields of the products there is a significant difference based on the electronic properties of the substituents.

| R$_1$ | R$_2$ | R$_3$ | Yield |
|---|---|---|---|
| 4-CH$_3$– | Ph | Ph | 93% |
| 4-COOCH$_3$ | Ph | Ph | 60% |
| H | Ph | CH$_3$ | 52% |

The potential for enatioselective intra-molecular hydroacylation to create an efficient pathway into dihydropyrroloindoles led to the study of rhodium catalyzed hydroacylations of N-vinylindole-2-carboxaldehydes **16**.[18] These dihydropyrrolindoles are core structures of indole alkaloids. The high level of enantiomeric excess of this reaction comes from the chirality of the biphenyl phosphine **17**. The reaction tolerated substitution throughout the indole and further reactivity of the substituted indole provided even more advanced intermediates used in synthetic chemistry.

| R | Yield | ee |
|---|---|---|
| 4-MeO-C$_6$H$_4$- | 99 | 99 |
| Me | 99 | 99 |
| C$_6$H$_{11}$ | 20 | 97 |

Transition-metal-catalyzed cross-coupling of indoles and aryl halides is one of the most sustainable pathways for the functionalization of indoles. Compared to the majority of reported palladium-catalyzed olefinations, the indolic C−H olefinations using rhodium catalysts often allow higher selectivity and broader substrate scope but is less documented. This method provides a direct and efficient process to the functionalization at C2 of protected indole **19** to provide the coupled product **20**[19].

**19**　　　　　　　　　　　　　　　　　　　　　**20**

## 4.4  HETEROATOM MANIPULATIONS

Heterogeneous rhodium catalysts do not have the same limitations that the homogenous catalysts do. Heterogeneous catalysis also helps to minimize waste during reaction work-up. These catalysts can be easily separated from the reaction mixture by simple filtration and possibly reused in successive reactions. The immobilization of the catalytic active species onto the solid support to produce the heterogeneous catalysts has attracted significant focus due to the ability to combine catalyst recovery with the high activity and selectivity of soluble catalyst.[4] Heterogeneous catalysis does have the limitation of low turnover rate and catalysis-leaching.

To present, silica-supported rhodium catalysts have been successfully used for hydrogenation, hydroformylation, and hydrosilylation reactions. Zhang and coworkers[4] developed a heterogeneous rhodium complexes **23** catalyzed carbon–heteroatom bond formation. The reaction couples disulfides **21** or diselenides with an alkyl or acyl halide to generate unsymmetrical sulfides (**24**) and selenides in good yields. The catalyst could be easily recovered and recycled by filtration of the reaction solution and re-used for five cycles without significant loss of activity (maintains over 90% yield).[4]

**21**

+

CH$_3$(CH$_2$)$_{11}$Br
**22**

**23**

**24**

From an economical and environmental point of view, one pot synthesis of amines from alkenes is a superior pathway over conventional multistep pathways.[20] Hydroaminomethylation is an attractive tool for the synthesis of amines and has been used in the pharmaceutical and chemical industries. Blum developed a heterogeneous system based on [Rh(cod)Cl]₂ immobilized within a sol-gel matrix as a highly efficient, regioselective, and recyclable catalyst for hydroaminomethylation reaction of vinylarenes with aniline derivatives (anilines or nitrobenzenes) under mild conditions.

The reaction advantages for environmental sustainability. The recyclable catalyst showed leaching rates of less than 0.063% and could be re-used three times without significant loss of selectivity. After the three cycles a significant loss of regioselectivity (branched versus linear amines **28:29**) due to the hydrolysis of Si–O bonds in the aqueous medium caused an inversion in the selectivity to provide a major product of the linear amine. The selectivity of the branched product to the linear product **28:29** was significantly higher at lower temperatures (60°C 32:1) compared to higher temperatures (90°C 5:1). This was also the first time that nitroarene was used as an amine precursor to provide the secondary amines in essentially the same selectivity and yields.

| | Product | **28** | **29** |
|---|---|---|---|
| Nitrobenzene | Yield | 91% | 3% |
| Aniline | Yield | 90% | 2% |

## 4.5   RING REACTIONS

The manipulation of highly reactive species is an attractive strategy for chemists due to the rapid generation of molecular complexity. Increasing the efficiency of the synthetic pathway can greatly minimize the waste stream of the sequence.

The tetrahydrofuran motif is common across many natural product classes and important bioactive compounds.[21] The formation of this important motif can be accessible through a number of synthetic pathways. Deska[22] developed a four-step procedure converting the achiral ether **30** to the 2,5-*trans*-disubstituted dihydrofuran-3-one **33** in modest enantiomeric excess and diastereomeric ratio. Temperature conditions for the cyclization were at room temperature or lower to achieve the highest diastereoselectivity. Yields from the synthetic pathway were not affected by the electronic properties on the substituents of the aromatic ring.

Another area that has been greatly exploited is the synthesis of 2,5-*trans*-disubstituted dihydrofuran-3-ones from α–diazoketones is carbenoid chemistry. Despite the value of products accessible from this reaction, the requirement for diazomethane to synthesize the substrates has limited its use to those with specialized training and equipment.[23] Using a rhodium(II) catalyst Boyer[23] developed a concise four-step procedure to develop a bicyclic stereotriad **36** in high diastereoselectivity from the allyl alcohol **34** eliminating the need for this difficult to handle substrate.

Sultams have attracted increasing attention due to their wide spectrum of bioactivities, such as anti-viral, anti-microbial, anti-leukemic, anti-cancer, and enzyme inhibition.[24] Regardless of the importance of the sultam there are only a limited number of synthetic approaches toward sultams such as **40** have been reported. The first approach, by Bunnett[25] involved the generation of the benzyne intermediate under strongly basic conditions using N-2′-chloroaryl alkanesulfonamides. This reaction provided moderate yields along with many byproducts. Later a process using the intramolecular nucleophilic substitution using N-fluoronitroaryl alkanesulfonamides[26] and another using intra-molecular oxidative substitution[27] under strongly basic conditions were developed both providing moderate yields. These approaches are not applicable to substrates with two aryl groups at the nitrogen of the sultam due to the facile N—S bond cleavage under strongly basic conditions. More pathways to the sultam have been developed, however, these approaches also suffer from the use of toxic reagents and harsh conditions.[24] Xu and coworkers developed a mild pathway using diazosulfonyl ester **39** with a rhodium (II) catalyst to provide the sultam in 99% yield. Under mild reaction conditions, this reaction pathway overcame the limitations of the previous pathways. Notably, multiple phenyl rings can be attached to the nitrogen of the sultam without cleavage of any bonds. It also eliminated the need for strongly harsh conditions and extremely toxic reagents.

The lactam is an important class of heterocycles and has been investigated due to its potent anti-bacterial activity. Among the multiple synthetic approaches, a [2 + 2] cycloaddition of a ketene with an imine, is one method to provide quick access to *cis* β-lactams.[28] The typical prerequisite for these protocols involves the generation of the ketene from an activated carboxylic acid derivative (often an acyl chloride), which limits the scope of this pathway.[29] Lee and coworkers developed a rhodium-catalyzed oxygenative addition reaction that furnishes the acyl chloride equivalent from a more stable terminal alkyne **41**. An intermediate rhodium vinylidene then undergoes a [2 + 2] cyclization with imines (**42**) to provide the β-lactam **4**. This reaction pathway enables more easily accessible alkynes to be employed as substrates for *trans*-β-lactam formation.

| R | R′ | Yield |
|---|---|---|
| 4-CH$_3$OC$_6$H$_4$ | Me | 80% |
| Ph | Bn | 86% |
| Ph | Ph | 32% |
| 4-NO$_2$OC$_6$H$_4$ | Me | 82% |

## 4.6  COUPLING REACTIONS

The Heck reaction is the combination of a C—H bond with a functional group capable of undergoing oxidative insertion with a transition metal such as palladium or rhodium to provide a new C—C sigma bond uniting two fragments into a new molecule. This coupling procedure has been used in numerous synthetic pathways.[30] Lautens and coworkers have developed a rhodium-catalyzed version of the Heck reaction using boronic acids in an aqueous media to unite vinyl arenes **45** with arylboronic acids **46** using a phase transfer catalyst such as sodium dodecyl sulfate (SDS) to provide the saturated coupled product **47** in high yields. Addition of organic cosolvents, sulfonated ligands, or triphenyl phosphines to the reaction mixture resulted in either no reaction after 15 h or less than 10% conversion. Unlike previous reports when organic solvents containing water was used, degradation of the boronic acid was kept at a minimum.[31]

A second generation rhodium catalyst capable of performing enatioselective coupling reactions was later developed by Lam.[32] Using a secondary amide (**49**) as a rhodium ligand, the coupling reaction provided **50** in 89% yield and 90% enantiomeric excess. This coupling process allowed for highly functionalized aromatic and heteroaromatic rings such as **48** with minimal impact on yields and selectivity. Hydrolytic deboronation was not reported as a challenge.

The use of aqueous solvents in lieu of organic solvents has many environmental and economic benefits.[33] Since normal transition-metal-catalysts have low solubility and reactivity in water, special catalysts needed to be developed to solve these challenges. Ligands **51** and **52** are the most popular water-soluble ligands used in transition-metal-catalyzed reactions.[43]

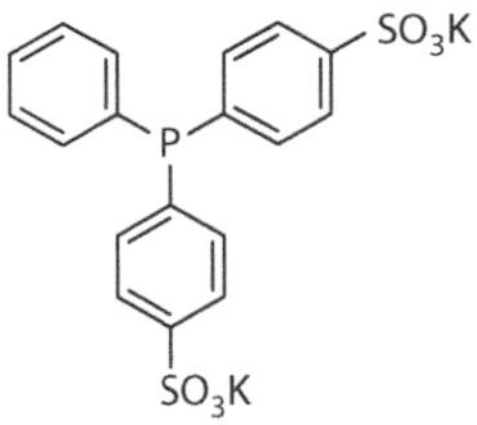

**51** 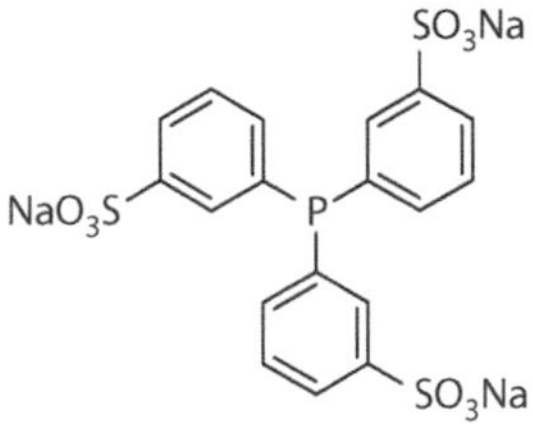                                                                                **52**

Water exhibits a combination of properties difficult to emulate with organic solvents and can be a means for catalyst recovery, reducing overall costs.[34] A common requirement for rhodium-catalyzed additions of arylboronic acids to olefins is that the olefin has to be activated by an electron-withdrawing group, which can facilitate the hydrolysis of an organorhodium intermediate. This constraint imposes a limitation on the scope of the substrate and leads to the addition of carbon and hydrogen across the pi bond.[35] Using heteroaromatic olefins (**53**), the addition product **54** was obtained. However styrenyl olefins **55** were used, 1,2diarylethylenes **56** were produced from the Heck type addition-β-hydride elimination process.[43]

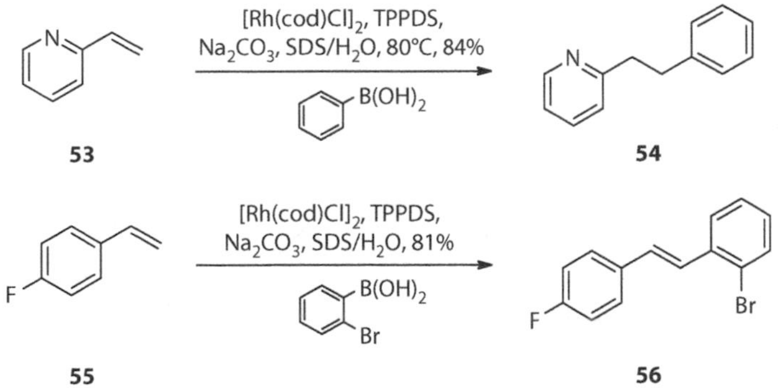

**53**                                                                                **54**

**55**                                                                                **56**

The aqueous addition of aryl boronic acids can be applied to the addition of alkenes as well. In a highly regioselective pathway, Lautens was able to add a boronic acids to the heteroaromatic alkyne **57** providing the trisubstituted alkene **59** in good yield. In optimizing this reaction and adding to the knowledge gained regarding water-soluble phosphine ligands, he discovered the coupling worked in the presence of pyridine. Taking advantage of this, he examined a novel pyridine substituted water-soluble ligand **58** improving the coupled yield (51%) of 1-pyridyl hexyne to phenyl boronic acid to provide the alkene **59** in 51%, as compared to the standard coupling ligands **51** (25%) and **52** (no reaction) as seen above.[36]

## 4.7  AMINE MANIPULATIONS

3,3-disubstituted oxindoles have been recognized as valuable structural motifs, forming the core structure of both natural products and pharmaceutical agents. This important structural framework has led to many synthetic pathways. Some of these pathways include harsh conditions such as perchloric acid,[37] sodium *tert*-butoxide,[38] or other strong bases,[39] which can inhibit functional group compatibility. Hu[40] has developed a pathway to condense 3-diazooxindole **60**, formaldehyde and amines to provide 3-amino-3-hydroxymethyloxindoles (**61**) or **60**, formaldehyde, and water to provide 3-hydroxymethyloxindoles (**62**) in a step- and atom-economical pathway. Many other transition metals were examined to test the scope of this reaction but only rhodium provided any appreciable yields even with expanded reaction times and elevated temperatures.

Highly substituted pyridines-based molecules represent an important class of heterocycles that are common in natural products and other functional materials. Due to the importance of the core structure, several methods for the synthesis of this motif have been recently developed.[41]

Rhodium-catalyzed C—H activation has been used as an efficient strategy for synthesis of heterocyclic scaffolds. In particular, pyridine rings were constructed through transformations of internal alkynes/alkenes with $\alpha$, $\beta$–unsaturated oximes and an internal oxidant.[42] Using the internal oxidant method, N—S bonds have not been reported to function as internal oxidants and more reactive electrophiles such as N–sulfonyliminimes 63 and internal alkynes 64 have rarely been developed as directing groups. Dong and coworkers developed a mild one-pot process that undergoes C—N bond formation and desulfonylation to form tetra-substituted pyridine rings 65 in good yields.[43]

Rhodium (III)-catalyzed cyclization of oxime derivatives with alkynes represent a useful procedure in isoquinolines and pyridine synthesis, however, the reactions with unsymmetrical or electron deficient alkynes suffer from low regioselectivity and reactivity. Glorius[44] was able to eliminate the standard ring formation/oxidation pathway for these N-oxides 68 by condensing oximes 66 with diazocarbonyls 67 to provide the target compounds in a single step. Traditional methods to the formation of isoquinolines and pyridine N-oxides have been reported using oxidants such as $m$CPBA, $H_2O_2$, and $CF_3CO_3H$ with the pyridine parent compound. One drawback to that is that the parent heterocycles need to be prepared in advance and suffer from the potential over oxidation of the highly functionalized substituents.

Recent pathways to five-membered acyl metal intermediates have been reported with nickel complexes through the elimination of small of small organic fragments such as carbon dioxide; however, this strategy conceptually lacks atom economy. Rhodium-based aldehyde C—H bond activation of benzaldehyde 69 followed by addition of the corresponding rhodium acyl hydride to internal alkyne 70 provided the substituted naphthalene bicycle 71 with 100% atom economy.[45]

## 4.8 QUATERNARY CARBON FORMATION

The stereoselective formation of quaternary stereogenic carbons remains both a challenging and important area of investigation. There have been numerous pathways to synthesize the carbon centers; among these attempts, enolate alkylation represents one synthetic strategy. The alkylation of the alpha carbon of a carbonyl remains one of the most fundamental reactions in organic chemistry. However, with this important reaction and the vast number of studies conducted on the generation of and substitution of the enolate, it still has a number of inherent limitations.[46]

These limitations include the ability to regioselectively form an enolate from an unsymmetrical ketone, polyalkylation via enolate equilibration, and the necessity to use simple electrophiles hinder both chiral and achiral enolate alkylations. Additionally, this alkylation usually requires specific enolate geometry for chirality of the product and since the origin of enatioselectivity is defined by a combination of enolate geometry and facial selectivity of the alkylation, chirality in this reaction is difficult while maintaining atom economical practices.

Evans and coworkers developed a rhodium-catalyzed allylic substitution of chiral acyclic tertiary allylic carbonates with acyl anion equivalents. This substitution pathway provides a route to undergo a stereo- and regioselective substitution to provide the quaternary carbon **73**. The enantioenriched product is isolated in greater than 95:5 selectivity and 87% yield. This pathway provides a product that could be obtained through enolate chemistry, only lacking the same atom economy obtained with rhodium-catalysis.

## 4.9 CONCLUSION

Rhodium-catalyzed C—H activation has gained attention and is being recognized for its potential to form many C—C and C–heteroatom bonds in an inherently more benign fashion.[47] With the success and wide scope of organic reactions conducted by rhodium-catalysis, organic chemistry continues to make strides in the reaction complexity and environmental sustainability. Further progress and study will continue to improve reactions and industrial use of rhodium.

## REFERENCES

1. Mishra, N. K.; Park, J.; Sharma, S.; Han, S.; Kim, M.; Shin, Y.; Jang, J.; Kwak, J. H.; Jung, Y. H.; Kim, I. S. Direct access to isoindoline through tandem Rh(III)-catalyzed alkenylation and cyclization of N-benzyltriflamides. *Chem. Commun.* **2014**, 50, 2350–2352.
2. Senthilkumar, N.; Gandeepan, P.; Jayaqkumar, J.; Cheng, C.-H. Rh(III)-catalyzed synthesis of 1-substituted isoquinolinium salts via a C—H bond activation reaction of ketimines with alkynes. *Chem. Commun.* **2014**, 50, 3106–3108.
3. Yu, S.; Li, X. Mild synthesis of chalcones via Rhodium(III)-Catalyzed C—C coupling of Arenes and cyclopropenones. *Org. Lett.* **2014**, 16, 1220–1223.
4. Zhang, H.; Hu, M.; Cai, M. Reductive coupling of disulfides and diselenides with alkyl halides catalyzed by a silica-supported phosphine rhodium complex using hydrogen as a reducing agent. *J. Chem. Res.* **2013**, 645–647.
5. Lyons, T. W.; Sanford, M. S. Palladium-catalyzed ligand-directed C—H functionalization reactions. *Chem. Rev.* **2010**, 110, 1147–1169.
6. Hofman, N.; Ackermann, L. meta-Selective C—H bond alkylation with secondary alkyl halides. *J. Am. Chem. Soc.* **2013**, 135, 5877–5884.
7. Mkhalid, I. A. I.; Barnard, J. A.; Marder, T. B.; Murphy, J. M.; Hartwig, J. F. C—H activation for the construction of C—B bonds. *Chem. Rev.* **2010**, 110, 890–931.
8. Cheng, C.; Hartwig, J. F. Rhodium-catalyzed intermolecular C—H silylation of arenes with high steric regiocontrol. *Science.* **2014**, 343, 853–857.
9. Buchner, E.; Curtius, T. Synthesis of β-keto esters from aldehydes and diazoacetic acid. *Chem. Ber.* **1885**, 18, 2371–2377.
10. Shin, K.; Ryu, J.; Chang, S. Orthogonal reactivity of acyl azides in C—H activation: Dichotomy between C—C and C—N amidations based on catalyst systems. *Org. Lett.* **2014**, 16, 2022–2025.
11. Hong, X.; Wang, H.; Qian, G.; Xu, B. Rhodium-catalyzed direct C—H bond cyanation of arenes with isocyanide. *J. Org. Chem.* **2014**, 79, 3228–3237.
12. Yang, C.; Williams, J. M. Palladium catalyzed cyanation of aryl bromides promoted by low-level organotin compounds. *Org. Lett.* **2004**, 6, 2837–2840.
13. Ushkov, A. V.; Grushin, V. V. Rational catalysis design on the basis of mechanistic understanding: Highly efficent Pd-catalyzed cyanation of aryl bromides with NaCN in recyclable solvents. *J. Am. Chem. Soc.* **2011**, 133, 10999–11005.
14. Buono, F. G.; Chidambaram, R.; Mueller, R. H.; Waltermire, R. E. Insights into palladium-catalyzed cyanation of bromobenzene: Additive effects on the rate-limiting step. *Org. Lett.* **2008**, 10, 5325–5328.
15. Xu, S.; Huang, X.; Hong, X.; Xu, B. Palladium-assisted regioselective C—H cyanation of heteroarenes using isonitrile as cyanide source. *Org. Lett.* **2012**, 14, 4614–4617.
16. Zhang, G.; Yu, H.; Qin, G.; Huang, H. Rh-Catalyzed oxidative C—H activation/annulation: Converting anilines to indoles using molecular oxygen as the sole oxidant. *Chem. Commun.* **2014**, 50, 4331–4334.
17. Shi, Z. Zhang, S.; Li, S.; Pan, D.; Ding, S.; Cui, Y.; Jiao, N. Indoles from simple anilines and alkynes: Palladium-catalyzed C—H activation using dioxygen as the oxidant. *Angew. Chem. Int. Ed.* **2009**, 48, 4572–4576.
18. Ghosh, A.; Stanley, L. M. Enantioselective hydroacylation of N-vinylindole-2-carboxaldehydes. *Chem. Commun.* **2014**, 50, 2765–2768.
19. Sharma, S.; Han, S.; Kim, M.; Mishra, N. K.; Park, J.; Shin, Y.; Ha, J.; Kwak, J. H.; Jung, Y. H.; Kim, I. S. Rh-Catalyzed oxidative C—C bond formation and C—N bond

cleavage: Direct access to C2-olefinated free (NH)-indoles and pyrroles. *Org. Biomol. Chem.* **2014**, 12, 1703–1706.

20. Nairoukh, Z.; Blum, J. Regioselective hydroaminomethylation of vinylarenes by a sol-gel immobilized rhodium catalyst. *J. Org. Chem.* **2014**, 79, 2397–2403.

21. Lorente, A.; Lamariano-Merketegi, J.; Albericio, F.; Álvarez, M. Tetrahydrofuran-containing macrolides: A fascinating gift from the deep sea. *Chem. Rev.* **2013**, 113, 4567–4610.

22. Skrobo, B.; Deska, J. Oxonium ylide rearrangement of enzymatically desymmetrized glutarates. *Org. Lett.* **2013**, 15, 5998–6001.

23. Boyer, A. Rhodium(II)-catalyzed stereocontrolled synthesis of dihydrofuran-3-imines from 1-Tosyl-1,2,3-triazoles. *Org. Lett.* **2014**, 16, 1660–1663.

24. Yang, Z.; Xu, J. Synthesis of benzo-γ-sultams via the Rh-catalyzed aromatic C—H functionalization of diazosulfonamides. *Chem. Commun.* **2014**, 50, 3616–3618.

25. Bunnett, J. F.; Kato, T.; Flynn, R. R.; Skorcz, J. Studies of ring closure via aryne intermediates. A. *J. Org. Chem.* **1963**, 28, 1–6.

26. Wojciechowski, K. Synthesis of nitrobenzophenones from nitro-α-sulfonyldiphenyl-methane derivatives. *Synth. Commun.* **1997**, 27, 135–144.

27. Wojciechowski, K. Synthesis of 2,1-benzisothiazoline 2,2-dioxide derivatives. *Pol. J. Chem.* **1992**, 66, 1121–1124.

28. Palomo, C.; Aizpurua, J. M.; Ganboa, I.; Oiarbide, M. Asymmetric synthesis of β-lactams by staudinger ketene-imine cycloaddition reaction. *Eur. J. Org. Chem.* **1999**, 12, 3223–3225.

29. Kim, I.; Roh, S. W.; Lee, D. G.; Lee, C. Rhodium-catalysed oxygenative [2 + 2] cyclo-addition of terminal alkynes and imines for the synthesis of β-lactams. *Org. Lett.* **2014**, 16, 2482–2485.

30. (a) Endo, A.; Yanagisawa, A.; Abe, M.; Tohma, S.; Kan, T.; Fukuyama, T. Total Synthesis of ecteinascidin 743. *J. Am. Chem. Soc.* **2002**, 124, 6552–6554. (b) Govek, S. P.; Overman, L. E. Total synthesis of asperazine. *J. Am. Chem. Soc.* **2001**, 123, 9468–9460.

31. Lautens, M.; Roy, A.; Fukuoka, K.; Fagnou, K.; Martin-Matute, B. Rhodium catalyzed coupling reactions of arylboronic acids to olefins in aqueous media. *J. Am. Chem. Soc.* **2001**, 123, 5358–5359.

32. Roy, I. D.; Burns, A. R.; Pattison, G.; Michel, B.; Parker, A. J.; Lam, H. W. A second-generation ligand for the enantioselective rhodium-catalyzed addition of arylboronic acids to alkenylazaarenes. *Chem. Commun.* **2014**, 50, 2865–2868.

33. Lautens, M.; Yoshida, M. Rhodium-catalyzed addition of arylboronic acids to alkynyl aza-heteroaromatic compounds in water. *J. Org. Chem.* **2003**, 68, 762–769.

34. Lautens, M.; Mancuso, J. Diastereoselective formation of indanes from arylboronate esters catalyzed by rhodium(I) in aqueous media. *Org. Lett.* **2002**, 4, 2105–2108.

35. Lautens, M.; Roy, A.; Fukuoka, K.; Fagnou, K.; Martin-Matute, B. Rhodium-catalyzed coupling reactions of arylboronic acids to olefins in aqueous media. *J. Am. Chem. Soc.* **2001**, 123, 5358–5359.

36. Lautens, M.; Yoshida, M. Regioselective rhodium-catalyzed addition of arylboronic acids to alkynes with a pyridine substituted water soluble ligand. *Org. Lett.* **2002**, 4, 123–125.

37. Zhou, F.; Ding, M.; Zhou, J. A catalytic metal-free Ritter reaction to 3-substituted 3-aminooxindoles. *Org. Biomol. Chem.* **2012**, 10, 3178–3181.

38. Marsden, S. P.; Watson, E. L.; Raw, S. A. Facile and general synthesis of quaternary 3-aminooxindoles. *Org. Lett.* **2008**, 10, 2905–2908.

39. Emura, T.; Esaki, T.; Tachibana, K.; Shimizu M. Efficient asymmetric synthesis of novel gastrin receptor antagonist AG-041R via highly stereoselective alkylation of oxindole enolates. *J. Org. Chem.* **2006**, 71, 8559–8564.

40. Wang, C.; Xing, D.; Wang, D.; Wu, X.; Hu, W. Synthesis of 3-Amino-3-hydroxymethyloxindoles and 3-Hydroxy-3-hydroxymethyloxindoles by Rh2(OAc)4-catalyzed three-component reactions of 3-diazooxindoles with formaldehyde and anilines or water. *J. Org. Chem.* **2014**, 79, 3908–3916.
41. (a) Zeni, G.; Larock, R. C. Synthesis of heterocycles via palladium-catalyzed oxidative addition. *Chem. Rev.* **2006**, 106, 4644–4680. (b) Gulevich, A. V.; Dudnik, A. S.; Chernyak, N.; Gevorgyan, V. Transition metal-mediated synthesis of monocyclic aromatic heterocycles. *Chem. Rev.* **2013**, 113, 3084–3213.
42. Liu, B.; Song, C.; Zhou, S.; Zhu, J. Rhodium(III)-catalyzed indole synthesis using N—N bond as an internal oxidant. *J. Am. Chem. Soc.* **2013**, 135, 16625–16631.
43. Zhang, Q.-R.; Huang, J. R.; Zhang, W.; Dong, L. Highly functionalized pyridines synthesis from N-sulfonyl ketimines and alkynes using the N—S bond as an internal oxidant. *Org. Lett.* **2014**, 16, 1684–1687.
44. Shi, Z.; Koester, D. C.; Boultadakis-Arapinis, M.; Glorius, F. Rh(III)-Catalyzed synthesis of multisubstituted isoquinoline and pyridine (N-oxides from oximes and diazocompounds. *J. Am. Chem. Soc.* **2013**, 135, 12204–12207.
45. Hojo, D.; Tanaka, K. Rhodium-catalyzed C—H bond activation/[4 + 2] annulation/aromatization cascade to produce phenol, naphthol, phenanthrenol, and triphenylenol derivatives. *Org. Lett.* **2012**, 14, 1492–1495.
46. Evans, P. A.; Oliver, S.; Chae, J. Rhodium catalyzed allylic substituion with an acyl anion equivalent: Stereospecific construction of acyclic quaternary carbon stereogenic centers. *J. Am. Chem. Soc.* **2012**, 134, 19314–19317.
47. Aïssa, C. Fürstener, A. A rhodium-catalyzed C—H activation/cycloisomerization tandem. *J. Am. Chem. Soc.* **2007**, 129, 14836–14837.

# 5 Nickel-Catalyzed C—H Activation

*Andrew C. Williams*

## CONTENTS

## 5.1 INTRODUCTION

Whilst the literature on nickel-catalyzed C—H activation is considerably less extensive than for, say, palladium or iridium, there has been a marked increase in the rate of publication since *ca.* 2005. Much of the earlier work is summarized in a small number of excellent reviews.[1–6] Recent examples of the use of nickel-catalyzed activation in natural product synthesis are summarized in a recent review by Yamaguchi et al.[7]

## 5.2 AROMATIC C(sp²)—H C—H ACTIVATION: ARYL–ARYL BOND FORMATIONS

Miura et al. have described the nickel-catalyzed direct arylation of azoles with aryl bromides, using nickel(II) bromide or a nickel(II) bromide diglyme complex, in the presence of 1,10-phenanthroline as ligand, and lithium $t$-butoxide as base, in diglyme at elevated temperature. The systems were optimized for benzthiazoles, benzoxazoles, and 5-phenyl oxazoles; benzthiazoles coupled efficiently without the addition of zinc to give 2-arylbenzthiazoles **3** (Scheme 5.1, Table 5.1).

Benzoxazoles and oxazoles required the addition of 0.5 equivalents of zinc powder, yields of **5**, **9**, and **10** were in the range 50%–96% (Schemes 5.2 and 5.3, Table 5.2).[8]

At the same time, Itami and coworkers described a very similar series of couplings of azoles (benzthiazoles, benzoxazoles, benzimidazoles, thiazoles, and imidazoles) with aryl and heteroaryl halides and triflates using nickel(II) acetate-bipyridyl and nickel(II) acetate-dppf catalysts, with lithium $t$-butoxide in dioxan. The Ni(OAc)$_2$. bipy catalyst system was optimal for aryl/heteroaryl bromides and iodides (Scheme 5.4, Table 5.3), and the Ni(OAc)$_2$.dppf system was effective for aryl/heteroaryl chlorides and triflates (Scheme 5.5, Table 5.4). Reaction temperatures were 85°C for the former and 140°C for the latter, and in both cases 1.5 equivalents of lithium $t$-butoxide was required, a considerably lower loading than required under Miura's conditions. Fortuitously, there are a number of final products common to both authors,

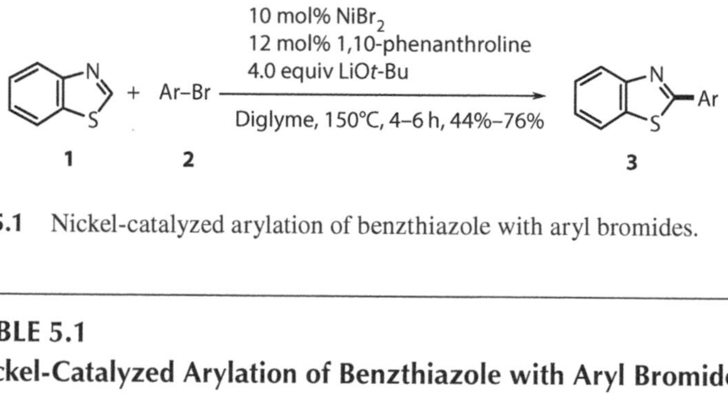

**SCHEME 5.1** Nickel-catalyzed arylation of benzthiazole with aryl bromides.

---

### TABLE 5.1
### Nickel-Catalyzed Arylation of Benzthiazole with Aryl Bromides

| Ar 2 | Time (h) | Yield 3 (%) |
|---|---|---|
| 4-MeC$_6$H$_4$ **2a** | 4 | 67 |
| 3-MeC$_6$H$_4$ **2b** | 4 | 55 |
| 2-MeC$_6$H$_4$ **2c** | 4 | 61 |
| 2,5-Me$_2$C$_6$H$_4$ **2d** | 4 | 69 |
| 1-Naphthyl **2e** | 4 | 76 |
| 4-MeOC$_6$H$_4$ **2f** | 4 | 56 |
| 4-FC$_6$H$_4$ **2g** | 4 | 60 |
| 4-F$_3$CC$_6$H$_4$ **2h** | 4 | 70 |
| 4-NCC$_6$H$_4$ **2i** | 4 | 44 |
| 4-MeC$_6$H$_4$ **2a**$^i$ | 6 | 50 |

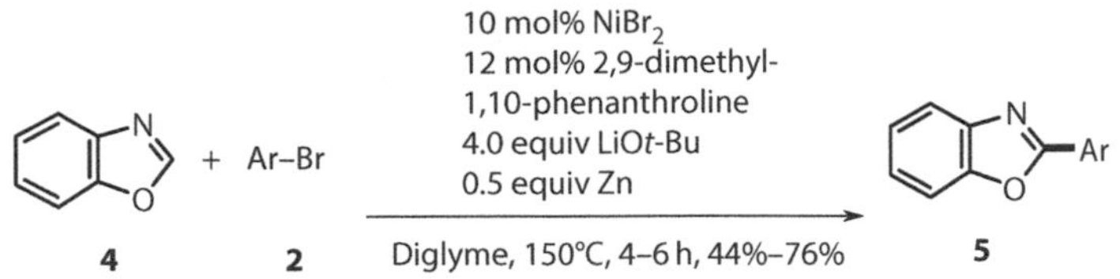

10 mol% NiBr$_2$
12 mol% 2,9-dimethyl-
1,10-phenanthroline
4.0 equiv LiO$t$-Bu
0.5 equiv Zn

Diglyme, 150°C, 4–6 h, 44%–76%

**4**    **2**    **5**

**SCHEME 5.2**   Nickel-catalyzed arylation of benzoxazole with aryl bromides.

10 mol% NiBr$_2$ · diglyme
12 mol% 1,10-phenanthroline
4.0 equiv LiO$t$-Bu
0.5 equiv Zn powder

Diglyme, 150°C, 4 h

**6**, X = H
**7**, X = CF$_3$    **8**    **9**, X = H 96%
**10**, X = CF$_3$ 89%

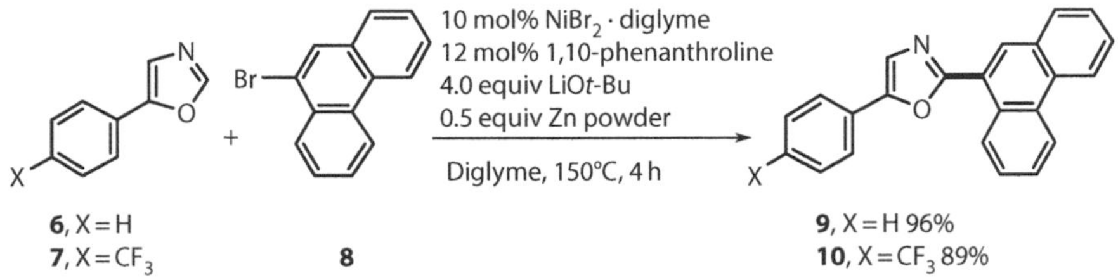

**SCHEME 5.3**   Nickel-catalyzed arylation of oxazole with aryl bromides.

### TABLE 5.2
### Nickel-Catalyzed Arylation of Benzthiazole with Aryl Bromides

| Ar 2 | Time (h) | Yield 5 (%) |
| --- | --- | --- |
| 2-MeC$_6$H$_4$ **2c** | 4 | 53 |
| 2,5-Me$_2$C$_6$H$_4$ **2d** | 4 | 66 |
| 1-Naphthyl **2e** | 4 | 62 |
| 4-MeO-2-MeC$_6$H$_4$ **2k** | 4 | 50 |
| 9-phenanthryl **2l** | 4 | 83 |
| 4-PhC$_6$H$_4$ **2m** | 4 | 60 |
| 9-anthracenyl **2n** | 4 | 68 |

allowing a direct comparison of the methods. Itami's conditions generally deliver higher yields than those of Miura, and the described scope of successful transformations is broader (Scheme 5.6).[9]

In a later extension of this methodology, Itami and coworkers extended the ranges of both sets of coupling partners, and examined the effect of replacement of lithium $t$-butoxide in dioxan with magnesium $t$-butoxide in dimethylformamide (DMF). Systematically comparing the two sets of conditions, they found that haloarenes **23**

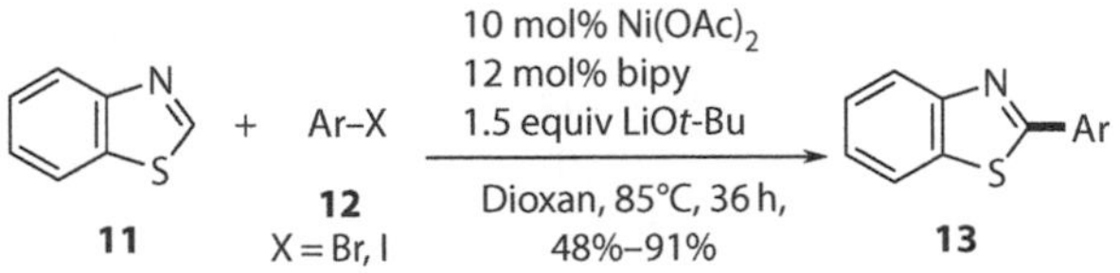

10 mol% Ni(OAc)$_2$
12 mol% bipy
1.5 equiv LiO$t$-Bu

Dioxan, 85°C, 36 h,
48%–91%

**11**    **12**
X = Br, I    **13**

**SCHEME 5.4**   Nickel-catalyzed arylation of benzthiazole with aryl bromides and iodides.

**TABLE 5.3**
**Nickel-Catalyzed Arylation of Benzthiazole with Aryl Bromides and Iodides**

| Ar 12 | X | Yield 13 (%) |
|---|---|---|
| Ph **12a** | I | 80 |
| Ph **12a** | Br | 62 |
| 2-MeC$_6$H$_4$ **12b** | I | 90 |
| 2-MeC$_6$H$_4$ **12b** | Br | 82 |
| 3-MeC$_6$H$_4$ **12c** | I | 71 |
| 3-MeC$_6$H$_4$ **12c** | Br | 63 |
| 4-MeC$_6$H$_4$ **12d** | I | 69 |
| 4-MeC$_6$H$_4$ **12d** | Br | 48 |
| 2-PhC$_6$H$_4$ **12e** | I | 91 |
| 2-MeOC$_6$H$_4$ **12f** | I | 65 |
| 4-MeOC$_6$H$_4$ **12g** | I | 70 |
| 3-F$_3$CC$_6$H$_4$ **12h** | I | 75 |
| 4-F$_3$CC$_6$H$_4$ **12i** | I | 68 |
| 4-NCC$_6$H$_4$ **12j** | I | 78 |
| 4-ClC$_6$H$_4$ **12k** | Br | 51 |
| 3,4,5-(MeO)$_3$C$_6$H$_2$ **12l** | Br | 61 |
| 1-Naphthyl **12m** | I | 91 |
| 3-Pyridyl **12n** | I | 57 |
| 3-Thienyl **12o** | I | 74 |
| 2-(3-Methylthienyl) **12p** | Br | 58 |

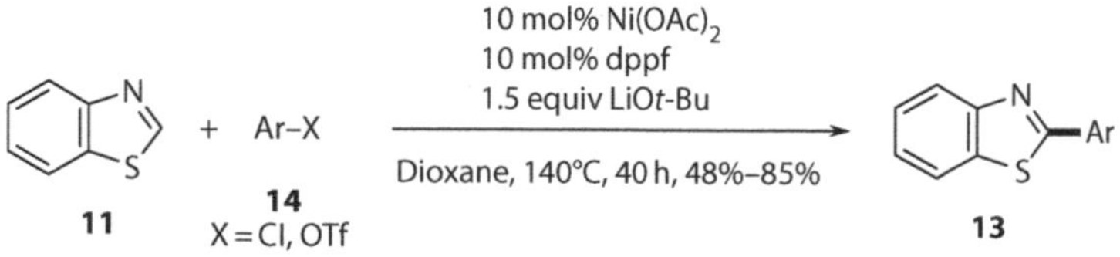

SCHEME 5.5   Nickel-catalyzed arylation of benzthiazole with aryl chlorides and triflates.

**TABLE 5.4**
**Nickel-Catalyzed Arylation of Benzthiazole with Aryl Chlorides and Triflates**

| Ar 14 | X | Yield 13 (%) |
|---|---|---|
| Ph **14a** | Cl | 74 |
| Ph **14a** | OTf | 48 |
| 2-MeC$_6$H$_4$ **14b** | Cl | 85 |
| 4-FC$_6$H$_4$ **14c** | Cl | 65 |
| 3-Thienyl **14d** | Cl | 58 |

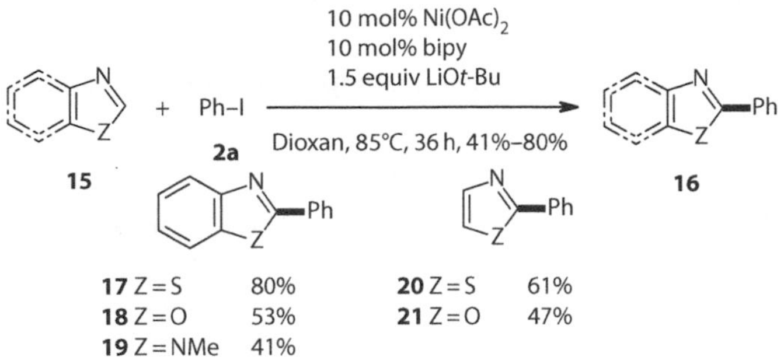

**SCHEME 5.6** Nickel-catalyzed arylation of azoles with aryl iodides.

were coupled with electron-neutral or electron-rich substituents, such as fluoride, bromide, methyl, and methoxy groups, with equal efficiency. In contrast, *ortho*-substituted haloarenes, chlorobenzene, and phenyl triflate were found to couple with inferior efficiency under the magnesium *t*-butoxide conditions. On the other hand, electron-deficient haloarenes, such as iodopyridine, iodonitrobenzene, and ester-substituted iodobenzene, were found to couple with superior arylating efficiency. Moreover, 4-nitroiodobenzene and ethyl 4-iodobenzoate were effectively coupled only under magnesium *t*-butoxide-mediated conditions (Scheme 5.7, Table 5.5). The authors concluded: "*Whereas the LiOt-Bu/1,4-dioxane system generally works well for robust substrates, the Mg(Ot-Bu)₂/DMF system is, in many cases, superior for substrates with base-sensitive functional groups. It should also be mentioned that Mg(Ot-Bu)₂ is less expensive than LiOt-Bu.*"[10]

More recently, Itami and coworkers described a new catalyst system for the coupling of phenol derivatives **26** (pivalates, carbamates, carbonates, sulfamates, triflates, tosylates, and mesylates) with azoles **25** in the presence of a Ni(cod)₂. dcype catalyst in generally excellent yields (80%–99%). Solvents used were dioxan, toluene, and DMF, compatible bases were caesium carbonate, potassium phosphate, and lithium *t*-butoxide, and reactions were generally run at 120°C (Scheme 5.8, Table 5.6).[11] Monocylic azoles behaved in a similar fashion to afford **28**, **29**, and **30** in good yield.

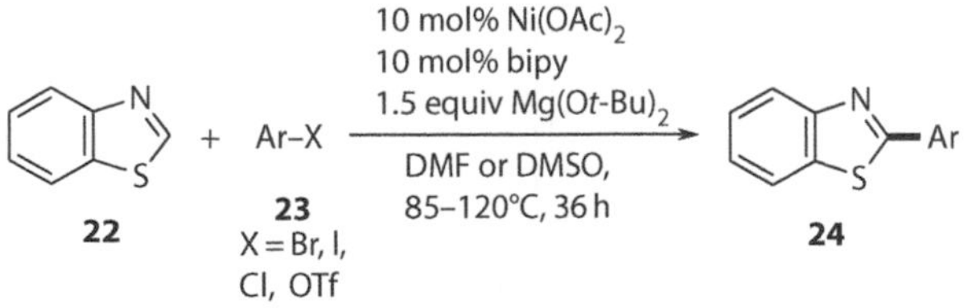

**SCHEME 5.7** Nickel-catalyzed arylation of azoles, comparison of LiOt-Bu and Mg(Ot-Bu)₂ protocols.

## TABLE 5.5
## Nickel-Catalyzed Arylation of Azoles, Comparison of LiO*t*-Bu and Mg(O*t*-Bu)$_2$ Protocols

| Superior | | | Equal | | | Inferior | | |
|---|---|---|---|---|---|---|---|---|
| Ar 23 | X | Yield 24 (%) | Ar 23 | X | Yield 24 (%) | Ar 23 | X | Yield 24 (%) |
| 4-F$_3$CC$_6$H$_4$ | I | 84 (68) | 3-F$_3$CC$_6$H$_4$ | I | 75 (75) | 2-F$_3$CC$_6$H$_4$ | I | 33 (72) |
| 3-Pyridyl | I | 85 (57) | Ph | Br | 57 (62) | 2-PhC$_6$H$_4$ | I | 26 (91) |
| 4-O$_2$NC$_6$H$_4$ | I | 29 (<1) | 4-FC$_6$H$_4$ | I | 61 (65) | 2-MeC$_6$H$_4$ | I | 58 (90) |
| 3-NCCC$_6$H$_4$ | I | 87 (47) | 4-MeC$_6$H$_4$ | I | 69 (69) | 1-Naphthyl | I | 63 (91) |
| 4-EtO$_2$CC$_6$H$_4$ | I | 48 (<1) | 3,5-Me$_2$C$_6$H$_4$ | I | 57 (65) | Ph | Cl | <5 (74) |
| | | | 3,4,5-(MeO)$_3$C$_6$H$_4$ | Br | 73 (61) | Ph | TfO | <5 (48) |

*Note:* Yields in parentheses are under the previously published conditions using lithium *t*-butoxide.

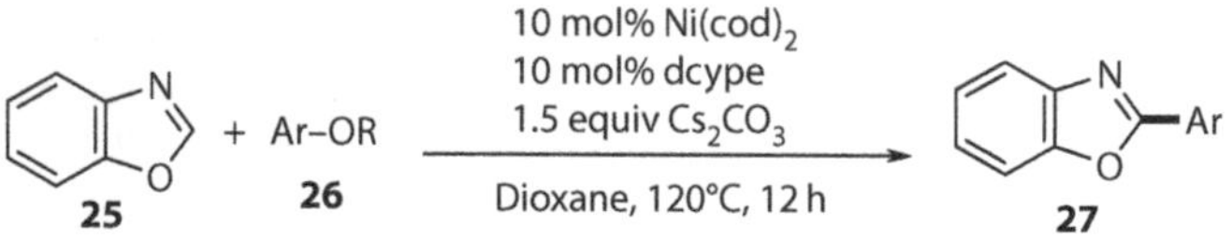

SCHEME 5.8   Nickel-catalyzed arylation of azoles with phenol derivatives.

## TABLE 5.6
## Nickel-Catalyzed Arylation of Azoles with Phenol Derivatives

| Ar 26 | X | Yield 27 (%) |
|---|---|---|
| 2-Naphthyl | CO$_2$*t*-Bu | 95 |
| 1-Naphthyl | CO$_2$*t*-Bu | 90 |
| Ph | CO$_2$*t*-Bu | 11 |
| Ph | Tf | 75 |
| 3-MeC$_6$H$_4$ | Tf | 85 |
| 4-MeOC$_6$H$_4$ | Tf | 55 |
| 4-F$_3$CC$_6$H$_4$ | Tf | 52 |
| 3-Pyridyl | CO$_2$*t*-Bu | 81 |
| 6-Quinolinyl | CO$_2$*t*-Bu | >99 |
| 6-Quinolinyl | CO$_2$*t*-Bu | 86 |

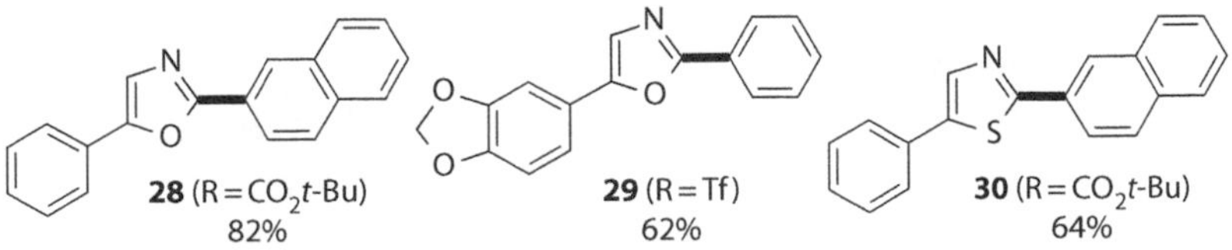

**28** (R = CO$_2$t-Bu)
82%

**29** (R = Tf)
62%

**30** (R = CO$_2$t-Bu)
64%

Iaroshenko et al. have recently described nickel-catalyzed arylations of 1-deazapurines **32** with aryl halides, using a [NiCl$_2$(PPh$_3$)$_2$] catalyst and potassium carbonate in DMF at 110°C. Yields were generally excellent (88%–97%) (Scheme 5.9, Table 5.7).[12]

Tobisu et al. have described very closely related C2 arylations of a range of electron-deficient heteroaromatics using organozinc reagents. This is conceptually related to the alkylation of arenes with Grignard reagents described below, Section 5.4.3. The reaction works with *in situ* formed organozinc reagents as well as preformed reagents such as **35** (Scheme 5.10).[13] Gartia et al. have reported a single example of the nickel-catalyzed arylation at the 2-position of furan with phenylmagnesium

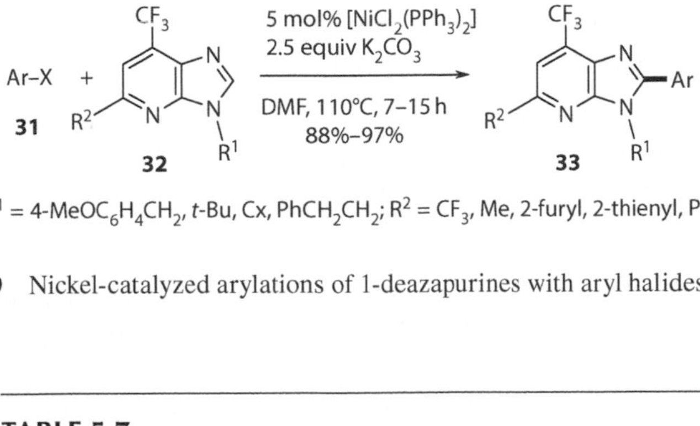

R$^1$ = 4-MeOC$_6$H$_4$CH$_2$, t-Bu, Cx, PhCH$_2$CH$_2$; R$^2$ = CF$_3$, Me, 2-furyl, 2-thienyl, Ph.

**SCHEME 5.9** Nickel-catalyzed arylations of 1-deazapurines with aryl halides.

**TABLE 5.7**
**Nickel-Catalyzed Arylations of 1-Deazapurines with Aryl Halides**

| Ar-X 31 | Time (h) | Yield 33 (%) |
|---|---|---|
| 4-MeC$_6$H$_4$I | 7–9 | 89–95 |
| 2,4-Me$_2$C$_6$H$_3$I | 7 | 92 |
| 4-EtC$_6$H$_4$I | 7–8 | 89–97 |
| PhBr | 12–13 | 94–95 |
| 2-FC$_6$H$_4$Br | 14–15 | 91–92 |
| 3-F$_3$CC$_6$H$_4$Br | 12 | 88 |
| 4-F$_3$CC$_6$H$_4$Cl | 7 | 93 |

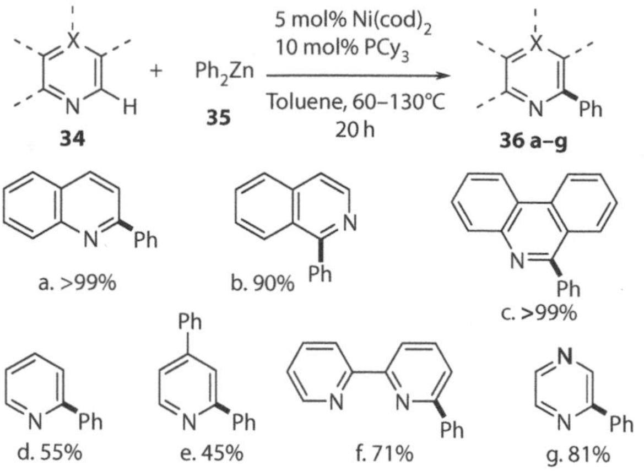

**SCHEME 5.10**   Nickel-catalyzed arylation of pyridines with organozinc reagents.

chloride. This is discussed in more detail in Section 5.4.3 in the context of nickel-catalyzed alkylation of aryl centers.[36]

## 5.3   AROMATIC C(sp²)—H C—H ACTIVATION: ARYL–VINYL BOND FORMATIONS

Recent developments in this area have been well summarized in three excellent reviews[3,5,14]: readers are directed toward these articles, and references cited therein. Some highlights from this area are summarized below.

Hiyama, Nakao and coworkers have described the direct alkenylation of heteroarenes **37** by reaction with alkynes **38** in the presence of a Ni(cod)$_2$/PCyp$_3$ catalyst in toluene at 35–100°C. The use of a bulky tri(sec-alkyl) phosphine ligand was crucial to the selective C—H activation over the competing C—CN insertion in nitrile-containing substrates, for example, **48**. The reaction was selective for aryl C—H over formyl C—H bond activation, was tolerant of a range of functional groups and even succeeded in the presence of substituents that might be susceptible to oxidative insertion and subsequent cross-coupling, such as Cl (Scheme 5.11).[15]

The same group later described the preparation of a bench-stable catalyst suitable for the hydroheteroarylation reaction.[16]

Miura and coworkers recently described an application of this system to the alkenylation of oxadiazoles: yields that ranged from 26% to 100%, with E/Z ratios from 79:21 to 99:1 (Scheme 5.12).[17]

Hiyama's group went on to show that the same catalyst system could be used to alkenylate pyridine-N-oxides. (Scheme 5.13) Yields were in the range 54–81%, E/Z ratios ranged from 93:7 to >99:1. For unsymmetrical alkynes, the regiochemistry was driven by arylation at the less-hindered terminal of the triple-bond.

They also demonstrated the efficient deoxygenation of the alkenylated pyridine-N-oxides to pyridines.[18]

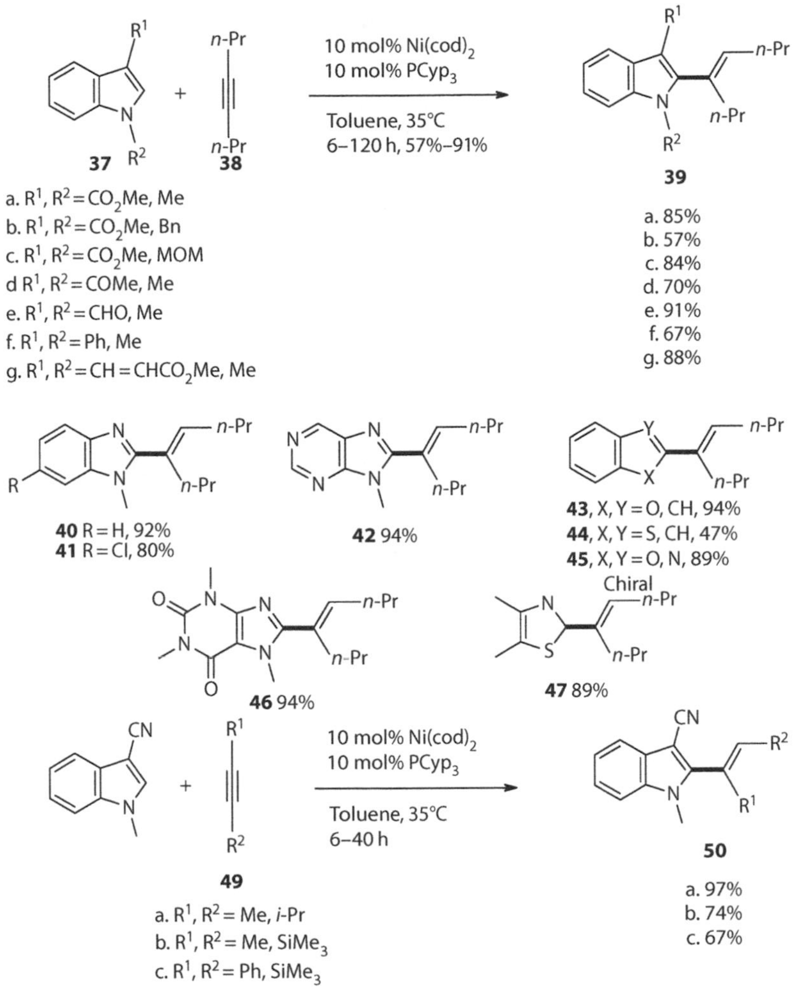
37
38
10 mol% Ni(cod)2
10 mol% PCyp3
Toluene, 35°C
6–120 h, 57%–91%
39
a. R1, R2 = CO2Me, Me
b. R1, R2 = CO2Me, Bn
c. R1, R2 = CO2Me, MOM
d R1, R2 = COMe, Me
e. R1, R2 = CHO, Me
f. R1, R2 = Ph, Me
g. R1, R2 = CH = CHCO2Me, Me
a. 85%
b. 57%
c. 84%
d. 70%
e. 91%
f. 67%
g. 88%
40 R = H, 92%
41 R = Cl, 80%
42 94%
43, X, Y = O, CH, 94%
44, X, Y = S, CH, 47%
45, X, Y = O, N, 89%
Chiral
46 94%
47 89%
10 mol% Ni(cod)2
10 mol% PCyp3
Toluene, 35°C
6–40 h
49
a. R1, R2 = Me, i-Pr
b. R1, R2 = Me, SiMe3
c. R1, R2 = Ph, SiMe3
50
a. 97%
b. 74%
c. 67%

**SCHEME 5.11** Nickel-catalyzed alkenylation of heteroarenes with alkynes.

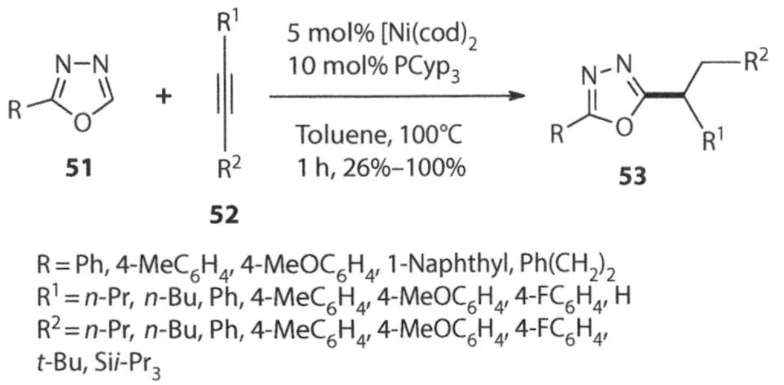
51
52
5 mol% [Ni(cod)2
10 mol% PCyp3
Toluene, 100°C
1 h, 26%–100%
53
R = Ph, 4-MeC6H4, 4-MeOC6H4, 1-Naphthyl, Ph(CH2)2
R1 = n-Pr, n-Bu, Ph, 4-MeC6H4, 4-MeOC6H4, 4-FC6H4, H
R2 = n-Pr, n-Bu, Ph, 4-MeC6H4, 4-MeOC6H4, 4-FC6H4,
t-Bu, Sii-Pr3

**SCHEME 5.12** Nickel-catalyzed alkenylation of 1,3,4-oxadiazoles with alkynes.

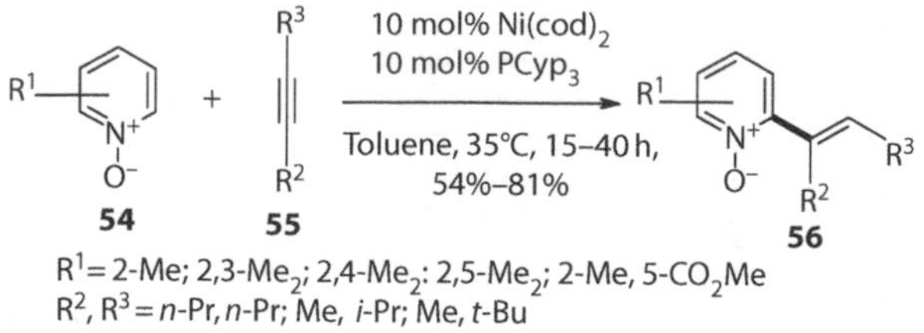

$R^1 = 2\text{-Me}; 2,3\text{-Me}_2; 2,4\text{-Me}_2: 2,5\text{-Me}_2; 2\text{-Me}, 5\text{-CO}_2\text{Me}$
$R^2, R^3 = n\text{-Pr}, n\text{-Pr}; \text{Me}, i\text{-Pr}; \text{Me}, t\text{-Bu}$

**SCHEME 5.13**   Nickel-catalyzed alkenylation of pyridine *N*-oxides with alkynes.

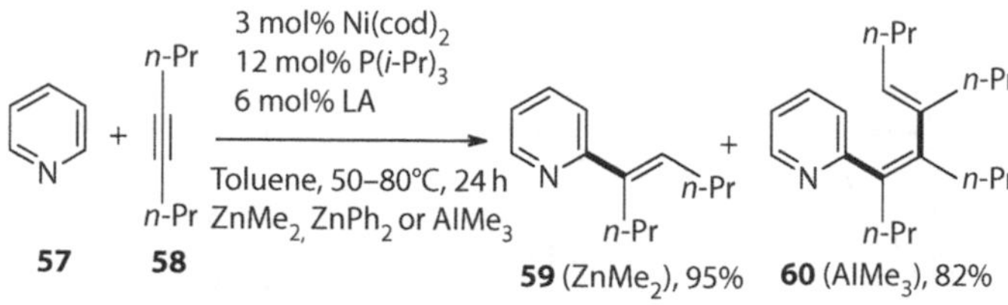

**SCHEME 5.14**   Nickel-catalyzed alkenylation of pyridines.

The direct C2 alkenylation of pyridines could be achieved by the use of a Lewis acid cocatalyst, such as trimethylaluminum, dimethyl-, or diphenylzinc. Trimethylaluminum favored formation of dienes **60**, zinc-based Lewis acids preferentially gave monoenes **59** (Scheme 5.14).[19]

These conditions are compatible with a range of functional groups and are not confined to pyridines (Scheme 5.15).

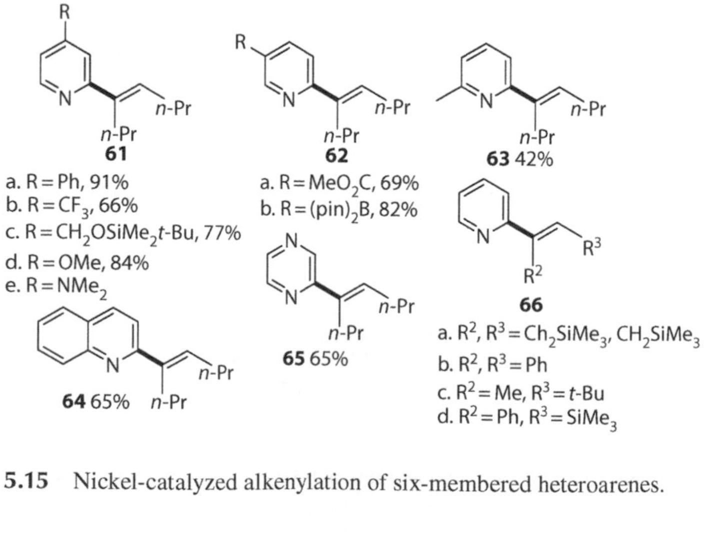

**SCHEME 5.15**   Nickel-catalyzed alkenylation of six-membered heteroarenes.

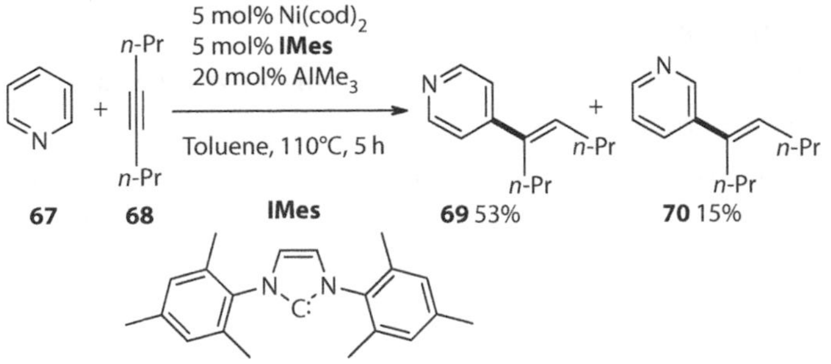

**SCHEME 5.16**   Nickel-catalyzed C3/C4 alkenylation of pyridines.

Hiyama, Nakao and coworkers were able to achieve direct C3/C4 alkenylation of pyridines **67** using the highly hindered NHC ligand **IMes** (1,3-(2,4,6-trimethylphenyl) imidazol-2-ylidene), and trimethylaluminum as Lewis acid cocatalyst (Scheme 5.16).[20]

Yap et al. have shed some light on the mechanistic details involved in these processes through their direct observation of η2,η1-Pyridine Ni(0)-Al(III) intermediates, using a combination of AlMe$_3$ and amino-NHC cocatalysts (Scheme 5.17, Table 5.8).[21]

Hiyama, Nakao, and coworkers have achieved the direct C6 alkenylation of 2-pyridones and related heterocycles, using Ni(cod)$_2$, P(i-Pr)$_3$, and AlMe$_3$ in toluene at 80°C. The reaction is limited to N-alkylpyridones but is otherwise of broad scope and synthetically useful yield (Scheme 5.18).[22]

Chatani have exploited the nickel-catalyzed chelation-assisted alkenylation of aryl C—H bonds in the synthesis of a wide variety of substituted isoquinolones **92–103** (Scheme 5.19).[23]

In a further extension of their earlier work, Hiyama, Nakao, and coworkers described the alkenylation of fluoroarenes **104** by reaction with alkynes **105** in the presence of a Ni(cod)$_2$/PCyp$_3$ catalyst in toluene at 80°C. More highly fluorinated arenes underwent coupling more quickly, and in higher yield, than less highly fluorinated substrates, which the authors ascribed to lower acidity of the C—H bond

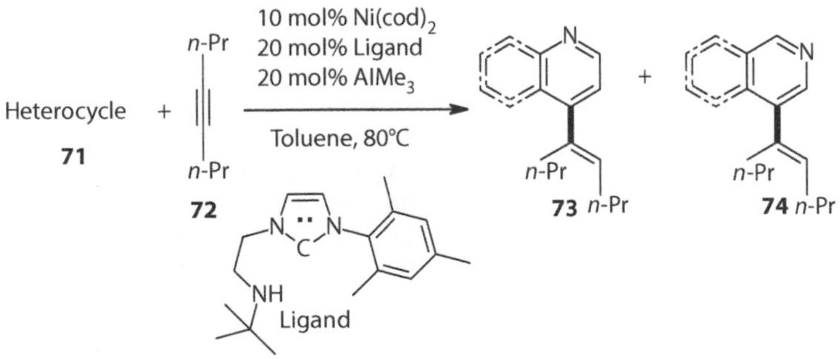

**SCHEME 5.17**   Nickel-catalyzed C3/C4 alkenylation of pyridines and fused pyridines.

**TABLE 5.8**
**Nickel-Catalyzed C3/C4 Alkenylation of Pyridines and Fused Pyridines**

| Heterocycle 71 | Yield (%) | Ratio 73:74 |
|---|---|---|
| Pyridine | 85 | 10:3 |
| 2-MePyridine | 56 | 1:0 |
| 3-MePyridine | 85 | 5:2 |
| 4-MePyridine | 34 | 0:1 |
| 3,5-$Me_2$Pyridine | 20 | 1:0 |
| 2-MeOPyridine | 49 | 1:0 |
| 3-MeOPyridine | 89 | 1:0 |
| 2-PhPyridine | 19 | 4:3 |
| 3-PhPyridine | 71 | 1:1 |
| Quinoline | 56 | 10:1 |
| 7-MeQuinoline | 82 | 10:1 |

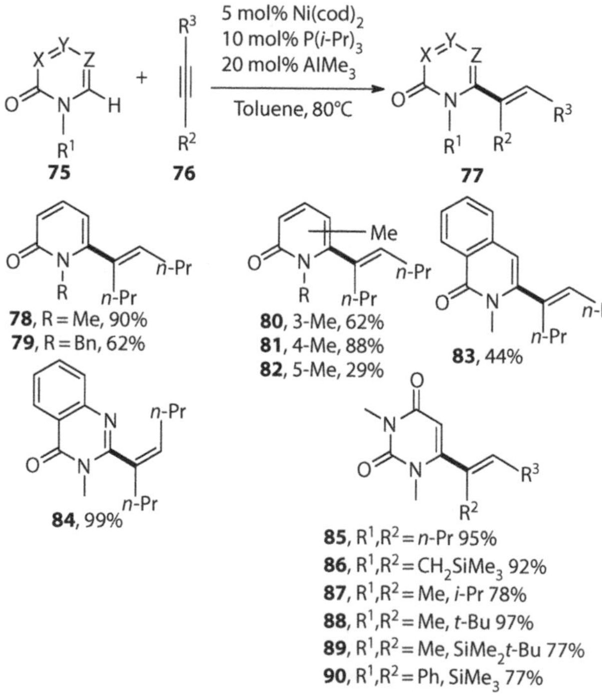

5 mol% Ni(cod)2
10 mol% P(i-Pr)3
20 mol% AlMe3
Toluene, 80°C
75
76
77
78, R = Me, 90%
79, R = Bn, 62%
80, 3-Me, 62%
81, 4-Me, 88%
82, 5-Me, 29%
83, 44%
84, 99%
85, R1,R2 = n-Pr 95%
86, R1,R2 = CH2SiMe3 92%
87, R1,R2 = Me, i-Pr 78%
88, R1,R2 = Me, t-Bu 97%
89, R1,R2 = Me, SiMe2t-Bu 77%
90, R1,R2 = Ph, SiMe3 77%


**SCHEME 5.18** C6 alkenylation of 2-pyridones and related heterocycles.

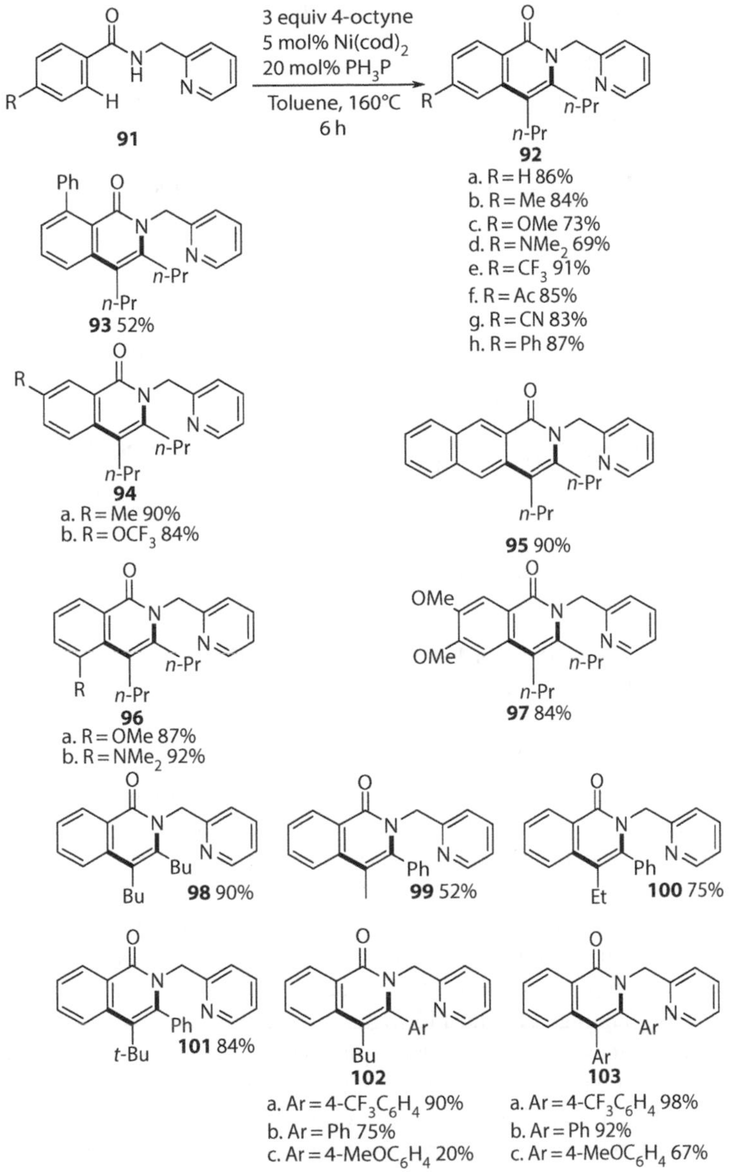

**SCHEME 5.19** Synthesis of substituted isoquinolines *via* nickel-catalyzed alkenylation of benzamides.

undergoing reaction. In unsymmetrical substrates the primary reaction site was generally at the most acidic C—H position (Scheme 5.20).[24] Perutz and coworkers have addressed the mechanism of Hiyama's nickel-catalyzed hydrofluoroarylation reactions *via* density functional theory (DFT) calculations and examined how the reactions are influenced by fluorine substitution.[25]

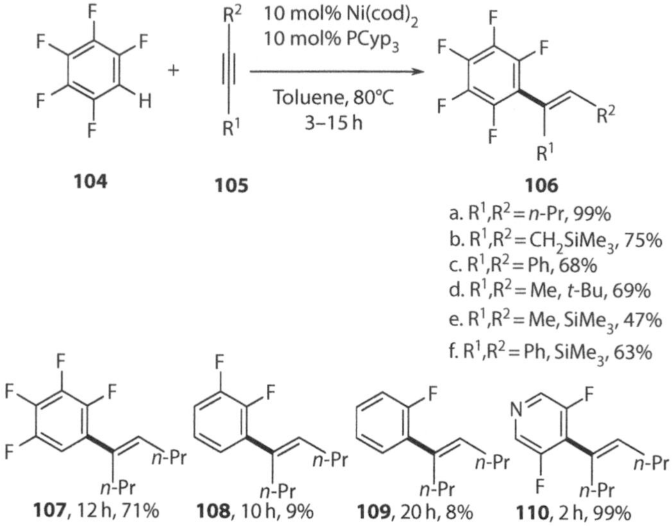
10 mol% Ni(cod)$_2$
10 mol% PCyp$_3$
Toluene, 80°C
3–15 h
104
105
106
a. R$^1$,R$^2$ = n-Pr, 99%
b. R$^1$,R$^2$ = CH$_2$SiMe$_3$, 75%
c. R$^1$,R$^2$ = Ph, 68%
d. R$^1$,R$^2$ = Me, t-Bu, 69%
e. R$^1$,R$^2$ = Me, SiMe$_3$, 47%
f. R$^1$,R$^2$ = Ph, SiMe$_3$, 63%
107, 12 h, 71%
108, 10 h, 9%
109, 20 h, 8%
110, 2 h, 99%

**SCHEME 5.20** Nickel-catalyzed alkenylation of fluoroarenes.

## 5.4 AROMATIC C(sp$^2$)—H C—H ACTIVATION: ARYL–ALKYL BOND FORMATIONS

### 5.4.1 Nickel-Catalyzed Hydroarylation of Alkenes

Under the same conditions, successfully deployed for the hydroarylation of alkynes, Hiyama, Nakao et al. were also able to achieve alkene hydroarylation to give alkylation products **111** and **112** in good yield, although the range of examples initially described was somewhat limited (Scheme 5.21).[24]

They later extended the utility of this reaction to the C6 alkylation of 2-pyridones, by the incorporation of trimethylaluminum as a Lewis acid cocatalyst. They were successful in accomplishing both inter-molecular couplings, to give **116–118** and intra-molecular couplings **114** and **115** (Scheme 5.22).[22]

Using NHC ligands instead of hindered phosphines, Hiyama's group was successful in alkylating a range of heteroarenes with vinylarenes (Scheme 5.23).[26]

Most recently, Hiyama, Nakao, and coworkers have been able to achieve direct C4 alkylation of pyridines **129** using the highly hindered Lewis acid cocatalyst (2,6-($t$-Bu)$_2$-4-MeC$_6$H$_2$O)$_2$AlMe (MAD) (Scheme 5.24).[20]

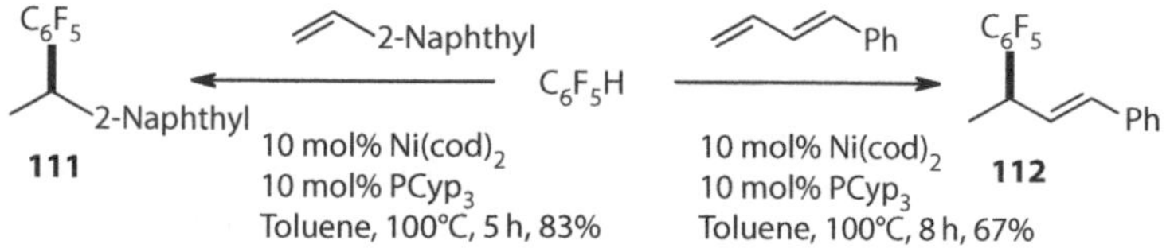
2-Naphthyl
C$_6$F$_5$H
Ph
111
112
10 mol% Ni(cod)$_2$
10 mol% PCyp$_3$
Toluene, 100°C, 5 h, 83%
10 mol% Ni(cod)$_2$
10 mol% PCyp$_3$
Toluene, 100°C, 8 h, 67%

**SCHEME 5.21** Nickel-catalyzed hydroarylation of alkenes.

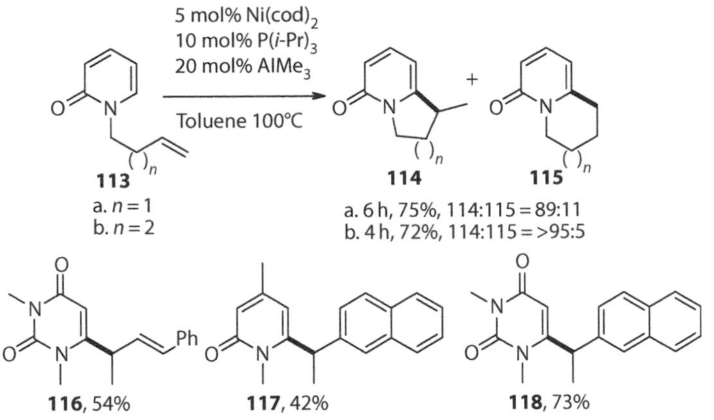

**SCHEME 5.22**   Nickel-catalyzed C6 alkylation of 2-pyridones.

Ong and coworkers have successfully applied the Ni(cod)$_2$/NHC catalyst combination to the tandem isomerization of allylarenes and subsequent regioselective hydroheteroarylation. Through appropriate choice of NHC it was possible to select for linear or branched products: IMes preferentially gave the branched products (Scheme 5.25), the more hindered NHC, IPr, preferentially gave the linear product, and this preference was amplified to near-complete selectivity by the addition of trimethylaluminum (Scheme 5.26).[27]

Doster and Johnson observed that the reaction of pentafluorobenzene with a trialkylvinyl stannane in the presence of a nickel catalyst derived from Ni(COD)$_2$ and an ancillary ligand such as MeNC$_5$H$_4$N$i$-Pr or $i$-Pr$_3$P produced either the aryl stannane

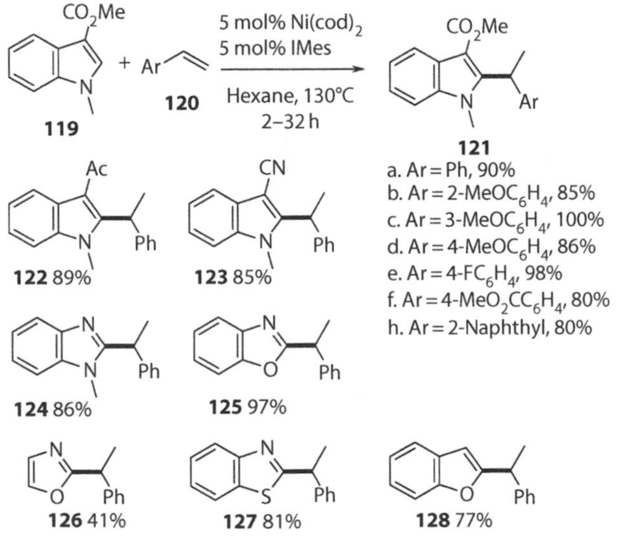

**SCHEME 5.23**   Nickel-catalyzed alkylation of five-membered heteroarenes.

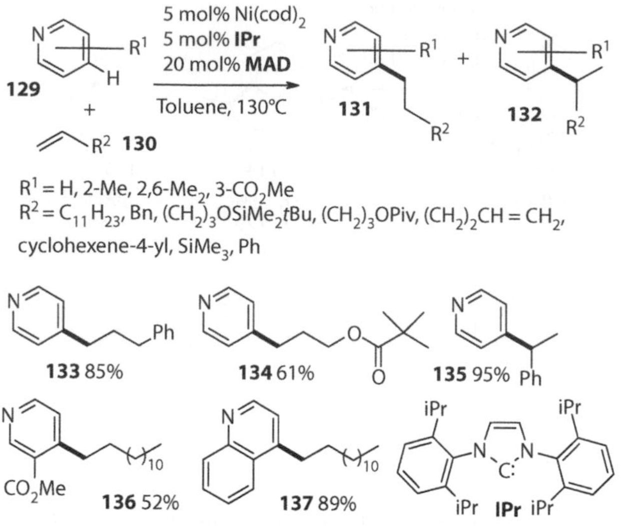

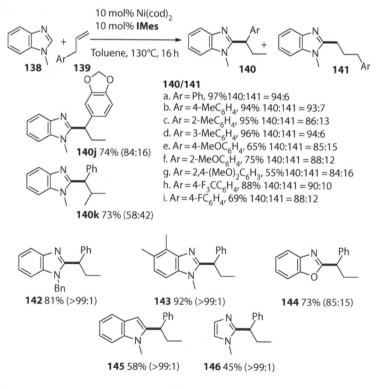

5 mol% Ni(cod)$_2$
5 mol% **IPr**
20 mol% **MAD**
Toluene, 130°C
129
130
131
132
R$^1$ = H, 2-Me, 2,6-Me$_2$, 3-CO$_2$Me
R$^2$ = C$_{11}$H$_{23}$, Bn, (CH$_2$)$_3$OSiMe$_2$tBu, (CH$_2$)$_3$OPiv, (CH$_2$)$_2$CH = CH$_2$, cyclohexene-4-yl, SiMe$_3$, Ph
**133** 85%
**134** 61%
**135** 95%
**136** 52%
**137** 89%
**IPr**

**SCHEME 5.24** Nickel-catalyzed direct C4 alkylation of pyridines.

10 mol% Ni(cod)$_2$
10 mol% **IMes**
Toluene, 130°C, 16 h
138
139
140
141
**140/141**
a. Ar = Ph, 97%140:141 = 94:6
b. Ar = 4-MeC$_6$H$_4$, 94% 140:141 = 93:7
c. Ar = 2-MeC$_6$H$_4$, 95% 140:141 = 86:13
d. Ar = 3-MeC$_6$H$_4$, 96% 140:141 = 94:6
e. Ar = 4-MeOC$_6$H$_4$, 65% 140:141 = 85:15
f. Ar = 2-MeOC$_6$H$_4$, 75% 140:141 = 88:12
g. Ar = 2,4-(MeO)$_2$C$_6$H$_3$, 55%140:141 = 84:16
h. Ar = 4-F$_3$CC$_6$H$_4$, 88% 140:141 = 90:10
i. Ar = 4-FC$_6$H$_4$, 69% 140:141 = 88:12
**140j** 74% (84:16)
**140k** 73% (58:42)
**142** 81% (>99:1)
**143** 92% (>99:1)
**144** 73% (85:15)
**145** 58% (>99:1)
**146** 45% (>99:1)

**SCHEME 5.25** Nickel-catalyzed alkylation of heteroarenes, branched selectivity.

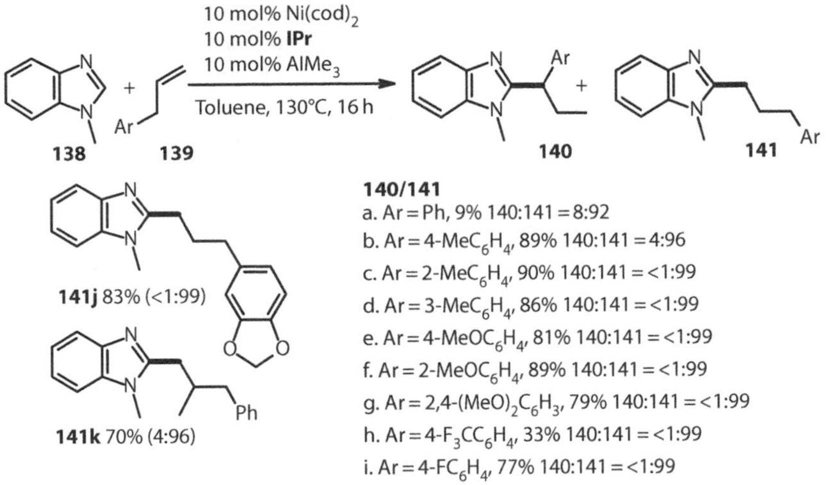

**SCHEME 5.26** Nickel-catalyzed alkylation of heteroarenes, linear selectivity.

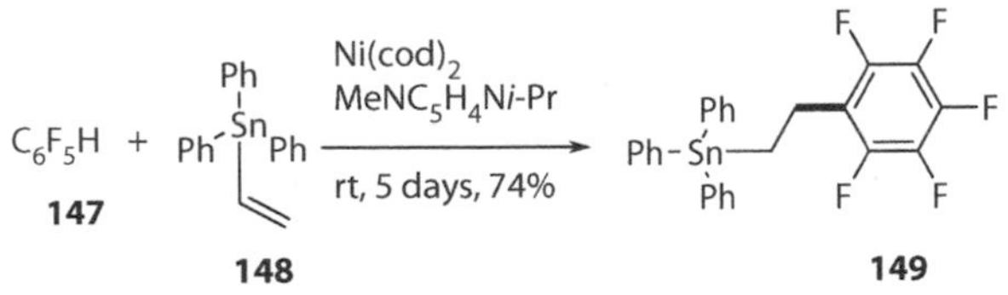

**SCHEME 5.27** Nickel-catalyzed hydroarylation of vinyl stannanes.

or the alkylated arene. The C—H stannylation pathway was favored for tributyl-stannanes; the alkylation pathway was favored by triphenylstannanes (Scheme 5.27). They have also optimized the stannylation route as a means to access otherwise inaccessible aryltributylstannanes (Section 5.5.2). They carried out extensive investigations and proposed a mechanistic manifold to account for their observations.[28]

Shi and coworkers have recently described their extensive computational investigation of the mechanistic origins of regioselectivity in nickel-catalyzed alkene hydroheteroarylation through C—H Activation.[29]

### 5.4.2 NICKEL-CATALYZED COUPLING OF ARENES WITH ALKYL HALIDES

Benzoxazoles and a range of other five-membered heterocycles have been successfully coupled with alkyl halides (chlorides, bromides, and iodides) using a nickel pincer complex, Nickamine **150**.

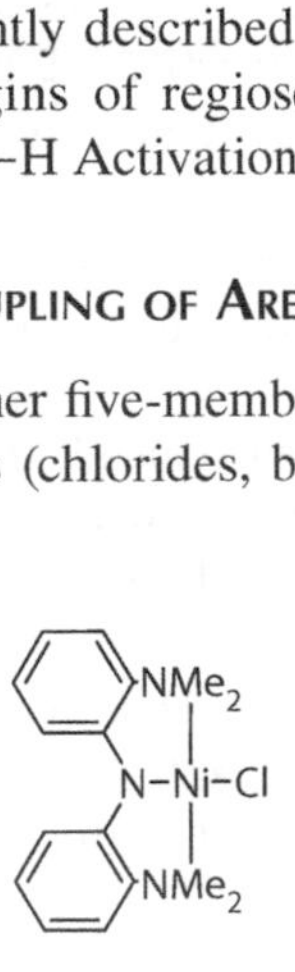

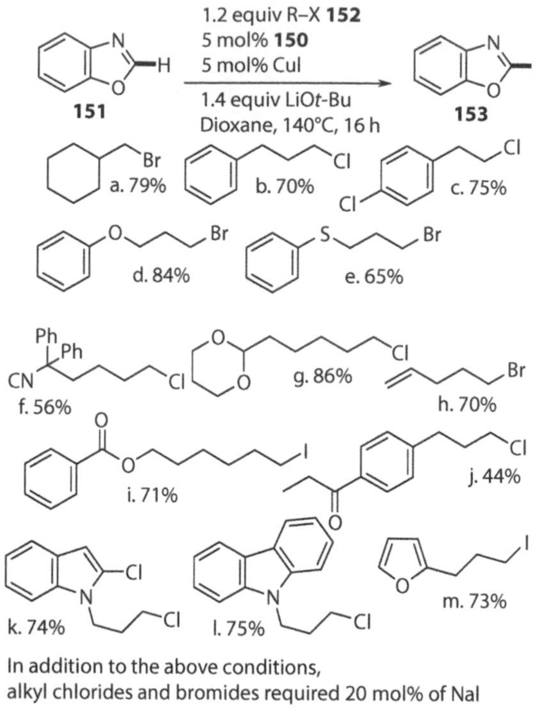

**SCHEME 5.28**  Nickel-catalyzed alkylation of benzoxazole with alkyl halides.

Copper iodide was required as a cocatalyst for successful coupling, and sodium iodide was required when coupling chlorides and bromides. Both electron-rich and -poor heterocycles were successful coupling partners (Scheme 5.28).[30]

The tolerated structural diversity in the heteroaryl coupling partner **154** is illustrated in Scheme 5.29.

Ackermann has recently described the direct alkylation of heteroarenes (benzoxazoles and benzthiazoles) with unactivated primary alkyl halides, using a user-friendly, nitrogen and phosphorus ligand-free catalytic system derived from inexpensive [(diglyme)NiBr$_2$]. CuI was found to be required as a cocatalyst and NaI was beneficial when using alkyl chlorides and bromides. However, the authors noted that the one secondary alkyl halide studied, 2-bromohexane, did not couple in an acceptable yield (Scheme 5.30).[31]

More recently, Ackermann has introduced the use of the chelating ligand BDMAE (bis(2-dimethylaminoethyl)ether), which makes possible the efficient coupling of secondary alkyl halides, even relatively unreactive secondary alkyl chlorides (Scheme 5.31).

Under these conditions, alkene C(sp$^2$)—H activation has been demonstrated; e.g., cyclohexene **166** was isolated in 58% yield. Alkylation in the 2-position of indole substrates, and previously unprecedented trifluoroethylations were also reported (Scheme 5.32).[32]

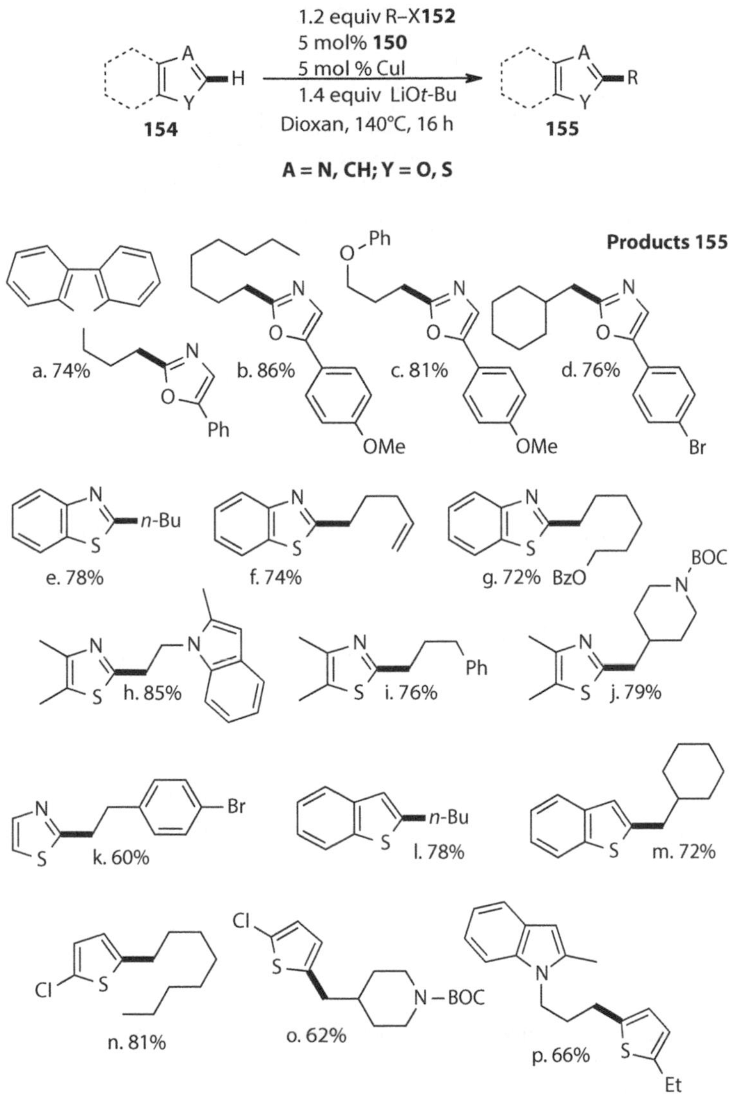

**SCHEME 5.29** Nickel-catalyzed alkylation of heteroarenes with alkyl halides.

Charette has demonstrated an intra-molecular alkylation–cyclization, promoted by Ni(PPh$_3$)$_4$ and NaHMDS. The mechanism was shown to be radical-based by addition of TEMPO or other radical scavengers (Scheme 5.33).[33,34]

### 5.4.3 Nickel-Catalyzed Coupling of Arenes with Grignard Reagents and Related Organometallics

Guo and coworkers have described a novel protocol for nickel-catalyzed alkylation of N-aromatic heterocycles using a Grignard reagent as the coupling partner. This

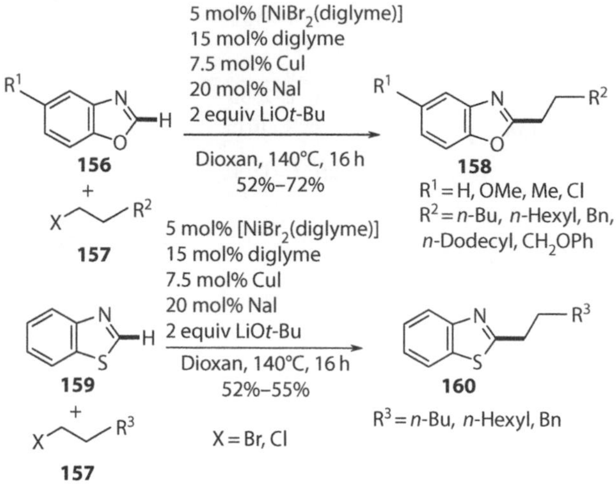

**SCHEME 5.30** Nickel-catalyzed direct alkylation of heteroarenes with primary alkyl halides.

new reaction proceeded efficiently at room temperature, providing access to a variety of alkylated heterocycles. Purines **182** and benzimidazoles **185** were particularly effective substrates; benzthiazoles **188** did not couple in such high yields (Scheme 5.34).[35]

Gartia et al. recently described a related coupling of furan with a small range of Grignard reagents using a nickel pincer catalyst **190** (Scheme 5.35).[36] The first example is more properly a case of C(sp$^2$)—C(sp$^2$) bond formation, and may be

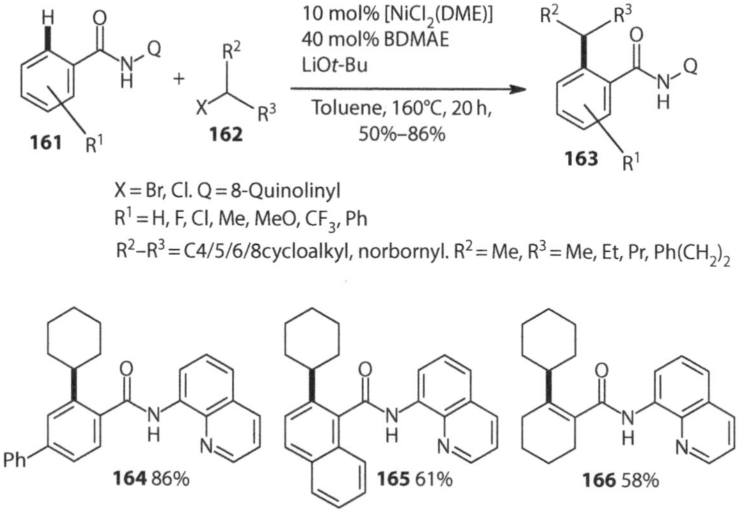

X = Br, Cl. Q = 8-Quinolinyl
R$^1$ = H, F, Cl, Me, MeO, CF$_3$, Ph
R$^2$–R$^3$ = C4/5/6/8cycloalkyl, norbornyl. R$^2$ = Me, R$^3$ = Me, Et, Pr, Ph(CH$_2$)$_2$

**SCHEME 5.31** Nickel-catalyzed alkylation of arenes with secondary alkyl halides.

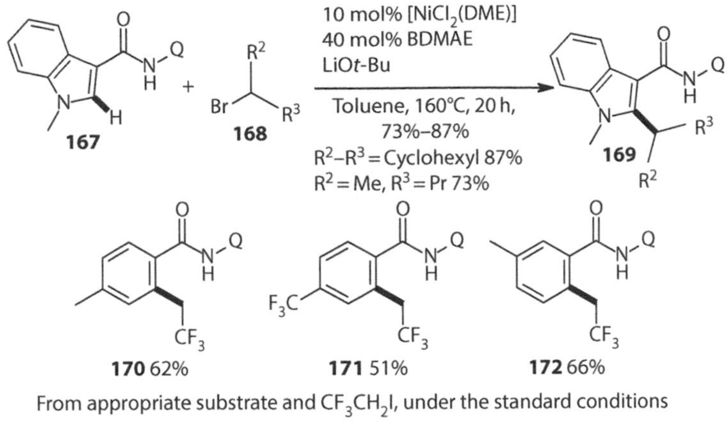

**SCHEME 5.32** Nickel-catalyzed alkylation of indoles and trifluoroethylation of arenes.

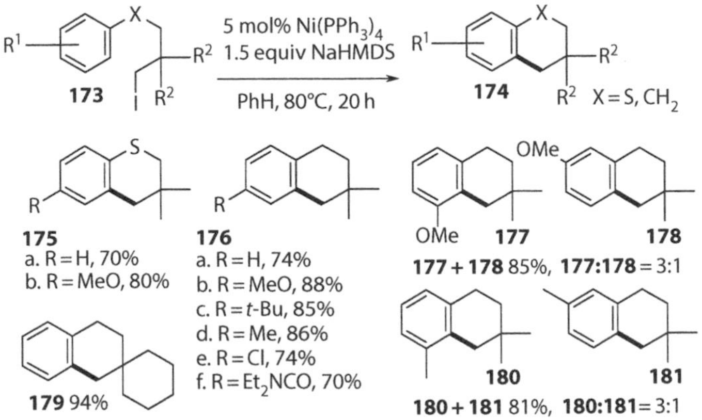

**SCHEME 5.33** Nickel-catalyzed intra-molecular alkylation–cyclization.

related to the nickel-catalyzed arylation of sp² centers with diphenylzinc described in Section 5.2.[13]

## 5.5 AROMATIC C(sp²)—H C—H ACTIVATION: ARYL–HETEROATOM BOND FORMATIONS

### 5.5.1 NICKEL-CATALYZED AMINATION OF ARENES AND HETEROARENES

In an alternative to Buchwald–Hartwig aminations, obviating the need for reactive aryl halides, sulfonates, expensive palladium catalysts and ligands, Li, Dean et al. have recently described the successful amination of substituted benzoxazoles in the 2-position, with a range of secondary amines. Their optimized conditions used $Ni(OAc)_2 \cdot 4H_2O$ as catalyst with TBHP as oxidant, propionic acid in aceto nitrile.

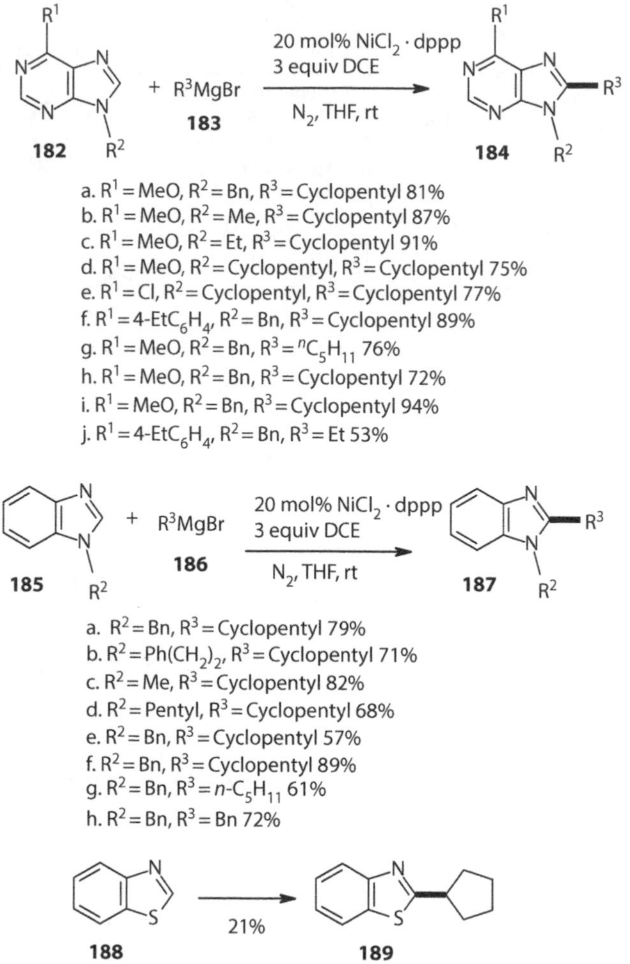

**SCHEME 5.34** Nickel-catalyzed alkylation of heterocycles with grignard reagents.

A mixture of the amine (1.2 equivalents), azole (1 equivalents) was heated with the catalyst system in a sealed tube at 70°C for 12 h. Yields ranged from 26% to 82%. Benzthiazoles and benzimidazoles failed to couple under these conditions (Scheme 5.36, Table 5.9).[37]

## 5.5.2 NICKEL-CATALYZED STANNYLATION OF ARENES AND HETEROARENES

As previously mentioned, Section 5.4.1, Doster and Johnson have demonstrated that the nickel-catalyzed reaction of fluorinated arenes and pyridines with tributylvinyl-stannane provides new aryltributylstannanes via C—H functionalization with the loss of ethylene. This represents a useful alternative to the existing methods for the preparation of these valuable Stille coupling reagents (Scheme 5.37).[38]

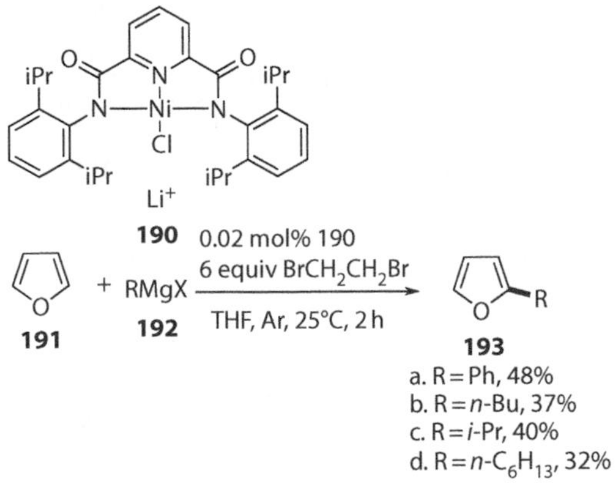

**SCHEME 5.35**  Nickel-catalyzed alkylation of furan with grignard reagents.

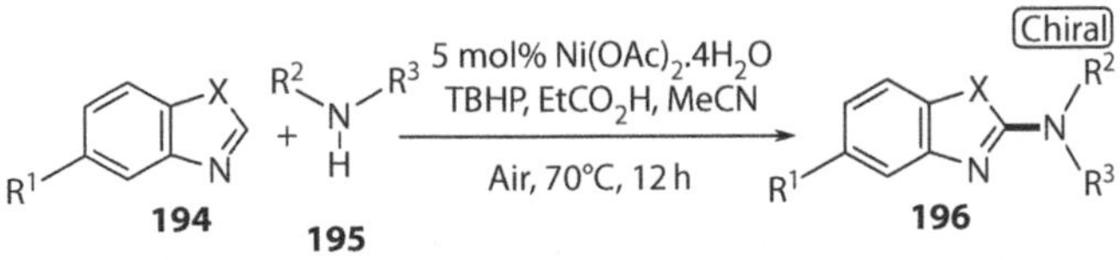

**SCHEME 5.36**  Nickel-catalyzed amination of azoles.

## TABLE 5.9
## Nickel-Catalyzed Amination of Azoles

|       | X  | R¹    | R²                    | R³           | Yield (%) |
|-------|----|-------|-----------------------|--------------|-----------|
| 196a  | O  | H     | n-Bu                  | n-Bu         | 68        |
| 196b  | O  | H     | n-Pr                  | n-Pr         | 73        |
| 196c  | O  | H     | Et                    | Et           | 66        |
| 196d  | O  | 5-Me  | n-Bu                  | n-Bu         | 70        |
| 196e  | O  | 5-Me  | n-Pr                  | n-Pr         | 78        |
| 196f  | O  | 5-Me  | Et                    | Et           | 75        |
| 196g  | O  | 5-Me  | $-(CH_2)_5-$          |              | 77        |
| 196h  | O  | 5-Me  | $-(CH_2)_2CHMe(CH_2)_2-$ |           | 72        |
| 196i  | O  | 5-Me  | Me                    | Bn           | 67        |
| 196j  | O  | 5-Me  | Me                    | 2-ClBn       | 67        |
| 196k  | O  | 5-Me  | $CH_2CH=CH_2$         | $CH_2CH=CH_2$ | 65       |
| 196l  | O  | 5-Me  | Me                    | Ph           | 35        |
| 196m  | O  | 5-tBu | n-Bu                  | n-Bu         | 69        |
| 196n  | O  | 5-Cl  | n-Bu                  | n-Bu         | 57        |
| 196o  | O  | 5-Cl  | $CH_2CH=CH_2$         | $CH_2CH=CH_2$ | 31       |
| 196p  | O  | 5-tBu | $CH_2CH=CH_2$         | $CH_2CH=CH_2$ | 64       |
| 1p6q  | O  | 5-tBu | Me                    | 2-ClBn       | 54        |
| 196r  | O  | 5-tBu | $-(CH_2)_5-$          |              | 70        |
| 196s  | S  | H     | n-Bu                  | n-Bu         | 0         |
| 196t  | NH | H     | n-Bu                  | n-Bu         | 0         |

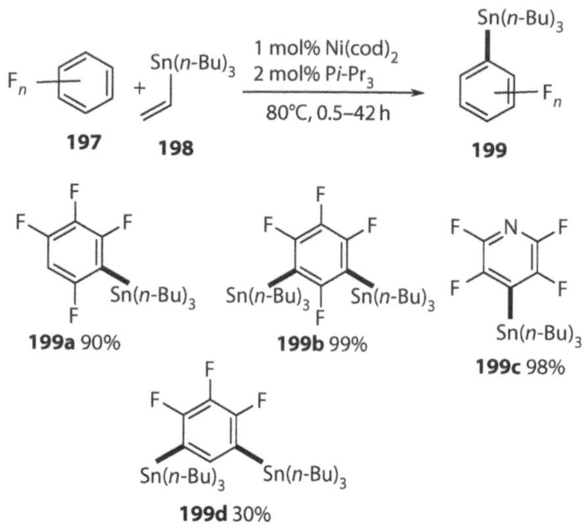

**SCHEME 5.37**  Nickel-catalyzed stannylation of fluoroarenes.

## 5.6   CARBONYL C(sp²)—H C—H ACTIVATION: ACYL–ALKENYL AND ACYL–ALKYL BOND FORMATIONS

Hiyama, Nakao and coworkers have demonstrated the efficient inter-molecular hydrocarbamoylation reactions of alkynes and 1,3-dienes with dialkyformamides, catalyzed by a nickel/Lewis acid combination, to allow direct atom-efficient access to a range of unsaturated amides. Intra-molecular hydrocarbamoylation of olefins also proceeds by the binary catalysis to give lactam derivatives (Scheme 5.38).[39]

In an extension of this approach, Hiyama, Nakao et al. demonstrated that $N,N$-bis(1-arylalkyl)formamides **203** underwent a previously unprecedented dehydrogenative [4 + 2] cycloaddition reaction with alkynes **204** using a nickel catalyst/AlMe$_3$ Lewis acid combination in a double functionalization of C(sp²)—H and C(sp³)—H

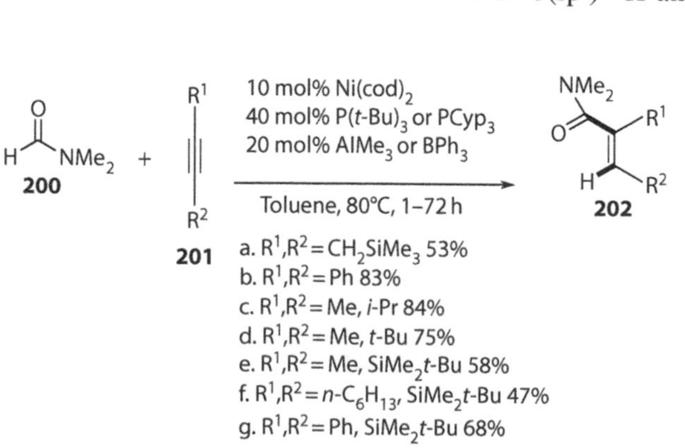

**SCHEME 5.38**  Nickel-catalyzed hydrocarbamoylation of alkynes.

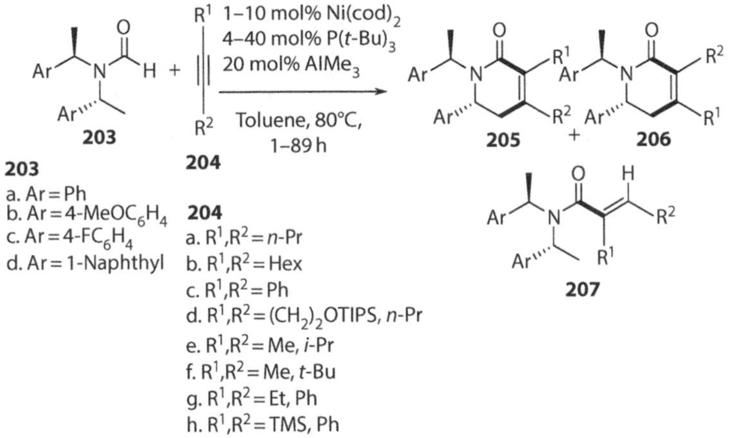

**SCHEME 5.39** Nickel-catalyzed dehydrogenative [4 + 2] cycloaddition of formamides with alkynes.

bonds, to give highly substituted dihydropyridone derivatives **205** and **206**, which are versatile synthetic precursors for nitrogen-containing six-membered heterocycles (Scheme 5.39, Table 5.10).[40,41]

The mechanism of this cooperative catalysis process has been studied using high level DFT calculations.[42] A similar transformation has been reported by Ogata and Fukuzawa; their system did not employ a Lewis acid cocatalyst, but did require a trialkylsilane for success.[43]

Donets and Cramer have developed an asymmetric version of these hydrocarbamoylation reactions using a diaminophosphine oxide ligand (Scheme 5.40, Table 5.11).[44]

## TABLE 5.10
## Nickel-Catalyzed Dehydrogenative [4 + 2] Cycloaddition of Formamides with Alkynes

| 203 | 204 | Yield (205 + 206 + 207, %) | 205:206 | (205 + 206):207 |
|-----|-----|----------------------------|---------|-----------------|
| a | a | 95 | | 96:4 |
| a | b | 85 | | 93:7 |
| a | c | 79 | | >95:5 |
| a | d | 86 | 76:24 | 96:4 |
| a | e | 79 | 62:38 | 80:20 |
| a | f | 63 | >95:5 | 80:20 |
| a | g | 80 | 57:43 | >95:5 |
| a | h | 23 | >95:5 | >95:5 |
| b | a | 99 | | 86:14 |
| c | a | 97 | | 92:8 |
| d | a | 74 | | >95:5 |

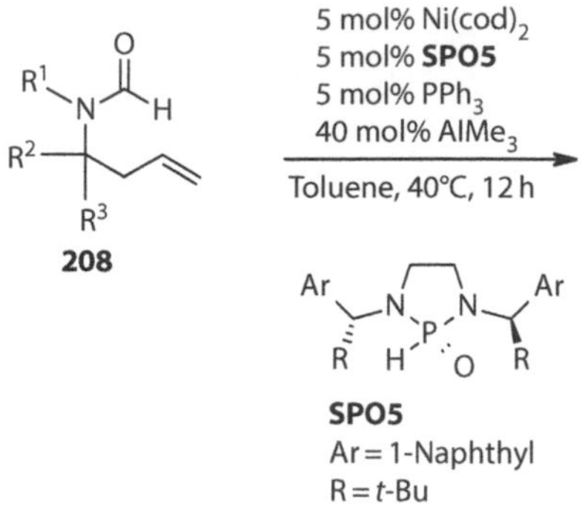

**SCHEME 5.40**   Asymmetric intra-molecular nickel-catalyzed hydroamination of alkenes.

**TABLE 5.11**
**Asymmetric Intra-Molecular Nickel-Catalyzed Hydroamination of Alkenes**

| R¹ | R² | R³ | Yield (%) | er |
|---|---|---|---|---|
| Bn | H | H | 89 | 94:6 |
| Ph(CH₂)₂ | H | H | 79 | 97:3 |
| Ph | H | H | 91 | 96.5:3.5 |
| EtO₂CCH₂ | H | H | 75 | 96:4 |
| Bn | Me | Me | 95 | 89.5:10.5 |
| Me | | –(CH₂)₆– | 82 | 92:8 |
| PMB | H | CH₂CH=CH₂ | 57:43 | >95:5 |
| –(CH₂)₃– | | CH₂CH=CH₂ | 98 | 94.5:5.5 |
| –(CH₂)₄– | | CH₂CH=CH₂ | 83 | 94:6 |
| –(CH₂)₂OCH₂– | | CH₂CH=CH₂ | 90 | 96:4 |
| –(CH₂)₂NMeCH₂– | | CH₂CH=CH₂ | 94 | 96:4 |
| –(CH₂)₅– | | CH₂CH=CH₂ | 77 | 93:7 |
| –(CH₂)₂O(CH₂)₂– | | CH₂CH=CH₂ | 81 | 96:4 |

## 5.7   ALKYNE C(sp)—H C—H ACTIVATION: ALKYNE–ALKYL AND ALKYNE–ALKYNE BOND FORMATIONS

### 5.7.1   Nickel-Catalyzed Alkylation of Alkynes with Alkyl Halides, Nickel-Catalyzed Sonogashira Reaction

Hu and coworkers have developed the first Ni-based methods for the Sonogashira coupling of unactivated, β-hydrogen-containing alkyl halides, including the first examples of the use of alkyl chlorides in Sonogashira reactions. The coupling, using Nickamine **150**, tolerates a wide range of functional groups in both coupling partners. Whilst the substituted alkyne products could also be prepared in comparable yields by the reaction of alkyl halides with alkynyl anions in liquid ammonia, this

5 mol% **150**; 0 or 20% NaI
1.4 equiv $Cs_2CO_3$

$R^1-X$ + ≡—$R^2$ ———————————→ $R^1$—≡—$R^2$

**210**        **211**     Dioxane, 100°C, 16 h        **212**

Coupling of bromides required NaI

5 mol% **150**; 3 mol% CuI; 20% mol% $n$-$Bu_4$NI
1.4 equiv $Cs_2CO_3$

$R^1-Cl$ + ≡—$R^2$ ———————————→ $R^1$—≡—$R^2$

**213**        **211**     Dioxane, 140°C, 16 h        **212**

**SCHEME 5.41**    Nickel-catalyzed alklation of alkynes with alkyl halides.

---

**TABLE 5.12**
**Nickel-Catalyzed Alklation of Alkynes with Alkyl Halides**

| $R^1$ | X | $R^2$ | Yield (%) | $R^1$ | X | $R^2$ | Yield (%) |
|---|---|---|---|---|---|---|---|
| $n$-$C_8H_{17}$ | I | $n$-$C_6H_{13}$ | 83 | $n$-$C_8H_{17}$ | Cl | $n$-$C_6H_{13}$ | 89 |
| $n$-$C_8H_{17}$ | I | TMS | 74 | $Ph(CH_2)_2$ | Cl | $-(CH_2)_4OAc$ | 66 |
| $Me_2CHCH_2$ | I | $-(CH_2)_4Cl$ | 84 | $EtO_2C(CH_2)_3$ | Cl | TIPS | 62 |
| $EtO_2C(CH_2)_3$ | I | $CH_2OTMS$ | 73 | $PhO_2C(CH_2)_6$ | Cl | $n$-$C_6H_{13}$ | 70 |
| $Et_2NCO(CH_2)_4$ | I | $n$-$C_6H_{13}$ | 68 | 4-$AcC_6H_4(CH_2)_3$ | Cl | $n$-$C_6H_{13}$ | 57 |
| 2-AcPyrrol-1-yl$(CH_2)_3$ | I | $n$-$C_6H_{13}$ | 61 | 1-Furyl$(CH_2)_3$ | Cl | $-(CH_2)_4OAc$ | 64 |
| 1-Furyl$(CH_2)_3$ | I | Ph | 84 | 1,3dioxan-2-yl$(CH_2)_5$ | Cl | $n$-$C_6H_{13}$ | 88 |
| $Ph(CH_2)_2$ | Br | $Ph(CH_2)_2$ | 89 | 4-$NCC_6H_4CO_2(CH_2)_3$ | Cl | $n$-$C_6H_{13}$ | 73 |
| $EtO_2C(CH_2)_3$ | Br | $n$-$C_6H_{13}$ | 73 | $Ph_2C(CN)(CH_2)_4$ | Cl | $n$-$C_6H_{13}$ | 66 |
| $EtO_2C(CH_2)_5$ | Br | $-(CH_2)_4Cl$ | 70 | Carbazol-1-yl$(CH_2)_3$ | Cl | $n$-$C_6H_{13}$ | 85 |
| $PhO(CH_2)_2$ | Br | $-(CH_2)_4Cl$ | 79 | 2-MeIndol-1-yl$(CH_2)_3$ | Cl | $Ph(CH_2)_2$ | 75 |
| 1,3 dioxan-2-yl$(CH_2)_2$ | Br | $n$-$C_4H_9$ | 76 | 3-$BrC_6H_4(CH_2)_3$ | Cl | $-(CH_2)_4OAc$ | 58 |
| $CH_2=CH(CH_2)_3$ | Br | $-(CH_2)_4OAc$ | 59 | 4-$ClC_6H_4(CH_2)_2$ | Cl | $Ph(CH_2)_2$ | 72 |
| 4-$BrC_6H_4(CH_2)_2$ | Br | $n$-$C_6H_{13}$ | 69 | | | | |

---

latter method has much lower functional-group tolerance for both alkyl halides and terminal alkynes. Thus, the new Sonogashira protocols are advantageous for the preparation of highly functionalized alkynes (Scheme 5.41, Table 5.12).

By appropriate choice of coupling conditions, different alkyl-X bonds could be differentiated, leading to orthogonal functionalization of alkyl iodides, bromides, and chlorides as demonstrated by the synthesis of **218** from 1-bromo-6-chlorohexane **214** (Scheme 5.42).[5,45]

## 5.7.2   NICKEL-CATALYZED COUPLING OF ALKYNES WITH ALKYNES, THE NICKEL-CATALYZED GLASER REACTION

A Ni/Cu-cocatalyzed oxidative coupling reaction between two different terminal alkynes which provided a variety of unsymmetrical conjugated diynes under mild

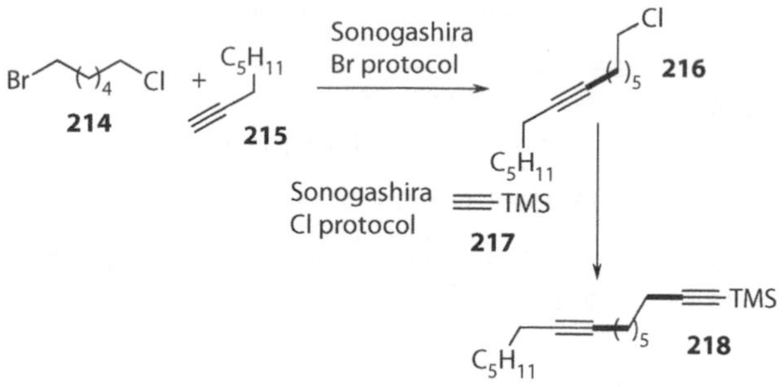

**SCHEME 5.42** Sequential selective nickel-catalyzed couplings of alkynes with difunctional alkyl halides.

conditions has recently been described by Lei and coworkers. Use of one of the alkynes in excess was required in order to obtain good yields of the hetero-coupled product and limit homo-coupling side reactions (Scheme 5.43).[46]

This methodology was used by Agrofoglio to prepare a range of anti-viral C5-substituted-(1,3-diyne)-2′-deoxyuridines **224** (Scheme 5.44).[47]

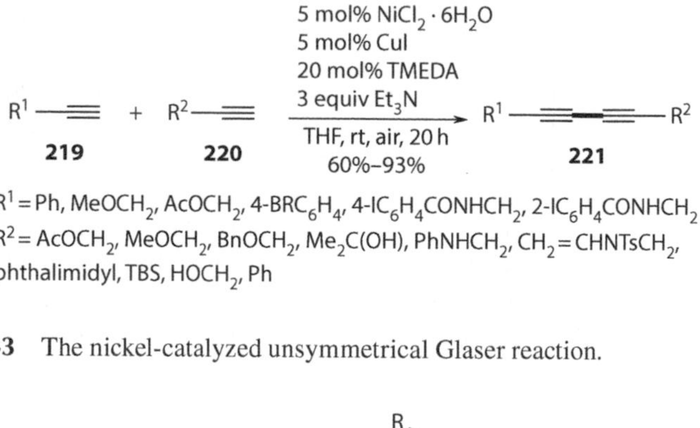

$R^1 = $ Ph, MeOCH$_2$, AcOCH$_2$, 4-BRC$_6$H$_4$, 4-IC$_6$H$_4$CONHCH$_2$, 2-IC$_6$H$_4$CONHCH$_2$
$R^2 = $ AcOCH$_2$, MeOCH$_2$, BnOCH$_2$, Me$_2$C(OH), PhNHCH$_2$, CH$_2$=CHNTsCH$_2$,
phthalimidyl, TBS, HOCH$_2$, Ph

**SCHEME 5.43** The nickel-catalyzed unsymmetrical Glaser reaction.

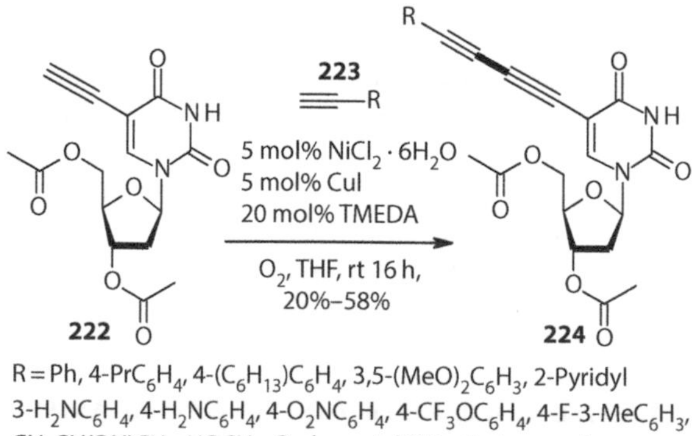

R = Ph, 4-PrC$_6$H$_4$, 4-(C$_6$H$_{13}$)C$_6$H$_4$, 3,5-(MeO)$_2$C$_6$H$_3$, 2-Pyridyl
3-H$_2$NC$_6$H$_4$, 4-H$_2$NC$_6$H$_4$, 4-O$_2$NC$_6$H$_4$, 4-CF$_3$OC$_6$H$_4$, 4-F-3-MeC$_6$H$_3$,
CH$_3$CH(OH)CH$_2$, HOCH$_2$, Cyclopentyl CH$_2$, Cyclopropyl

**SCHEME 5.44** The nickel-catalyzed unsymmetrical Glaser reaction in the synthesis of anti-viral agents.

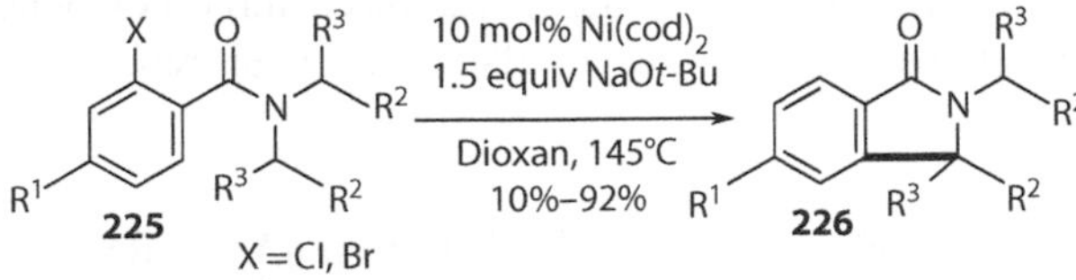


X = Cl, Br
$R^1$ = H, F, MeO
$R^2$,$R^3$ = H, H; H, Me; Me, Me; -$(CH_2)_5$-

**SCHEME 5.45** Nickel-catalyzed intra-molecular arylation of $sp^3$ C—H bonds.

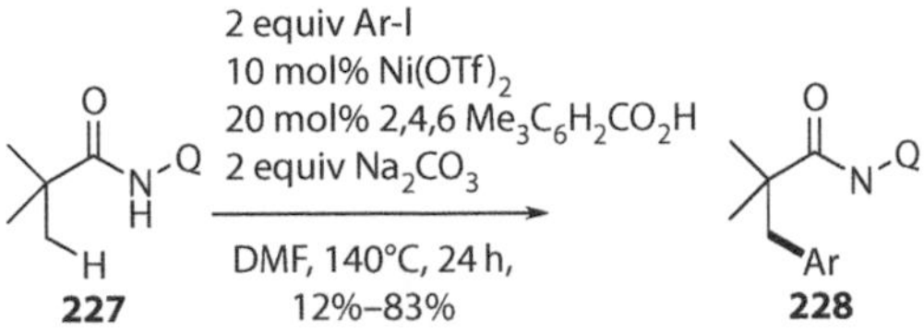


Q = 8-Quinolinyl
$R^1$,$R^2$ = Ph,Ph; Bu,Bu; Bn,Et; Ph,Me; –$(CH_2)_5$–
–$(CH_2)_6$–; Cx,H; Ph,$C_5H_{11}$

Ar = 4-$MeOC_6H_4$; 4-$F_3C_6H_4$; 4-$MeO_2CC_6H_4$; 4-$AcC_6H_4$
4-$IC_6H_4$; 4-$ClC_6H_4$; 4-$MeC_6H_4$; 4-$H_2NC_6H_4$; 4-$Me_2NC_6H_4$
2-$MeC_6H_4$; 5-indolyl; 2-thienyl

**SCHEME 5.46** Nickel-catalyzed inter-molecular arylation of C($sp^3$)—H bonds.

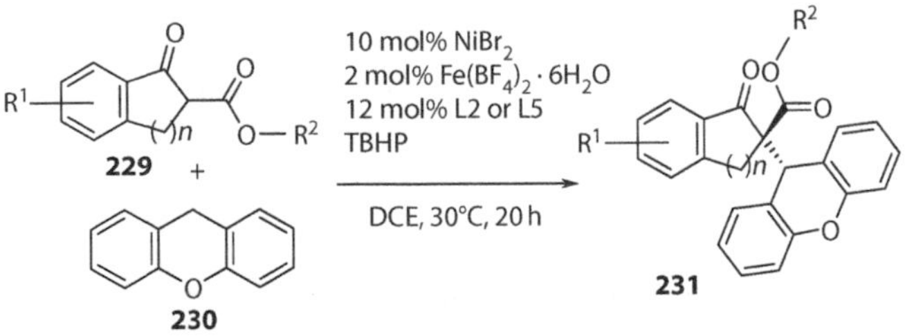


*n* = 1: $R^1$ = H, 6-F, 6-Cl, 6-Br, 5-F, 5-Cl, 5-I; $R^2$ = *t*-Bu, 1-adamantyl
yields 68%–90%, *ee* 97%–99%
*n* = 2: $R^1$ = H, 7-Br, 7-MeO, 5-MeO, 5,7-$Me_2$; $R^2$ = *t*-Bu, 1-adamantyl
yields 34%–72%, *ee* 99%

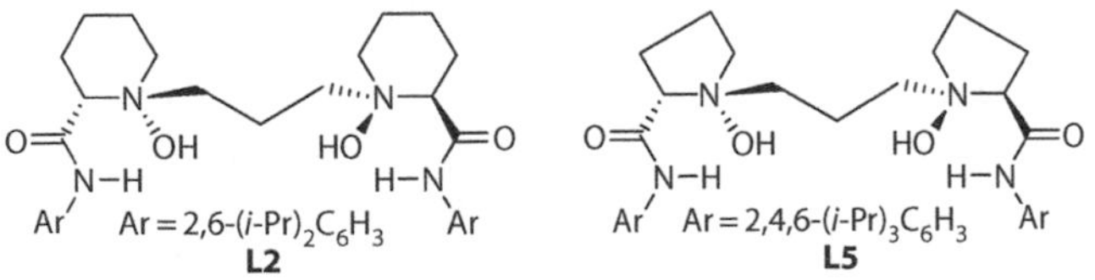


**SCHEME 5.47** Nickel-catalyzed enantioselective oxidative cross-dehydrogenative coupling of β-ketoesters with xanthenes.

Extension of this methodology to nickel-catalyzed three component couplings of three alkynes, or two alkynes with an alkene has been reported by Ogata and Fukuzawa.[43]

## 5.8 ALKYL C(sp³)—H C—H ACTIVATION: ALKYL–ARYL, ALKYL–ALKENYL, AND ALKYL–ALKYL BOND FORMATIONS

The dehydrogenative [4 + 2] cycloaddition of formamides with alkynes described by Hiyama, Nakao et al., and summarized in Section 5.6, as well as representing an example of C($sp^2$)—H activation, also represents an example of C($sp^3$)—H activation alkyl–alkenyl bond formation.[40,41]

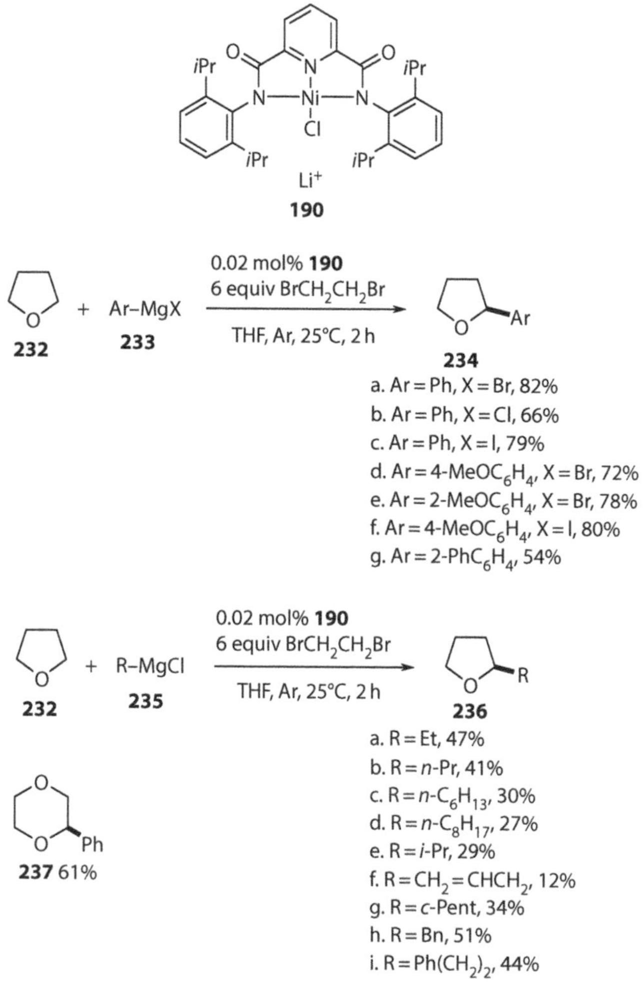
iPr
O
N
O
iPr
N—Ni—N
Cl
iPr
iPr
Li+
190
0.02 mol% 190
6 equiv BrCH2CH2Br
O
+ Ar–MgX
THF, Ar, 25°C, 2 h
O
Ar
232
233
234
a. Ar = Ph, X = Br, 82%
b. Ar = Ph, X = Cl, 66%
c. Ar = Ph, X = I, 79%
d. Ar = 4-MeOC6H4, X = Br, 72%
e. Ar = 2-MeOC6H4, X = Br, 78%
f. Ar = 4-MeOC6H4, X = I, 80%
g. Ar = 2-PhC6H4, 54%
0.02 mol% 190
6 equiv BrCH2CH2Br
+ R–MgCl
THF, Ar, 25°C, 2 h
R
232
235
236
a. R = Et, 47%
b. R = n-Pr, 41%
c. R = n-C6H13, 30%
d. R = n-C8H17, 27%
e. R = i-Pr, 29%
f. R = CH2 = CHCH2, 12%
g. R = c-Pent, 34%
h. R = Bn, 51%
i. R = Ph(CH2)2, 44%
Ph
237 61%

**SCHEME 5.48** Nickel-catalyzed alkylation of cyclic ethers with grignard reagents.

Kalyani and coworkers have described the development of intra-molecular arylation of C(sp³)—H bonds adjacent to an amide nitrogen, using aryl chlorides and bromides with a Ni(COD)-based catalyst system (Scheme 5.45).[48]

Aihara and Chatani have reported the first example of the nickel-catalyzed direct arylation of otherwise unactivated C(sp³)—H bonds in aliphatic amides. The presence of a bidentate directing group, such as an 8-aminoquinoline moiety is required for the reaction to proceed. In comparison to other methods, their methodology shows considerably wider functional group tolerance (Scheme 5.46).[49]

In an example of C(sp³)—H activation alkyl–alkyl bond formation, Feng and coworkers have developed a one-pot, highly enantioselective oxidative cross-dehydrogenative coupling reaction of α-substituted β-ketoesters with xanthenes. These were catalyzed by a $NiBr_2/Fe(BF_4)_2$ catalyst system in the presence of a chiral ligand. Various optically active xanthene derivatives bearing a quaternary stereogenic center were obtained in moderate to good yields (up to 90%) with excellent *ee* values (up to 99% *ee*) under mild conditions (Scheme 5.47).[50]

In addition to the C—H activation and Grignard cross-coupling of furans, Gartia et al. have demonstrated the utility of their nickel pincer catalyst system for the cross-coupling of various tetrahydrofurans and dioxans. The nickel (II) complex **190** enabled cross-couplings at room temperature in 1–2 h (Scheme 5.48).[36]

# REFERENCES

1. Koester, D. C.; Werz, B. *Nachr. Chem.* **2011**, *59*, 40–43.
2. Hirano, K.; Miura, M. *Yuki Gosei Kagaku Kyokaishi* **2011**, *69*, 252–265.
3. Becker, R.; Jones, W. D. *Catalysis without Precious Metals* Bullock, M. Ed.; Wiley: Hoboken, NJ, **2010**, pp 143–164.
4. Nakao, Y. *Shokubai* **2008**, *50*, 705–709.
5. Hu, X. *Chimia* **2012**, *66*, 154–158.
6. Johnson, S. A.; Hatnean, J. A.; Doster, M. E. *Prog. Inorg. Chem.* **2012**, *57*, 255–352.
7. Yamaguchi J.; Itami, K.; Yamaguchi, A. D. *Angew. Chem. Int. Ed.* **2012**, *51*, 8960–9009.
8. Hachiya, H.; Hirano, K.; Satoh, T.; Miura M. *Org. Lett.* **2009**, *11*, 1737–1740.
9. Canivet, J.; Yamaguchi, J.; Ban, I.; Itami, K. *Org. Lett.* **2009**, *11*, 1733–1736.
10. Yamamoto, T.; Muto, K.; Komiyama, M.; Canivet, J.; Yamaguchi, J.; Itami, K. *Chem. Eur. J.* **2011**, *17*, 10113–10122.
11. Muto, K.; Yamaguchi, J.; Itami, K. *J. Am. Chem. Soc.* **2012**, *134*, 169–172.
12. Iaroshenko, V. O.; Ali, I.; Mkrtchyan, S.; Semeniuchenko, V.; Ostrovskyi, D.; Langer, P. *Synlett* **2012**, *23*, 2603–2608.
13. Tobisu, M.; Hyodo, I.; Chatani, N. *J. Am. Chem. Soc.* 2009, 131, 12070–12071.
14. Nakao, Y. *Chem. Rec.* **2011**, *11*, 242–251.
15. Nakao, Y.; Kanyiva, K. S.; Oda, S.; Hiyama, T. *J. Am. Chem. Soc.* **2006**, *128*, 8146–8147.
16. Kanyiva, K. S.; Nakao, Y.; Hiyama, T. *Heterocycles* **2007**, *72*, 677–680.
17. Mukai, T.; Hirano, K.; Satoh, T.; Miura, M. *J. Org. Chem.* **2009**, *74*, 6410–6413.
18. Kanyiva, K. S.; Nakao,Y.; Hiyama, T. *Angew. Chem. Int. Ed.* **2007**, *46*, 8872 –8874.
19. Nakao, Y.; Kanyiva, K. S.; Hiyama, T. *J. Am. Chem. Soc.* **2008**, *130*, 2448–2449.
20. Nakao, Y.; Yamada, Y.; Kashihara, N.; Hiyama, T. *J. Am. Chem. Soc.* **2010**, *132*, 13666–13668.
21. Tsai, C.-C.; Shih, W.-C.; Fang, C.-H.; Li, C.-Y.; Ong, T.-G.; Yap, G. P. A. *J. Am. Chem. Soc.* **2010**, *132*, 11887–11889.

22. Nakao, Y.; Idei, H.; Kanyiva, K. S.; Hiyama, T. *J. Am. Chem. Soc.* **2009**, *131*, 15996–15997.
23. Shiota, H.; Ano, Y.; Aihara, Y.; Fukumoto, Y.; Chatani, N. *J. Am. Chem. Soc.* **2011**, *133*, 14952–14955.
24. Nakao, Y.; Kashihara, N.; Kanyiva, K. S.; Hiyama, T. *J. Am. Chem. Soc.* **2008**, *130*, 16170–16171.
25. Guihaume, J.; Halbert, S.; Eisenstein, O.; Perutz, R. N. *Organometallics* **2012**, *31*, 1300–1314.
26. Nakao, Y.; Kashihara, N.; Kanyiva, K. S.; Hiyama, T. *Angew. Chem. Int. Ed.* **2010**, *49*, 4451–4454.
27. Lee, W.-C.; Wang, C.-H.; Lin, Y.-H.; Shih, W.-C.; Ong, T.-G. *Org. Lett.* **2013**, *15*, 5358–5361.
28. Doster, M. E.; Johnson, S. A. *Organometallics* **2013**, *32*, 4174–4184.
29. Jiang, Y.-Y.; Li, Z.; Shi, J. *Organometallics* **2012**, *31*, 4356–4366.
30. Vechorkin, O.; Proust, V.; Hu, X. *Angew. Chem. Int. Ed.* **2010**, *49*, 3061–3064.
31. Ackermann, L.; Punji, B.; Song, W. *Adv. Synth. Catal.* **2011**, *353*, 3325–3329.
32. Song, W.; Lackner, S.; Ackermann, L. *Angew. Chem. Int. Ed.* **2014**, *53*, 2477–2480.
33. Beaulieu, L.-P. B.; Roman, D. S.; Vallee, F.; Charette, A. B. *Chem. Commun.* **2012**, *48*, 8249–8251.
34. Mousseau, J. J.; Charette, A. B. *Acc. Chem. Res.* **2013**, *46*, 412–424.
35. Xin, P.-Y.; Niu, H.-Y.; Qu, G.-R.; Ding, R.-F.; Guo, H.-M. *Chem. Commun.* **2012**, *48*, 6717–6719.
36. Gartia, Y.; Ramidi, P.; Jones, D. E.; Pulla, S.; Ghosh, A. *Catal. Lett.* **2014**, *144*, 507–515.
37. Li, Y.; Liu, J.; Xie, Y.; Zhang, R.; Jin, K.; Wang, X.; Duan, C. *Org. Biomol. Chem.* **2012**, *10*, 3715–3720.
38. Doster, M. E.; Hatnean, J. A.; Jeftic, T.; Modi, S.; Johnson, S. A. *J. Am. Chem. Soc.* **2010**, *132*, 11923–11925.
39. Nakao, Y.; Idei, H.; Kanyiva, K. S.; Hiyama, T. *J. Am. Chem. Soc.* **2009**, *131*, 5070–5071.
40. Nakao, Y.; Idei, H.; Morita, I.; Hiyama, T. *J. Am. Chem. Soc.* **2011**, *133*, 3264–3267.
41. Kumar, P.; Louie, J. *Angew. Chem. Int. Ed.* **2011**, *50*, 10768–10769.
42. Anand, M.; Sunoj, R. B. *Org. Lett.* **2012**, *14*, 4584–4587.
43. Ogata, K.; Fukuzawa, S. *Yuki Gosei Kagaku Kyokaishi* **2012**, *70*, 2–10.
44. Donets, P. A.; Cramer, N. *J. Am. Chem. Soc.* **2013**, *135*, 11772–11775.
45. Vechorkin, O.; Barmaz, D.; Proust, V.; Hu, X. *J. Am. Chem. Soc.* **2009**, *131*, 12078–12079.
46. Yin, W.; He, C.; Chen, M.; Zhang, H.; Lei, A. *Org. Lett.* **2009**, *11*, 709–712.
47. Sari, O.; Roy, V.; Balzarini, J.; Snoeck, R.; Andrei, G.; Agrofoglio, L. A. *Eur. J. Med. Chem.* **2012**, *53*, 220–228.
48. Wertjes, W. C.; Wolfe, L. C.; Waller, P. J.; Kalyani, D. *Org. Lett.* **2013**, *15*, 5986–5989.
49. Aihara, Y.; Chatani, N. *J. Am. Chem. Soc.* **2014**, *136*, 898–901.
50. Cao, W.; Liu, X.; Peng, R.; He, P.; Lin, L.; Feng, X. *Chem. Commun.* **2013**, *49*, 3470–3472.

# 6 Iron-Catalyzed C—H Activation

*Narendra B. Ambhaikar*

## CONTENTS

## 6.1 INTRODUCTION: C—H ACTIVATION BEYOND PRECIOUS TRANSITION METALS

In the face of constant challenges, the fine chemical and pharmaceutical industries continue to require greener processes by way of synthetic strategies that easily provide substituted or functionalized carbons in aliphatic, aromatic, and heterocyclic systems.[1] Transition-metal-catalyzed functionalization of unactivated C—H bonds avoiding the use of pre-functionalized starting material is a valuable and straightforward synthetic strategy. It is considered a "green" synthetic approach owing to the fact that pre-functionalization or pre-programming and protection of substrates or starting materials required for the transformations are minimized or avoided. The "greenness" is essentially a consequence of a shorter and superior synthesis resulting from a direct or an indirect reduction in time, cost, and hazardous waste. It is, therefore, not surprising that C—H functionalization of heterocycles has been recognized by the ACS Green Chemistry Roundtable as one of the most valuable new reactions for the pharmaceutical industry.[2]

A variety of transition metals have played a dominant role in catalyzing organic transformations. The recent decades have witnessed breakthrough work in late-stage transition metal catalysis, with metals such as palladium,[3] platinum,[4] ruthenium,[5] rhodium,[6] and others at the forefront of new catalytic technologies. These organometallic strategies have led to the introduction of novel bond-forming and oxidation state-altering strategies, thus adding new reactions to the toolbox of the organic chemist. Unfortunately, the transition metals involved here suffer from concerns such as high cost and toxicity, which preclude their applicability beyond a certain point or at the commercial scale. As a result, some first row transition metals, notably copper[7] and iron[8] have begun to emerge as favorable catalysts with a rich potential due to their relatively low cost and abundance. Iron has gained, surprisingly, limited attention as a catalyst despite the fact that this metal offers a multitude of advantages. Despite the commercial availability of many iron salts and complexes, their utility is relatively unexplored in comparison to other transition metals. Some noteworthy facts about this metal as a catalyst are as follows:

1. With its ubiquitous presence in the geosphere and its 4.7% weight natural abundance, iron is inexpensive and environmentally friendly as a catalyst.
2. Iron also displays interesting electrochemical properties, thus being a much desired element in catalysis.
3. Iron is one of the most abundant metals around, also being inexpensive and environmentally friendly; it is the second-most abundant metal in the Earth's crust; more than 1 trillion kg of iron is produced each year. In the human body, iron is the most abundant transition metal (~4 g/person).
4. Iron is often found in the biosphere as part of catalytic systems.

As a result of these attractive features, the recent years have witnessed the use of iron as a catalyst in a multitude of organic syntheses, including oxidations and cross-coupling reactions.[9] Quite clearly, development of iron-catalyzed chemistry for applications toward efficient synthetic reactions deserves greater emphasis. The reasons will be evident from the examples in the sections that follow. This chapter concentrates on some selected literature reports of innovative C—H activation chemistry utilizing iron as a catalyst.[10]

## 6.2   HISTORIC PERSPECTIVE AND INSPIRATION FROM NATURE

The field of iron-catalyzed C—H bond oxidation is over a hundred years old. Yet significant breakthrough applications in organic synthesis and promising synthetic methodologies have emerged only in the past two decades. More than a century ago in 1896, the reaction of hydrogen peroxide ($H_2O_2$) and ferrous ions ($Fe^{2+}$) in water with tartaric acid was discovered by British chemist H. J. H. Fenton.[11] The environmentally clean transformation[12] was found to oxidize most organic compounds, albeit without any substantial specificity. Phenol yields were generally poor, with isomeric mixtures obtained from substituted arenes. Here, $Fe^{2+}$ is the catalyst and $H_2O_2$ is the oxidant. Due to its lack of specificity, it is not well suited for specific local oxidation toward the synthesis of complex organic molecules in the presence of multiple reactive sites. Yet, the Fenton reaction, as it has come to be known, has, over the years, found some practical applications such as oxidation of contaminants and waste water.[13] Forty years later, Haber and Weiss[14] indicated hydroxyl radical to be the actual oxidant in the Fenton system, which was further studied by Merz and

Waters in the 1940s.[15] Many more recent methods discussed in this chapter, involving C—H oxidation catalyzed by iron have originated from this century-old reaction and the importance of hydroxyl radicals has since grown.[16]

An area that became widely studied was the iron-based porphyrin system of the P450 enzymes.[17] As part of biomimetic studies being carried out for alkane oxidation, iron porphyrin systems were conceived to mimic Cytochrome P450 enzymes. The transformations involving iron catalysis for alkane and arene hydroxylation were low-yielding, with limited selectivity and substrate scope. From 1978 onwards until his death, Sir D. H. R. Barton carried out extensive studies on the Gif chemistry while in France,[18] through which a series of "Gif reagents" were developed. The studies were aimed at "mimicking non-enzymatically" Nature's ubiquitous oxygenated saturated unreactive carbons, with the ultimate goal of gaining insight into biochemical oxygenation and developing reagents that would be applicable generally in organic synthesis.[19] The work of Que at Minnesota is also noteworthy and relevant where focus has been on the mechanisms of dioxygen activation by non-heme iron enzymes, on the design and synthesis of iron complexes that mimic enzyme function, stabilize key oxidizing intermediates, or serve as bio-inspired green oxidation catalysts.[20] With this stage set, remarkable growth has taken place over the past 10 years with major contributions from Nakamura, White and others on C—H activation and functionalization of "not activated" aliphatic and aromatic substrates. These new developments seek a high level of site differentiation and selectivities but they have made limited impact on the world of pharmaceutical synthesis so far. However, there is enormous potential for growth in this area of research owing to the overall possibly "green chemistry" features of the methodologies.

This chapter focuses on some of the prominent developments that have recently taken place in the field of C—H activation or functionalization catalyzed by iron. Iron-catalyzed reactions conventionally require high temperatures since iron possesses low catalytic reactivity. New methodologies have involved development of low temperature and mild reaction conditions. Many more developments continue and will continue. While several reports have been published dealing with basic reports of methodologies, their application to pharmaceutically useful molecules or complex natural products are rather limited. While iron catalysis is an old and well established field through Gif and Fenton chemistry, it is only recently that its application to C—H functionalization is being viewed as an opportunity for applications. Thus, this area of research has tremendous potential, especially when one considers it from the perspective of green chemistry and industry. The following discussion is based on types of substrates that are employed and the bond-forming processes that are involved, that lead to interesting transformations.

## 6.3 NITROGEN-DIRECTED ARYL C—H ACTIVATION

Nitrogen-directed C—H activation to form C—C bonds like aryl–aryl is a relatively recent class of reactions. In 2008, Nakamura and coworkers presented the first report of an iron-catalyzed C—H arylation reaction—a homogeneous iron catalysis featuring C—C bond formation via C—H activation.[21] As an overall synthetic transformation, it was a formal nucleophilic displacement of the *ortho*-hydrogen atom by an aryl zinc nucleophile. *In situ* generated organozinc derived from aryl Grignard reagent and zinc chloride was involved in the presence of TMEDA ligand. Especially interesting was the fact that the reaction took place at 0°C, instead of the usual >80°C temperature that is required for C—H activation reactions.[22] Upon screening of reaction conditions and the reagent, a rather specific blend of the metals iron, zinc, and magnesium; plus 1,10-phenanthroline, TMEDA and 1,2-dichloro-2-methylpropane was thus deemed necessary to favor product formation. Here, the phenanthroline is presumed to coordinate to iron, and TMEDA to zinc; with the reaction involving a redox cycle of iron with the electron-accepting dichloride.

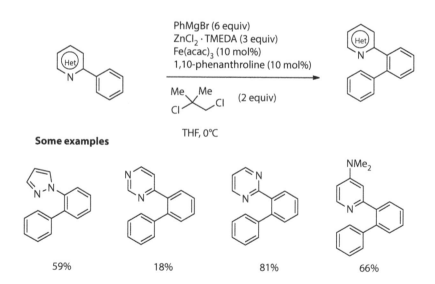

2-Phenylpyridine derivatives with substituent on the 4-position of the phenyl ring react to give the corresponding phenylated products in excellent yields. The reaction is relatively fast when the phenyl group is equipped with an electron-donating group.

While the reaction is somewhat insensitive to the electronic effect of the substituent on the aryl group, it is very sensitive to any steric effects. The reaction of 2-*o*-tolylpyridine is very slow and yields little product due to restricted conformation. Nitrogen-containing heterocycles other than pyridine also undergo this arylation reaction. Some examples are shown above. In this interesting case of iron catalysis involving C—C bond formation reaction with C—H bond activation, the overall reaction is a formal nucleophilic displacement of the *ortho*-hydrogen atom by an arylzinc nucleophile. While potentially useful, aryl magnesium bromide and zinc chloride–TMEDA are required to be used in excess.

Later on, it has been observed by Nakamura et al. that the expensive organodichloride can also be replaced with dioxygen as an oxidant. This variant of the C—H arylation involves the slow introduction of dioxygen into the reaction mass, which allows the desired oxidative cross-coupling to take place smoothly. An example of this is shown below.[23]

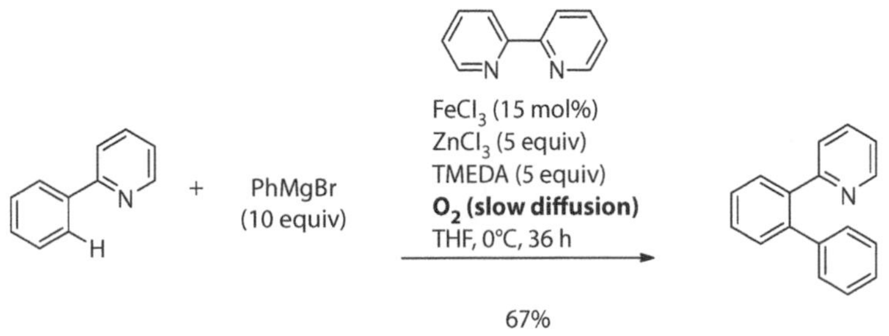

FeCl$_3$ (15 mol%)
ZnCl$_3$ (5 equiv)
TMEDA (5 equiv)
**O$_2$ (slow diffusion)**
THF, 0°C, 36 h

PhMgBr
(10 equiv)

67%

Nakamura et al. further extended this chemistry to aromatic imines to activate aryl C—H. They have reported an iron-catalyzed coupling of an aromatic imine and a diarylzinc reagent to yield a direct displacement product of the *ortho* C—H bond of the imine by the aryl group of the zinc reagent.[24] Aryl imines can also allow for directed *ortho* C—H bond activation chemoselectively enabling *ortho*-arylation *via* iron catalysis in reasonably good yields. The *ortho*-aryl group is thus introduced by starting with an acetophenone and then its conversion to an appropriate imine. Di-*tert*-butyl-2,2′-bipyridine was found to be a ligand of choice in case of an aryl imine as substrate unlike palladium, which catalyzes the conventional substitution reaction while iron catalyzes an imine-directed C—H bond activation to cause *ortho*-arylation from an acetophenone-derived imine using a diarylzinc reagent.

With these reaction conditions that involve PhMgBr, ZnCl$_2$ · TMEDA, cat. Fe(acac)$_3$, dtbpy and 1,2-dichloroisobutane in tetrahydrofuran, mild and selective C—H bond activation and functionalization in the presence of aryl bromide, chloride, or sulfonate groups can be achieved in good yields. Alongside, 1,2-dichloroisobutane remains essential to attain such selectivity. Remarkably, the reaction takes place even in the presence of a diverse group of substituents thus demonstrating chemoselectivity.[25] The imine group appears to be a very favorable directing group for the bond activation, particularly (4-methoxyphenyl)imine of the ketone because of the ease of preparation and handling. It is worth noting that methoxime and *N,N*-dimethylhydrazone, unlike imines, do not react to induce any C—H activation. This is a practical method for the synthesis of *ortho*-substituted aromatic ketones under mild conditions (0°C).

The substitution of *ortho*-hydrogen atom occurs preferentially over that of the electrofugal leaving groups on the same aryl ring, where dtbpy happens to be ligand of choice among the various ones screened. Good tolerance is exhibited by this reaction for halide and pseudohalide groups, which is particularly significant, given that reaction conditions are similar to those of iron-catalyzed cross-coupling of aryl halides and pseudohalides.[26] Selective C—H bond activation occurs even in the presence of good leaving groups including halides. In fact, the reaction also tolerates the presence of triflate and tosylate groups. In one interesting case, complete conversion of the starting material is achieved by using a near stoichiometric amount (2.5 equiv) of the diphenylzinc reagent for a shorter reaction time (9 h).

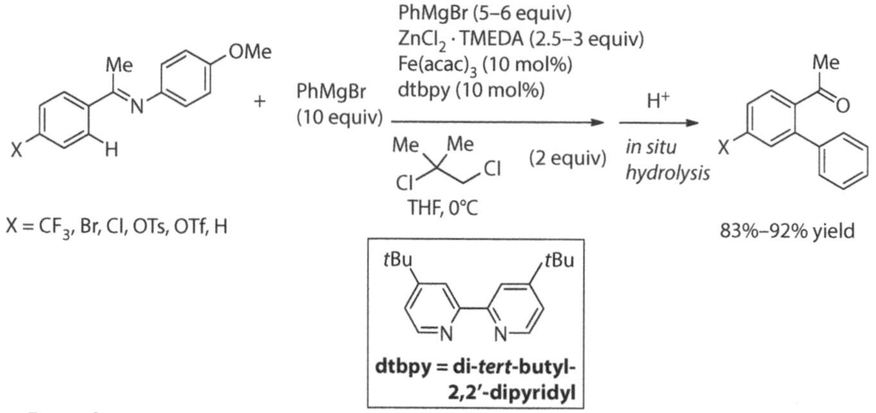

X = CF$_3$, Br, Cl, OTs, OTf, H

83%–92% yield

**Examples**

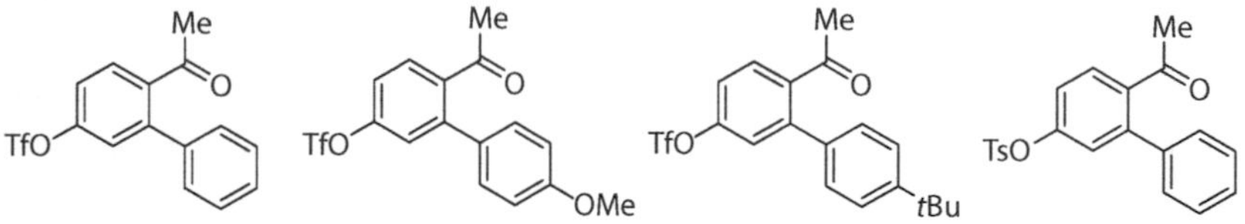

DeBoef and coworkers have taken this concept a step further by reporting iron-catalyzed arylation of heterocycles via directed C—H bond activation using an imine directing group.[27] The imine group also allows for the ortho functionalization of heterocycles such as pyridines, thiophenes, and furans with complete conversion in just 15 min at 0°C. Thus, iron-catalyzed arylation via C—H bond action can be carried out on many different substrates containing N, S, and O heterocycles at mild temperatures in short reaction times. The reaction is clean—with only the starting materials, the desired biaryl product and the Grignard homo-coupling derived biphenyl product as part of the reaction mix.

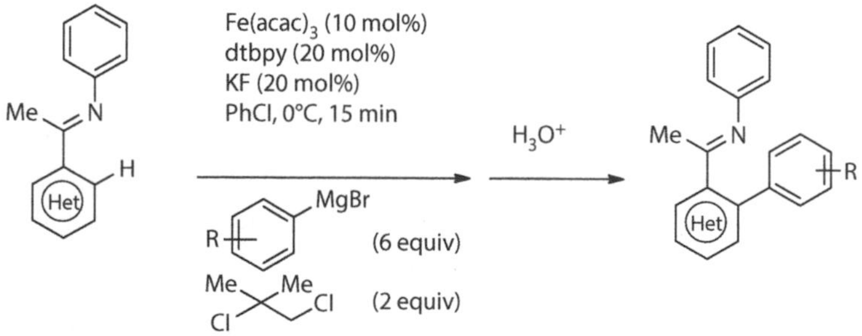

**Some examples**

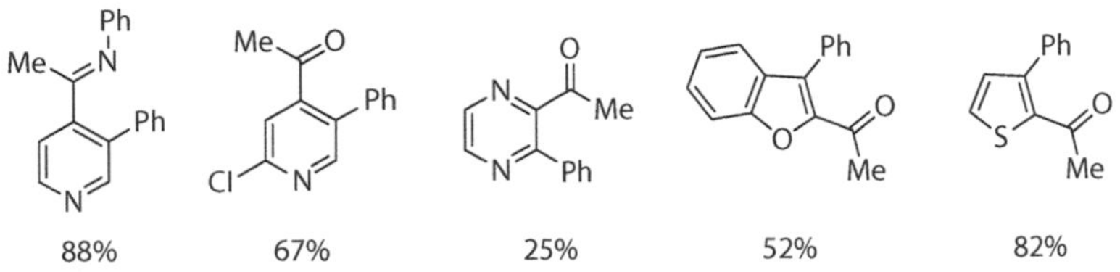

| 88% | 67% | 25% | 52% | 82% |

The Nakamura group has later demonstrated the utility of carboxamides bearing a uniquely effective directing group, namely 8-quinolinamide that enables iron-catalyzed C—H activation. Interesting cases of C—H activation chemistry have been reported employing this directing auxiliary. For example, conditions for replacing a C($sp^3$)—H bond with a new C—aryl bond at the β-position of a 2,2-disubstituted carboxamide. The overwhelmingly higher reactivity of a methyl group over a benzylic group excludes a radical mechanism, and the high sensitivity of the yield to the structure of the substrate and the ligand suggests involvement of organoiron intermediates in some crucial steps.[28]

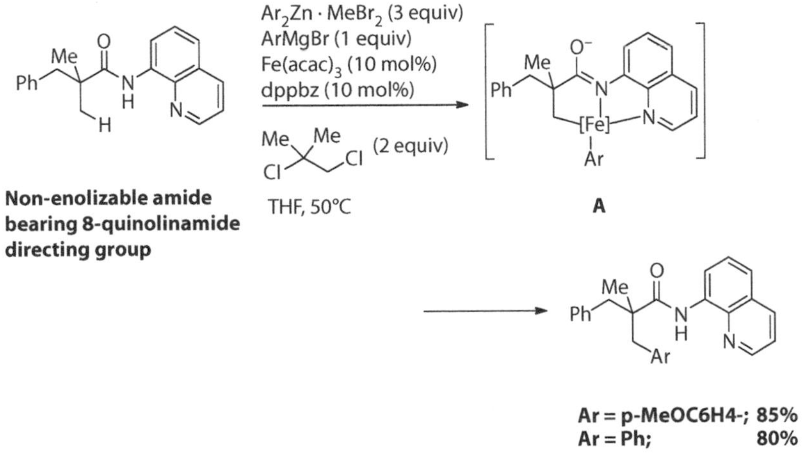

**Non-enolizable amide bearing 8-quinolinamide directing group**

THF, 50°C

**A**

Ar = p-MeOC6H4-; 85%
Ar = Ph;  80%

## 6.4  IRON(II)-CATALYZED DIRECT RADICAL C—H ARYLATION

Iron(II)-catalyzed radical arylation of arenes with aryl and heteroaryl iodides has been reported by Charette's group. This is a relatively versatile methodology involving low amounts of iron catalyst (5 mol%).[29]

As an alternative to the aryl iodides, aryl bromides also undergo this reaction, although they generate the desired product in lower yields. A prerequisite for this transformation is that the symmetric arenes be applied in order to attain chemoselectivity. Mechanistically, the reaction presumably proceeds via an iron-catalyzed radical process. Steric bulk does not appear to impact the product formation, thereby making it a reliable arylation methodology. A base is required in order to quench the generated hydrogen iodide or bromide and maintain the appropriate acidity. Unfortunately, while the iron catalyst may be inexpensive, bathophenanthroline and iodide appear to be cost-contributing factors in this reaction.

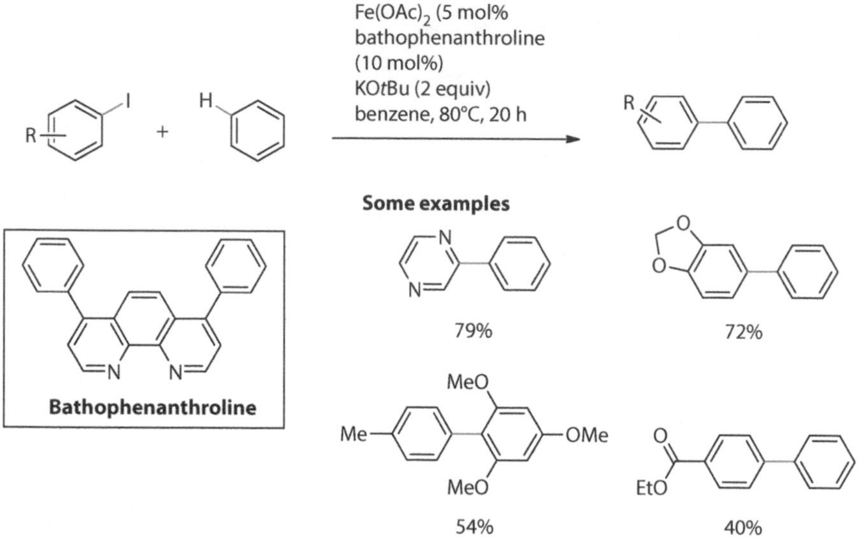

**Some examples**

79%                    72%

54%                    40%

Lei's group has also independently worked on this transformation and found FeCl₃ to be more efficient than Fe(II) salts as the catalyst, with dimethylethylenediamine (DMEDA) being the ligand of choice in combination with a base. While aryl iodides reacted well, the team found that with LiHMDS as a base, even aryl bromides were found to react with electron-donating groups on aryl halides enhancing reactivity and steric bulk decreasing it.[30]

## 6.5   IRON-CATALYZED CHEMOSELECTIVE OXIDATIVE COUPLING OF TERTIARY AMINES

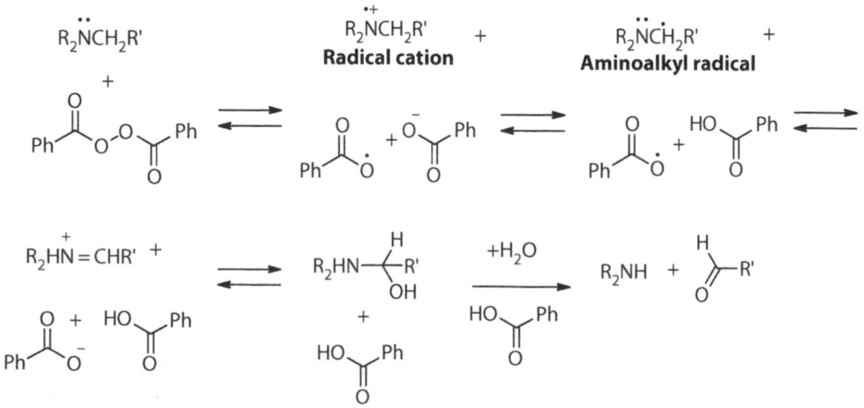

The activation of C—H bonds in tertiary amines is a reasonably well understood reaction. It has its origins in the dealkylation chemistry reported by dePaolini in 1932[31] and the mechanistic explanation put forth by Horner et al.[32] It is similar to the oxidative dealkylation of tertiary amines with peroxides catalyzed by peroxidases and P450 enzymes.

Later, with Murahashi's report in 1988 that ruthenium salts could catalyze the same reaction, nucleophiles like allylsilanes and hydrogencyanide have led to potentially useful compounds, breaking new ground and leading to interesting insights.[33] Miura and coworkers in 1989 reported the generation of N-propargyl amines via $O_2$ oxidation of N,N-dimethylaniline catalyzed by metal salts with terminal alkynes among other products due to reactions of radical intermediates.[34] Based on this work, Vogel and Volla, in 2009, reported the first iron-catalyzed oxidative C—C cross-coupling of tertiary amines with terminal alkynes yielding propargylamines. Aromatic and nonaromatic[35] amines and alkynes were employed as coupling partners in a reaction that was chemoselective for steric reasons, where the methylamino group reacted faster than other alkylamino groups. In this case tert-butyl hydroperoxide (t-BuO)$_2$ was used as the oxidant. The reaction can be applied to aromatic as well as aliphatic amines and alkynes without solvent. Steric factors contribute to the high chemoselectivity for aminomethyl groups.

Iron-catalyzed oxidative α-cyanation of tertiary amines via direct functionalization of C($sp^3$)—H bonds has been reported to provide rapid access to α-aminonitriles.[36] Recently, important contributions have come from the research groups of Sain and Ofial. The α-aminonitriles, some of which are traditionally generated using the Strecker reaction[37] are versatile synthetic intermediates that have found wide utility in synthetic strategies.[38] While other metals such as RuCl$_3$ and V$_2$O$_3$ were earlier reported to facilitate this transformation, the Sain group came up with an iron-catalyzed version where hydrogen peroxide was the oxidant in the presence of sodium cyanide using polymer supported iron phthalocyanines as catalysts in methanol—acetic acid as solvent mixture.[39]

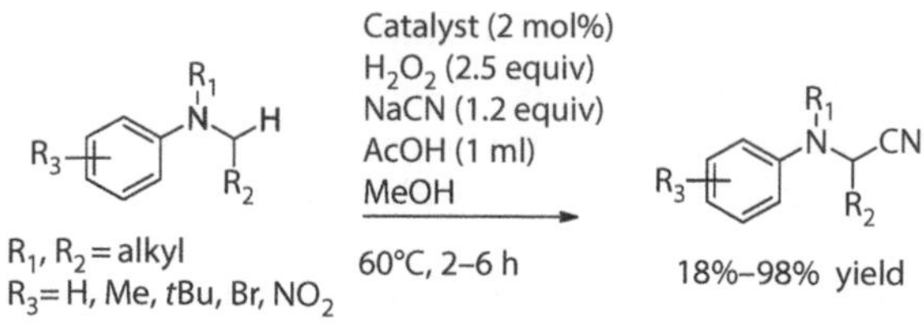

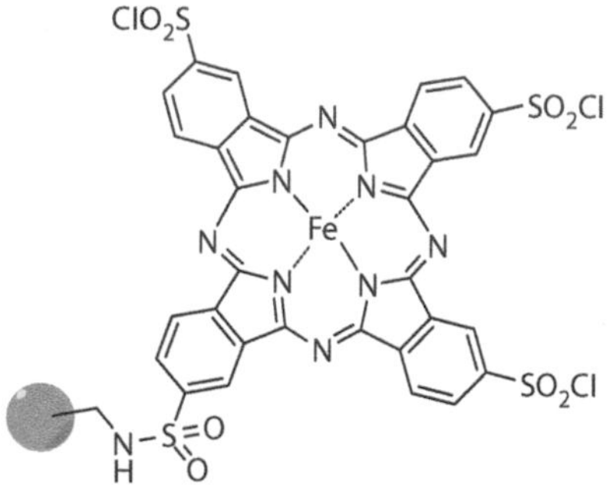

Catalyst (2 mol%)
H$_2$O$_2$ (2.5 equiv)
NaCN (1.2 equiv)
AcOH (1 ml)
MeOH

R$_1$, R$_2$ = alkyl
R$_3$ = H, Me, *t*Bu, Br, NO$_2$

60°C, 2–6 h          18%–98% yield

**Polymer supported phthalocyanine catalyst**

A little later, Armin Ofial's group reported an acid-free FeCl$_2$-catalyzed synthesis of α-aminonitriles under mild and acid-free conditions by oxidation of tertiary amines in the presence of "ligandless" iron salts.[40] Relatively environmentally friendly conditions were developed so that under oxidative conditions, an iron salt could selectively activate C(*sp*$^3$)—H bonds adjacent to nitrogen in tertiary amines to allow for cross-coupling reactions with cyanide even at room temperature. During this work, a variety of copper and iron (II and III) salts as well as cyanide sources such as K$_3$[Fe(CN)$_6$], *n*-Bu4N$^+$CN$^-$ and Me$_3$SiCN (TMSCN) were screened, keeping methanol as the solvent, under nitrogen atmosphere. The α-cyanation product could be obtained with several iron (II) and iron (III) salts. A combination of FeCl$_2$ with di-*tert*-butylperoxide [(*t*-BuO)$_2$], which is typically known to cause C(*sp*$^3$)—H bond activation adjacent to nitrogen, does not yield product when TMSCN is used as the cyanation reagent. The most favorable conditions for oxidative cyanation involve a combination of FeCl$_2$, *t*-BuOOH, MeOH, and TMSCN, as shown below.

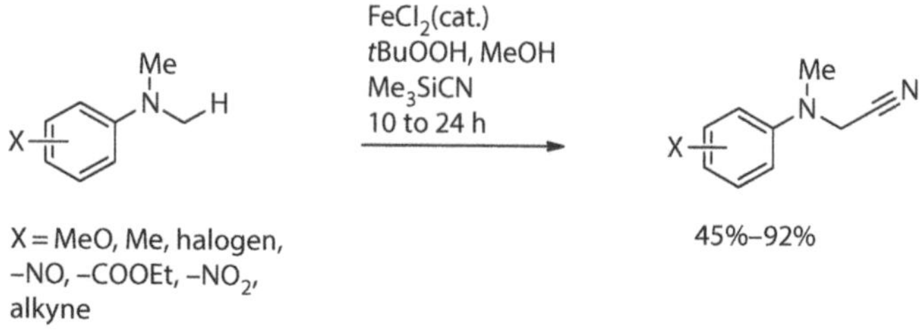

FeCl$_2$(cat.)
*t*BuOOH, MeOH
Me$_3$SiCN
10 to 24 h

X = MeO, Me, halogen,
–NO, –COOEt, –NO$_2$,
alkyne

45%–92%

**Some examples**

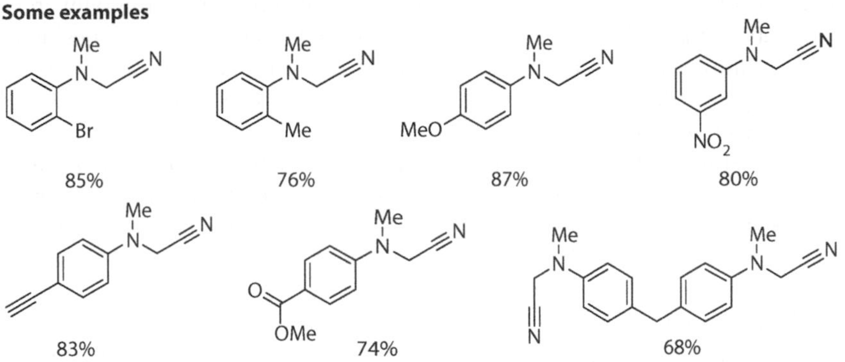
Me
N
N
Br
85%
Me
N
N
Me
76%
Me
N
N
MeO
87%
Me
N
N
NO2
80%
Me
N
N
83%
Me
N
N
O
OMe
74%
Me
N
N
Me
N
N
68%

Remarkably, chemoselectivity is observed even in the presence of an alkynyl group on the phenyl ring, thus favoring the cyanation of the *N*-methyl group and complementing the Vogel protocol described in earlier in this section. While most substrates react well at room temperature, the ones with electron-withdrawing substituents, especially at the *para* position, tend to require reflux temperatures. When aromatic rings possess multiple *N,N*-dimethyl groups in the substrate, cyanation can take place at each of these groups. For the most part, environmentally benign iron salts under oxidative conditions selectively activate C($sp^3$)—H bonds adjacent to nitrogen in tertiary amines. This activation results in efficient cross coupling reactions with cyanide at room temperature.[41] *N*-Phenyl substituted cyclic amines could also be obtained in high yields when similar conditions were employed. An interesting case is that of *N*-phenyl pyrrolidine that gets double cyanated to yield *trans*-2,5-dicyanophenyl pyrrolidine,[42] a compound that was evaluated as part of class of dipeptidyl peptidase IV inhibitors in type 2 diabetes.[43]

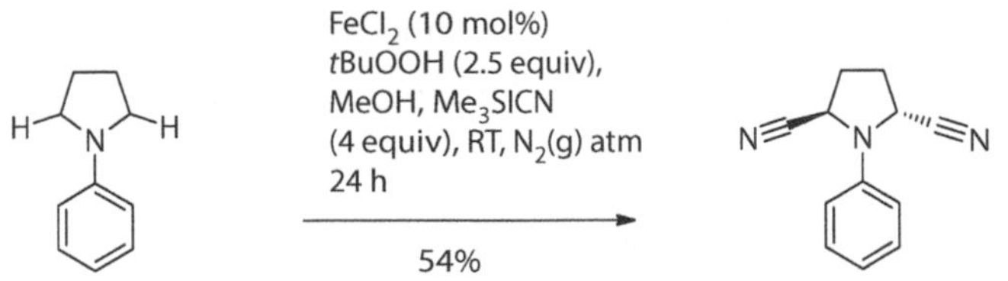
H
N
H
FeCl2 (10 mol%)
tBuOOH (2.5 equiv),
MeOH, Me3SICN
(4 equiv), RT, N2(g) atm
24 h
54%
N
N
N

**N-phenyl pyrrolidine**

**Dipeptidyl peptidase IV inhibitor**

Offial and coworkers have also devised a similar strategy to functionalize C—H bond in *N,N*-dialkylanilines via iron-catalyzed "dehydrogenative phosphonation."[44]

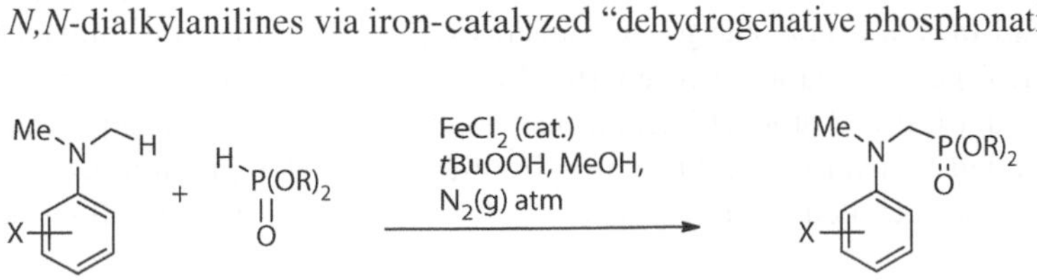
Me
N
H
X
+
H
P(OR)2
O
FeCl2 (cat.)
tBuOOH, MeOH,
N2(g) atm
Me
N
P(OR)2
O
X

The use of the same iron salt $FeCl_2$ in combination with *tert*-butylhydroperoxide enables the efficient activation of $C(sp^3)$—H bonds in the α-position in *N,N*-dialkylanilines and generates a C—P bond in the product. A variety of functional groups on the substrate compounds are tolerated to yield acylic or cyclic α-aminophosphonates.

An interesting C—H activation strategy involving α-arylation of aliphatic amines at the $sp^3$ C—H has been reported by Nakamura et al. Here, acyclic and aliphatic amines bearing *N*-(2-iodophenyl methyl group, undergo arylation with a Grignard or diorganozinc reagents yielding the corresponding α-arylated amine. This reaction proceeds via an intra-molecular 1,5-hydrogen transfer and then a reductive elimination. The preferred catalyst used was $Fe(acac)_3$ although the reaction works comparably with $FeCl_3$ or $FeCl_2$. The reaction involves the use of an ethereal solvent such as *tert*-butyl methyl ether or diethyl ether. The use of THF can possibly lead to competitive reactions and is therefore not desirable.[45]

X = Br, I
$R_1 = R_2 = alkyl$

Fe(acac)₃ (2.5–5.0 mol%)
PhMgBr (1.2–2.0 equiv)
MTBE, 50°C, 15–30 min

40%–84% yield

This reaction is presumed to be initiated by *N*-(2-iodobenzyl)-group on the amine so that cleavage of the C—H bond adjacent to the nitrogen amine takes place. Iron promotes de-iodination or de-bromination, creating an aryl radical intermediate, which then goes through a 1,5-hydrogen transfer causing an organoiron intermediate to form. The final mechanistic step ends with the formation of $C(sp^3)$—$C(sp^2)$ bond via reductive elimination of the iron intermediate.

### 6.5.1  OXIDATIVE MANNICH REACTIONS

Iron (II) and iron (III) salts have been investigated for oxidative Mannich reactions of *N,N*-dialkylanilines and tetrahydroisoquinolines with carbon nucleophile, they are effective with only *tert*-butylhydroperoxide (TBHP) or di-*tert*-butylperoxide as oxidants. Recently, with the goal of finding the most atom economic conditions to oxidatively functionalize aliphatic C—H bond adjacent to nitrogen in *N,N*-dialkylanilines, a simple and green transformation involving iron-catalyzed *aerobic* C—H functionalization was reported by Doyle and coworkers.[46] The simplicity of this ligandless and atom efficient transformation conveys a strong potential for it to be applied to a variety of molecules, in order to develop green processes. Their earlier work had demonstrated that 4-substituted *N,N*-dialkylanilines undergo oxidation in methanol to the corresponding methoxy hemiaminals by oxygen ($O_2$)-catalyzed by CuBr, $FeCl_3$, or $Co(OAc)_2$, with later results indicating better performance by $FeCl_3$.[47]

Historically, oxidative Mannich reactions have usually been developed for catalysis with ruthenium (II) and dirhodium (II) complexes,[48] depending on an oxidant like *tert*-butyl hydroperoxide. Earlier cases of Fe(II) and Fe(III) salts required reactions of *N,N*-dialkylanilines and tetrahydroisoquinolines with carbon nucleophiles to be run in the presence of di-*tert*-butylperoxide. This work involves the absence of ligands and dioxygen as a green, atom efficient and inexpensive oxidant. Vinylogous oxidative Mannich-type reactions as well as aza-Henry reactions with carbon nucleophiles catalyzed by inexpensive ferric chloride can take place efficiently. An example of this $FeCl_3$-catalyzed aerobic oxidative vinylogous Mannich reaction of *N,N*-dialkylanilines with siloxyfuran, a known precursor for γ-butyrolactones, affords a variety of racemic butenolide products. Electron-rich *N,N*-dialkylanilines lead to the formation of Mannich adducts in high yields, while moderately electron-deficient substrates form products in moderate amounts. Unsymmetrically substituted anilines show a preference to undergo selective reaction at only the *N*-methyl position. The oxidative protocol involving procedural simplicity enables facile functionalization of an aliphatic C—H bond adjacent to nitrogen with siloxyfuran, nitroalkanes, and other nucleophiles such as indoles, pyrroles, nitroalkanes, and methyl malonate.

1.0 equiv          1.5 equiv
R = Me, Et

40%–84% yield

In pursuit of exploring generality of this chemistry, other nucleohpiles were investigated. Typically *N*-aryltetrahydroisoquinolines can undergo functionalization with the aid of TBHP or photolytic catalysis when nitromethane or nitroethane is employed (aza-Henry reaction) however dialkylanilines appear mostly insensitive.[49] Not surprisingly, the above conditions involving $FeCl_3$ and $O_2(g)$ with silyloxyfuran do not enable dialkylanilines to react with nitromethane with the product forming only in trace amounts, presumably due to catalyst inhibition. A well optimized version of this transformation shown in the scheme below involves the use of weakly acidic components such as 1.0 equiv of 1,1,1,3,3,3-hexafluoro-2-propanol (HFIP) rather than organic bases.

1.0 mmol                                              **Major**          **Minor (double addition)**

**R = H, 71% yield; major:minor = 98:2**
**R = Me, 77% yield; major:minor = >99:1**

Dioxygen bound to iron (II) does not initiate the oxidative Mannich reaction, since the reaction does not proceed well in the presence of FeCl$_2$, showing only <5% product.

The reaction likely proceeds through an electron transfer between ferric chloride and amine producing a radical cation and iron (II) chloride. Hydrogen abstraction from the radical cation is proposed to lead to iminium species either directly or *via* iron(II)-bound dioxygen.[50]

## 6.6 IRON-CATALYZED C—H ACTIVATION VARIANT OF THE SUZUKI–MIYAURA REACTION

Replacing a halide or the leaving group in the electrophilic component of a cross-coupling reaction with C—H bonds can be a significant step to develop a green synthetic transformation. It avoids the need to pre-modify the substrate and consequently makes the reaction atom-economical. C—H activation and functionalization in $sp^2$–$sp^2$ cross-coupling reactions has been demonstrated with various transition metals like Pd, Ru, and Rh.[51] Extending iron-catalysis to C—C bond-formation through C—H bond-functionalization has remained a challenge. Yu and coworkers reported in 2008 their investigative work on an iron-mediated cross-coupling reaction generating biaryl compounds through C—H bond activation.[52] According to their report, the reaction employed "easily handled reagents with low toxicity." While the reaction involved stoichiometric levels of iron, several substituted phenylboronic acids were demonstrated to react with simple unactivated arenes yielding the desired product. Later, the same group in 2010 described the first iron-mediated direct Suzuki–Miyaura reaction between *N*-heteroaryl compounds and arylboronic acids as a new method to *ortho*-arylate pyrrole and pyridine (shown below) using modified reaction conditions.[53] Both electron-rich and electron-deficient heteroarenes were demonstrated to productively undergo the reaction. Subsequent investigations from the same laboratory led to the development of a catalytic version of this reaction. Here the iron-macrocyclic polyamine (MCPA) complex plays a key in governing the outcome of the reaction. Oxygen is an indispensable oxidant for the coupling since its absence yields only a trace amount of the product. Substituted pyrroles can be synthesized using this protocol. A typical procedure involves heating pyrrole, the appropriate boronic acid, MCPA ligand (0.2 mmol), $FeC_2O_4 \cdot 2H_2O$ (0.2 equiv), **L2** (0.2 equiv), 130°C, 10 h, in the presence of air.

Product of 2-arylation is formed preferentially with little or no formation of 3-arylation product. Electron-withdrawing substituents seem the favor the reaction while the electron-donating groups may inhibit it. Substituted pyrroles get phenylated at the 2-position with excellent regioselectivity yielding desired products in moderate to good yields.

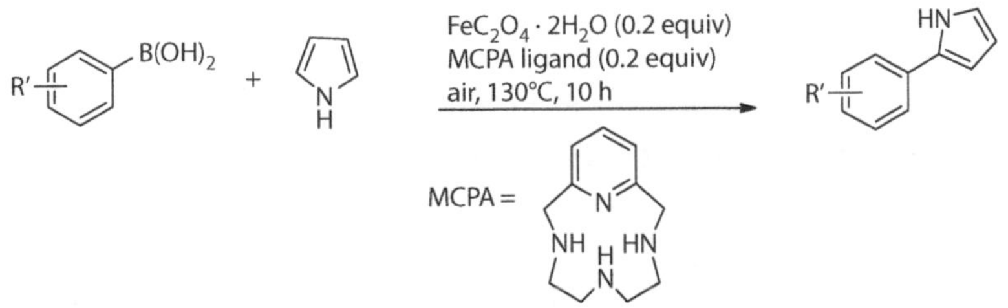

Pyridine, an electron deficient heteroarene is difficult to arylate through direct and selective C—H arylation. The presence of the Fe–MCPA combination enables products to form with high regioselectivity in this case. In other words, the methodology offers an opportunity to directly arylate electron-deficient heteroarenes.

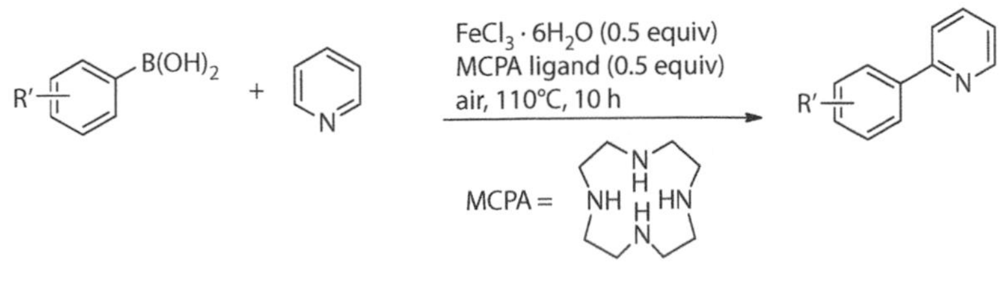

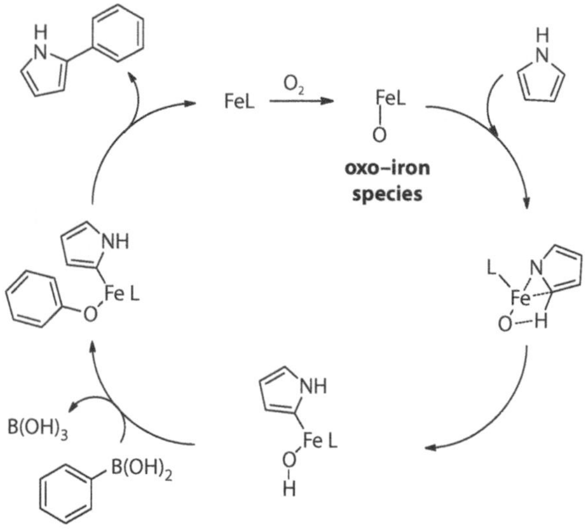

Proposed catalytic cycle for iron mediated direct Suzuki–Miyaura reaction

A catalytic cycle shown above has been proposed. The reaction begins with the oxidation of the iron complex to form the oxo-iron species by reaction with oxygen. An electrophilic attack of this oxo-iron complex on C-2 position of heteroarenes is aided by the heteroatom. Following a consequential deprotonation, a tertiary complex comprising iron, heterocycle, and MCPA ligand is formed. With the introduction of arylboronic acid, transmetalation and then elimination of the iron complex can yield the desired product.

## 6.7  CONSTRUCTION OF C−N BONDS VIA IRON-CATALYZED DIRECT C−H BOND ACTIVATION

While the formation of C−O bond in Fenton and Gif chemistry is well documented, the formation of C−N bonds through the activation of unactivated $sp^3$ C−H bonds has been challenging. The primary nitrogen source in most reported procedures involve the use of nitrenes or their derivatives.[54] Much of this chemistry employs the expensive late transition metals. It would be beneficial to employ inexpensive first row transition metals as catalysts to generate C−N bonds. An early example that demonstrates this is an efficient, inexpensive, and air-stable catalyst/an oxidant (FeCl$_2$ and N-bromosuccinimide or NBS) system that promotes amidation of benzylic $sp^3$ C−H bonds in ethyl acetate under mild conditions.[55]

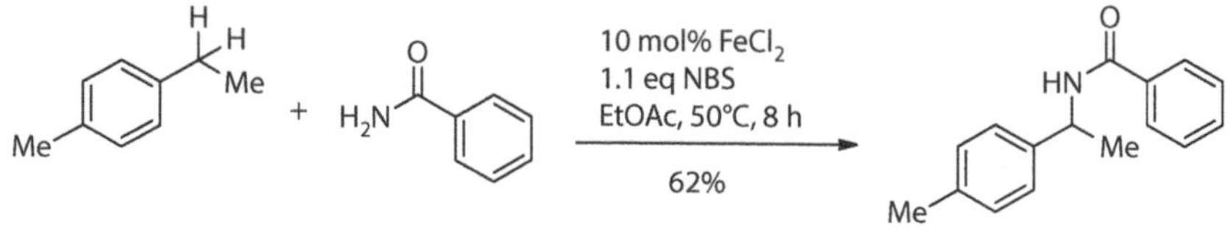

A variety of sulfonamides and carboxamides have been introduced at the benzylic position using this methodology in reasonably good yields. NBS plays the role of an oxidant as well as a free radical initiator. The reaction is proposed to involve the generation of an *N*-bromocarboxamide or *N*-bromosulfonamide as the initial intermediate that then reacts with iron salt eventually forming the reactive iron–nitrene complex that can aminate the benzylic position.

In another report, Li and coworkers have reported the synthesis of azole derivatives by iron-catalyzed oxidative reactions of azoles and ethers.[56] The typical *N*-alkylation of azoles demanding a deprotonation step and then a nucleophilic substitution reaction often suffers from the use of a strong base and over-alkylation. Substituted triazoles, imidazoles, and benzimidazoles have been shown to undergo reaction with a variety of cyclic and acyclic ethers in the presence of $FeCl_3 \cdot 6H_2O$ as catalyst and *tert*-butyl hydroperoxide (TBHP) as an oxidant in reasonably good yields. The mechanism of this reaction involves a key role being played by $Fe^{2+}$—$Fe^{3+}$ redox processes in the C—N bond formation. The main steps in the reaction are, specifically, the reductive heterolytic cleavage of the O—O bond in the peroxide and the addition of the ether carbon radical to the oxonium species.

The Nakamura research group has also reported an iron-catalyzed directed amination of a $C(sp^2)$—H bond with chloroamines to produce anthranilic acid derivatives in up to quantitative yield.[57] The structure of the directing group and the control of the electronic properties of the diphosphine ligand bring about selective formation of the aminated product. Several anthranilic acid derivatives involving ortho amination of aromatic carboxamides with *N*-chloroamines have been prepared. A combination of the quinolin-8-yl group (Q) and the diphosphine ligand slows down the undesired C—C bond formation pathway, thus enabling the C—N pathway. This reaction provides rapid access to anthranilic acid derivatives in up to quantitative yields. The structure of the directing Q and the control of the electronic properties of the diphosphine ligand allow selective formation of the aminated product.

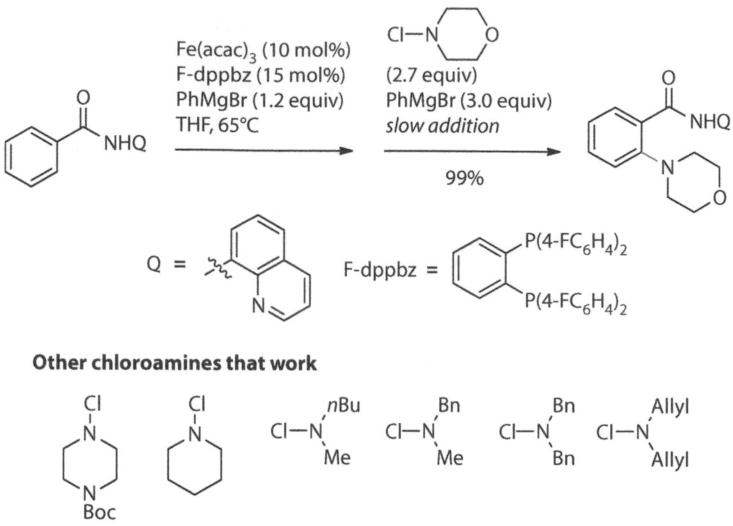

**Other chloroamines that work**

The reaction is understood to proceed through the iron intermediate **A**, shown below, from the starting material amide comprising the directing $N$-(quinolin-8-yl)-benzamide auxiliary.[58] In the presence of the earlier mentioned dihalo oxidant, namely 1,2-dichloroisobutant, intermediate **A** undergoes oxidation to form the *ortho*-phenylated product. However with $N$-chloromorpholine around, **A** reacts much faster than the oxidative C—C bond-formation, yielding the *ortho*-aminated product as shown in the scheme above.

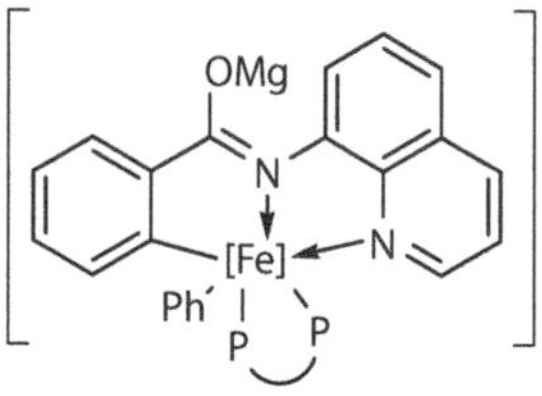

**Intermediate A**

It is noteworthy that there are several factors that influence the desired product against the other potential pathways. For example, the unique bidentate quinolin-8-yl (Q) group and less electron-rich diphosphine ligand together effectively retard the C—C bond-forming pathway; PhMgBr being the ideal Grignard reagent. Organozinc reagents described in the earlier Nakamura report are ineffective and only lead to *ortho*-phenylation. A high level of purity of chloroamine is necessary for larger (gram scale) reaction, with any decomposed chloroamine impeding the catalytic cycle. Interestingly, no diaminated product is observed in a variety of examples studied where the aryl group ranged from "non-*ortho*" substituted phenyl, naphthyl, and heteroaryl systems including thiophene and indole; the yields ranging from 54% to quantitative.

Betley et al. have recently described unique iron-dipyrrinato catalysts that can selectively aminate $sp^3$ C—H bonds.[59] This is a synthetic strategy inspired by nature where cytochrome P450 enables the conversion of C—H bonds into functional group handles. It is based on the 30-year-old mechanistic precedent involving the reaction of dioxygen with heme iron in cytochrome P450 generating strong oxidant comprising a reactive iron–oxygen multiple bond (iron–oxo) and possessing radical character, causing the entire unit to be a reactive functionality.[60] Subsequently, this electronic configuration causes the activation of substrate aliphatic C—H bonds via H-atom abstraction, overcoming any orbital spatial restrictions responsible for possibly prohibitive oxidative addition pathways.[61] The overall result is substrate functionalization due to a recombination of the organic radical generated in the activation step with the open-shell iron-hydroxyl to produce an alcohol product with corresponding reduction of the iron. Thus, ideally, any metal–ligand multiple bond can replicate the the electronic structure of the cytochrome P450 reactive iron–oxo intermediate, thereby enabling a general approach to convert unactivated C—H bonds to a C–heteroatom bond containing product. Lately, there have been several examples that prove this point, notable among which would be the metal-stabilized carbene and nitrene transfer reactions.[62] For example, the C—H amination Rh$_2$-dicarboxylate catalysts generate cyclic carbamates, guanidines, sulfamides, and, more recently, azides that yield indolines through intra-molecular $sp^3$ C—H amination.[63] Betley and coworkers have extended this concept by designing an iron–dipyrrinato catalyst that selectively aminates $sp^3$ C—H bonds. Here, only single azide functionality in the simple linear aliphatic substrate is required to bring about the synthesis of complex cyclic amine structures as shown.

**Some examples**

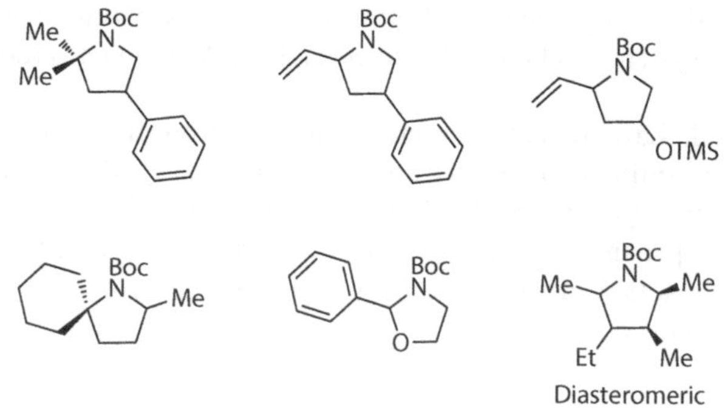
Boc
Me
Me
N
Boc
N
Boc
N
OTMS
Boc
N
Me
Boc
N
O
Boc
N
Me
Me
Et
Me
Diasteromeric

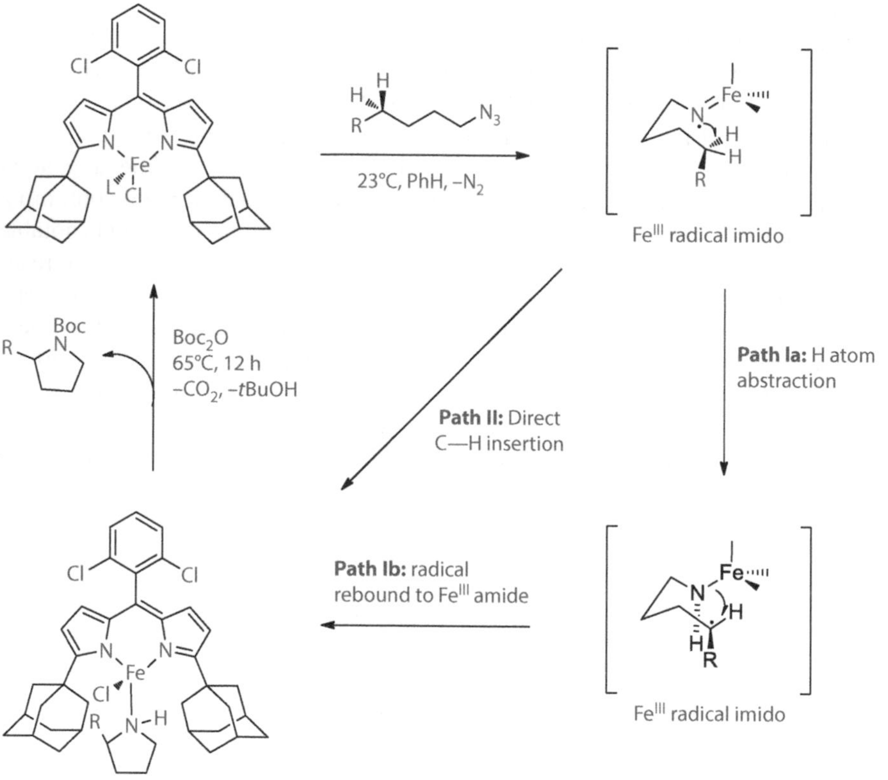
Cl
Cl
N
Fe
N
L
Cl
H
H
R
N3
23°C, PhH, –N2
Fe
N
H
H
R
FeIII radical imido
Boc
R
N
Boc2O
65°C, 12 h
–CO2, –tBuOH
Path II: Direct
C—H insertion
Path Ia: H atom
abstraction
Cl
Cl
N
Fe
N
Cl
R
N–H
Path Ib: radical
rebound to FeIII amide
Fe
N
H
H
R
FeIII radical imido

Benzylic, allylic, and even the less-reactive tertiary and secondary C—H bonds can be functionalized using this methodology. An attempt to functionalize primary C—H leads to the formation of the corresponding amine and imine. Under catalytic conditions, the resultant free heterocyclic product is not obtained because of a possible tight Lewis acid/base pair between dipyrrinato iron and the heterocyclic nitrogen causing product inhibition. $Boc_2O$ therefore works out to be an *in situ* protection reagent, thereby reducing the nucleophilicity of the product *N*-heterocycle while avoiding the generation of by-products that might retard or prevent catalysis. By-products $Boc_2O$ (*t*BuOH, $CO_2$) do not inhibit catalyst turnover, enabling the synthesis of heterocycle with catalytic amounts of the iron catalyst. Secondary and primary C—H bonds can be activated and functionalized as well, albeit in low yields.

The reaction presumably involves the C—H functionalization via a three-step sequence consisting of (i) oxidation of the Fe(II) catalyst to an Fe(III) imido radical by the alkyl azide substrate, (ii) intra-molecular H-atom abstraction causing the formation of an alkyl radical and an Fe(III) amide (Path 1a), and finally (iii) radical recombination to form the observed *N*-heterocyclic product. An alternative direct C—H bond insertion (Path II) by the Fe(III) imido radical cannot be ruled out. Both hypothesized mechanisms necessitate the alignment of substrate C—H bond to be in close proximity to the reactive Fe-imido radical.[64] A conformation requiring the C—H bond substrate to approach the imido radical opposite the chloride ligand is expected to enable rapid C—H bond functionalization. With this potentially promising synthetic strategy, azetidines, pyrrolidines, and piperidines can be prepared.

## 6.8 IRON-CATALYZED C—H ACTIVATION OF UNACTIVATED ALKANE *sp³* C—H BONDS

Probably, the most challenging area in C—H activation is that of strong $sp^3$ C—H bonds. Highly selective C—H activation of such unactivated systems is a much desired goal. There has been an immense development in this area through the use of a variety of metal catalysts.[65] In the field of iron catalysis to activate and functionalize $sp^3$ C—H bonds, the focus has been on the development of reactions using green oxidants such as oxygen, air, or hydroperoxides. Nature provides inspiration to the chemist through iron or copper containing metalloenzymes. Formally, oxoiron(V) oxidants are postulated in the catalytic cycles of several iron enzymes, such as the ones in cytochromes P450 that carry out difficult oxidation including the hydroxylation of strong C—H bonds in methane. Extensive work in this field by Que Jr., led to the first example describing a non-heme catalyst: iron tris(2-pyridylmethyl)amine complex involving hydrogen peroxide as the oxidant.[66] It involved stereospecific alkane hydroxylation using this multidentate nitrogen-containing ligand responsible for excellent reactivity.[67] This in-depth study indicated that an $Fe^V = O$ species plays a key role in the oxygen-activation mechanisms postulated for non-heme iron enzymes. Since then, the design of multidentate nitrogen ligands in iron catalysis has gained wide interest and acceptance by the organic chemist, leading to significant developments in the field of activation or functionalization of strong hydrocarbon C—H bonds, catalyzed by iron.[68] The ability of Fe(mep) complexes [mep ≡ *N,N'*-dimethyl-*N,N'*-bis(2-pyridylmethyl)-ethane, 1,2-diamine] to operate via an electrophilic metal oxidant with the presence of a bulky yet adaptable ligand framework and their indicated usage in preparative epoxidations of olefins containing functionality make them attractive as catalysts in C—H activation.[69] Despite much development in the use of high-valence metal oxos for C—H hydroxylation, a general approach in the form a synthetic methodology for directed C—H hydroxylation has been much awaited. Recently, ground-breaking research in this challenging field has led to remarkable applications in the total syntheses of a few complex natural products, notably from the White group. In 2007, they reported the use of an electrophilic catalyst called Fe($S$,$S$-PDP) with a bulky ligand framework using inexpensive $H_2O_2$ oxidant to generate highly selective oxidations on a variety of substrates.[70] The design of the catalyst was based on earlier studies suggesting a correlation between the flexibility of the mep ligand and liability of the corresponding iron complex. Fe($S$,$S$-PDP)[71] was seen as a rigidified form of the Fe(mep) complex that would lead to improved selectivities.

The invention of the new methodology by White and coworkers was based on the hypothesis that site-selective oxidations of unactivated $sp^3$ C—H bonds could be predictably controlled using a suitably reactive metal catalyst capable of discriminating subtle electronic and steric differences between C—H bonds in complex substrates; much like Sharpless' seminal work on site-selective olefin oxidations using electrophilic metal catalysts.[72] White and coworkers demonstrated that the site of oxidation with the Fe($S$,$S$-PDP) catalyst could be predicted in complex organic substrates on the basis of electronic and steric environments of the C—H bonds. Preexisting carboxylate functionality could direct oxidation toward five-membered lactone formation.

**Some examples**

60% (18% rsm)   52% (20% rsm)   57% (27% rsm)   43% (33% rsm)

There is an important relationship between electronic and steric factors in determining the selectivities of C—H oxidations with the Fe($S$,$S$-PDP) catalyst. Hydroxylation occurs favorably at the most electron-rich tertiary (3°) C—H bond in many cases examined. Oxidation occurs (i) preferentially at the most electron-rich 3°C—H bonds and (ii) at the least sterically hindered most electron-rich methylene site. Hydroxylation is also directed to the sterically hindered 3° C—H site by free carboxylic acid. Sterics can provide a second handle for selectivity in molecular structures where C—H bonds of similar electron densities are present. When electron-withdrawing groups are present in the α or β positions, C—H oxidations with the Fe(S,S-PDP) catalyst are subject to electronic deactivation.

**(+)-arteminisinin**

Anti-malarial compound (+)-artemisinin, with its five 3° C—H bonds along its tetracyclic skeleton and the sensitive endoperoxide moiety, is prone to Fe(II)-mediated cleavage.[73] The electron-rich and sterically unencumbered 3° C—H bond at C-10 would be expected to be oxidized preferentially over the remaining 3° C—H bonds in α or β with respect to electron-withdrawing ester and endoperoxide. Diastereomerically (+)-10b-Hydroxyartemisinin, was generated as the major product in 54% yield considering recovery of the starting material. Compared to the earlier reported biocatalytic hydroxylation reaction (47% yield) with microbial cultures of *Cunninghamella echinulata*, the Fe(S,S-PDP) method is superior in terms of short reaction time of a few hours versus four days *via* biocatalysis. A few other groups have also studied this methodology and have come up with improvements. For example, Costas et al. have reported a redesigned catalyst that shows improved performance. The redesign was carried out at the remote position of the pyridine ring with a bulky hydrocarbon group, forming a robust cavity for iron, thereby increasing the selectivity and efficiency compared with the results from the Fe(PDP) catalyst.[74]

Gibberellic acid derivative

As part of a detailed and exhaustive study, the same bulky, electrophilic Fe(S,S-PDP) catalyst was shown by White et al. in 2010 to be capable of site-selective oxidation of isolated, unactivated secondary C—H bonds yielding mono-oxygenated products without resorting to directing or activating groups.[75] The chemoselectivity advantages for secondary C—H bond oxidation presumably stem from intermediate properties of secondary C—H bonds. Thus, increased steric accessibility compared to tertiary C—H bonds and a greater electron-rich character compared to primary C—H bonds, make secondary C—H bonds amenable to tuning by both electronic activation and deactivation. For example, a terminally attached electron-withdrawing group (EWG) such as an ester through its inductive effects deactivates the proximal sites creating differential electronic environments and thus biasing the oxidation sites. With an EWG around, the distal sites therefore show high selectivity for oxidation and typically the major product is a result of ketonization at the methylene site furthest away from EWG, provided there is a careful balance between reactivity and selectivity. Similarly, hyperconjugative activation caused by electronic activating groups (EAGs) like the cyclopropyl group can enable attainment of site selectivities with Fe(S,S-PDP) orthogonal to those based on inductive or steric effects. Likewise, a lone-pair electrons of an ethereal oxygen with adjacent secondary C—H (cyclic ethers, for example, tetrahydrofuran) can yield lactone with over-oxidation products.

These steric, electronic, and stereoelectronic effects on selectivity of secondary C—H bond oxidation would be expected to be more prominent on complex substrates such as natural products, than seen on the above relatively simple products. This is indeed the case, as evidenced in the examples that follow. With several C—H bonds, lack of selectivity would indeed be a valid concern. Yet, trends of selectivity are observed and seem pronounced in these cases.

In the example below there are nine possible sites of oxidation. However, oxidation has been observed at only the C2 and C3 sites. Consistent with the analysis, C2 is the major site of oxidation and yields 53% of the isolated C2 product.

The explanation is as follows. Two tertiary sites in the bicyclic substrate above are electronically and sterically deactivated toward oxidation: C9 is alpha to a ketone, and C5 is both in an axial orientation and adjacent to a *gem*-dimethyl group. The two ketones (EWGs) electronically deactivate the side chain and the entire B ring, with only 2° C—H bonds of A ring as the probable sites for methylene oxidation. C-2 is furthest away from the sterically bulky quaternary centers on the A ring. Also, C-2 oxidation may possibly relieve repulsive 1,3-diaxial interactions with two methyl substituents (C15 and C16). Thus, distinct selectivity factors can mutually reinforce in a complex molecular construction, leading to predictable and highly selective oxidation of secondary C—H bonds.

*Secondary C—H sites: 5*
*Tertiary C—H sites: 3*

**Fe(*R,R*-PDP)** (25 mol%)
AcOH (0.5 equiv)
H$_2$O$_2$ (5.0 equiv)

45%

Another example above shows site-selectivity through hyper-conjugative activation in the presence of the Fe(PDP)-catalyst in case of terpenoid substrate comprising sensitive cyclopropane and cyclobutane annulated nine-membered ring. The cyclopropane component effectively overrides steric effects and activates the two adjacent methylene groups at C-6, thus favoring C-3 for oxidation for its remoteness from ketone (EWG). C-3 oxidation is thus a major product with other oxidation products formed in minor amounts, where the Fe(*R,R*-PDP) enantiomeric version of the catalyst shows improved performance over the earlier Fe(*S,S*-PDP) due to a better match of the catalyst with substrate topology.

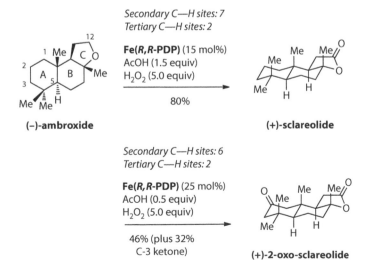

*Secondary C—H sites: 7*
*Tertiary C—H sites: 2*

**Fe(*R,R*-PDP)** (15 mol%)
AcOH (1.5 equiv)
H$_2$O$_2$ (5.0 equiv)

80%

**(–)-ambroxide**

**(+)-sclareolide**

*Secondary C—H sites: 6*
*Tertiary C—H sites: 2*

**Fe(*R,R*-PDP)** (25 mol%)
AcOH (0.5 equiv)
H$_2$O$_2$ (5.0 equiv)

46% (plus 32%
C-3 ketone)

**(+)-2-oxo-sclareolide**

Using the Fe(PDP) catalyst, selective and sequential oxidations have also been investigated and demonstrated. In an otherwise electronically unbiased hydrocarbon structure, an electron-activating group (cyclic ether) was employed to direct the initial site of oxidation. Following this early oxidation, the newly installed oxo group, as part of the lactone, now acts as an EWG, to inductively bias the molecule during the next oxidation thus imparting selectivity. This unique and powerful strategy is shown above through the sequential oxidation of (–)-ambroxide into (+)-2-oxo-sclareolide. It is quite evident that this catalysis provides the organic chemist with a unique way to propose highly efficient and green syntheses of complex organic molecules. While the yields reported can undergo further improvement, the strategy itself is designed to bring in efficiency in synthesis. There is no doubt that many more examples with improved variations will be the subject of future reports.

The reactions of this non-heme iron catalyst whether directed or non-directed proceed with complete retention of stereochemistry. The activation of unactivated C—H bonds as methodology demonstrating applications at the laboratory scale represents a landmark development. In terms of practicality this process is rather clean, since the oxidant is hydrogen peroxide—a cheap, easily available reagent commonly used in traditional iron-catalyzed Gif and Fenton chemistry.[76] The mechanism of catalysis as studied earlier by Que et al. in 2001[77] indicates a high-valence iron–oxo species being formed during the transformation. Non-heme iron enzymes such as methane monooxygenase and Rieske dioxygenases catalyze similar alkane C—H oxidations inspiring the development of such synthetic models. The reaction of the catalyst with excess $H_2O_2$ generates an Fe(III)-OOH intermediate. In the stereospecific alkane hydroxylation by non-heme iron catalysts, Que et al. have alluded to mechanistic evidence for an $Fe^V=O$ active species and the formation of alkyl radical forming from the abstraction of hydrogen by this active iron species (Ref: Chemical Reviews C—H activation). Fe(PDP)-type complexes catalyze stereospecific alkane hydroxylation by a mechanism involving both a low-spin Fe(III)-OOH intermediate and an $Fe^V=O$ species derived from O—O bond heterolysis.

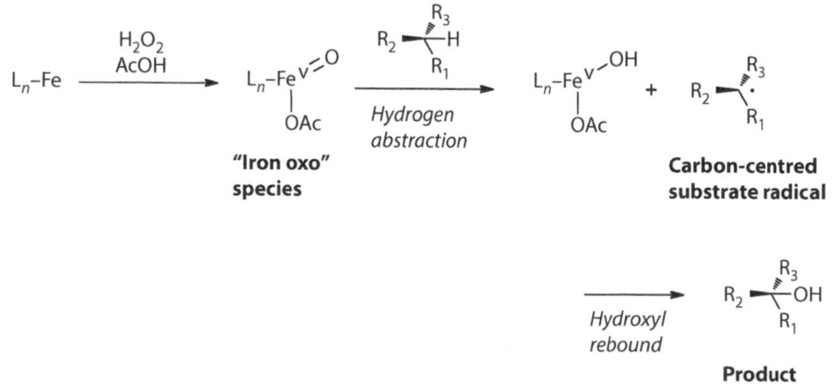

"Iron oxo" species

Carbon-centred substrate radical

Product

Thus, the Fe(PDP) with $H_2O_2$ and acetic acid (or alternatively a carboxylic acid substrate) generates an iron oxo species as the active oxidant. Intermolecular or intramolecular hydrogen abstraction by the iron oxo species leads to a short-lived aliphatic carbon-centered substrate radical. Hydroxyl rebound finally yields the hydroxylated product.[78]

## 6.9  CONCLUSION

The field of C—H activation and functionalization catalyzed by iron has seen substantial development during the past decade. As a catalyst, iron offers the distinct advantage of being cost-effective; selective activation of C—H bonds that does not require redundant pre-programming of substrates leans toward green chemistry by avoiding environmental waste. With increased understanding of the Fenton as well as Gif chemistry and "mimicking non-enzymatically" nature's oxygenated saturated unreactive carbons, the field has been going through an impressive revival. As result, new methodologies have emerged that are beginning to be applied in total syntheses. It is quite evident from the foregoing sections that this area of research will continue to experience growth. However, one of the key challenges is to develop reaction conditions that are practical to the point of being applied (eventually) on large scale in manufacturing, which is the ultimate testing ground for any synthetic reaction.

## REFERENCES

1. (a) Dyker, G. *Handbook of C—H Transformations: Applications in Organic Synthesis*; Wiley-VCH: Weinheim, NY, 2005. (b) Godula, K.; Sames, D. *Science* **2006**, *312*, 67. (c) Alberico, D.; Scott, M. E.; Lautens, M. *Chem. Rev.* **2007**, *107*, 174. (d) Bellina, F.; Rossi, R. *Tetrahedron* **2009**, *65*, 10269.
2. Constable, D. J. C.; Dunn, P. J.; Hayler, J. D.; Humphrey, G. R.; Leazer, J. L., Jr.; Linderman, R. J.; Lorenz, K. et al. *Green Chem.* **2007**, *9*, 411.
3. (a) Beccalli, E. M.; Broggini, G.; Martinelli, M.; Sottocornola, S. *Chem. Rev.* **2007**, *107*, 5318. (b) Browning, A. F.; Greeves, N. In *Palladium-Catalyzed Carbon-Carbon Bond Formation*; Beller, M., Bolm, C., Eds.; Wiley-VCH: Wienheim, NY, 1997; p. 35.
4. (a) Chianese, A. R.; Lee, S. J.; Gagné, M. R. *Angew. Chem. Int. Ed.* **2007**, *46*, 4042. (b) Fürstner, A.; Davies, P. W. *Angew. Chem. Int. Ed.* **2007**, *46*, 3410. (c) Albrecht, M.; van Koten, G. *Angew. Chem. Int. Ed.* **2001**, *40*, 3750.
5. (a) Trost, B. M.; Toste, F. D.; Pinkerton, A. B. *Chem. Rev.* **2001**, *101*, 2067. (b) Naota, T.; Takaya, H.; Murahashi, S.-I. *Chem. Rev.* **1998**, *98*, 2599.
6. (a) Fagnou, K.; Lautens, M. *Chem. Rev.* **2003**, *103*, 169. (b) Doyle, M. P.; Duffy, R.; Ratnikov, M.; Zhou, L. *Chem. Rev.* **2010**, *110*, 704.
7. Allen, S. E.; Walvoord, R. R.; Padilla-Salinas, R.; Kozlowsk, M. C. *Chem. Rev.* 2013, 113*(8)*, 6234.
8. Bolm, C.; Legros, J.; Le Paih, J.; Zani, L. *Chem. Rev.* **2004,** *104*, 6217.
9. (a) Fürstner, A. *Angew. Chem. Int. Ed.* **2009**, *48*, 1364. (b) Sherry, B. D.; Fürstner, A. *Acc. Chem. Res.* **2008**, *41*, 1500. (c) Correa, A.; Mancheño, O. G.; Bolm, C. *Chem. Soc. Rev.* **2008**, *37*, 1108. (d) Enthaler, S.; Junge, K.; Beller, M. *Angew. Chem. Int. Ed.* **2008**, *47*, 3317.
10. For a more detailed review on the subject: Sun, C.-L.; Li, B.-L.; Shi, Z.-J. *Chem. Rev.* **2011**, *111*, 1293.
11. Fenton, H. J. H. *Chem. News* **1876**, *33*, 190.

12. (a) Bishop, D. F.; Stern, G.; Fleischman, M.; Marshall, L. S. *Ind. Eng. Chem. Des. Dev.* **1968**, *7*, 110. (b) Pérez, M.; Torrades, F.; García-Hortal, J. A.; Domènech, X.; Peral, J. *Appl. Catal. B: Environ.* **2002**, *36*, 63.
13. (a) Sawyer, D. T.; Sobkowiak, A.; Matsushita, T. *Acc. Chem. Res.* **1996**, *29*, 409. (b) Walling, C. *Acc. Chem. Res.* **1998**, *31*, 155. (c) MacFaul, P. A.; Wayner, D. D. M.; Ingold, K. U. *Acc. Chem. Res.* **1998**, *31*, 159. (d) Goldstein, S.; Meyerstein, D. *Acc. Chem. Res.* **1999**, *32*, 547.
14. Haber, F.; Weiss, J. J. *Proc. Roy. Soc. London, Ser. A.* **1934**, *147*, 332.
15. (a) Merz, J. H.; Waters, W. A. *Discuss. Faraday. Soc.* **1947**, *2*, 179. (b) Merz, J. H.; Waters, W. A. *J. Chem. Soc.* **1949**, *S15*, 2427.
16. Walling, C. *Acc. Chem. Res.* **1975**, *8*, 125.
17. Meunier, B.; de Visser, S. P.; Shaik, S. *Chem. Rev.,* 2004, 104, 3947–3980.
18. Gif chemistry was named after the location Gif-sur-Yvette in France. http://www.encyclopedia.com/doc/1G2-2830905471.html.
19. (a) Barton, D. H. R.; Doller, D. *Acc. Chem. Res.* **1992**, *25*, 504. (b) Barton, D. H. R. *Chem. Soc. Rev.* **1996**, *25*, 237. (c) Barton, D. H. R. *Tetrahedron* **1998**, *54*, 5805.
20. Costas, M.; Mehn, M. P.; Jensen, M. P.; Que, Jr., L. *Chem. Rev.* **2004**, *104*, 939.
21. Norinder, J.; Matsumoto, A.; Yoshikai, N.; Nakamura, E. *J. Am. Chem. Soc.* **2008**, *130*, 5858.
22. (a) Oi, S.; Fukita, S.; Inoue, Y. *Chem. Commun.* **1998**, 2439. (b) Kakiuchi, F.; Kan, S.; Igi, K.; Chatani, N.; Murai, S. *J. Am. Chem. Soc.* **2003**, *125*, 1698. (c) Kakiuchi, F.; Matsuura, Y.; Kan, S.; Chatani, N. *J. Am. Chem. Soc.* **2005**, *127*, 5936. (d) Chen, X.; Li, J.-J.; Hao, X.-S.; Goodhue, C. E.; Yu, J.-Q. *J. Am. Chem. Soc.* **2006**, *128*, 78. (e) Chen, X.; Goodhue, C.; Yu, J.-Q. *J. Am. Chem. Soc.* **2006**, *128*, 12634. (10) (f) Oi, S.; Fukita, S.; Hirata, N.; Watanuki, N.; Miyano, S.; Inoue, Y. *Org. Lett.* **2001**, *3*, 2579. (g) Kalyani, D.; Deprez, N. R.; Desai, L. V.; Sanford, M. *J. Am. Chem. Soc.* **2005**, *127*, 7330. (h) Ackermann, L. *Org. Lett.* **2005**, *7*, 3123. (d) Shabashov, D.; Daugulis, O. *Org. Lett.* **2005**, *7*, 3657. (i) Ackermann, L.; Althammer, A.; Born, R. *Angew. Chem. Int. Ed.* **2006**, *45*, 2619–2622. (j) Hull, K. L.; Sanford, M. S. *J. Am. Chem. Soc.* **2007**, *129*, 11904.
23. Yoshikai, N.; Matsumoto, A.; Norinder, J.; Nakamura, E. *Synlett* **2010**, 313.
24. Yoshikai, N.; Matsumoto, A.; Norinder, J.; Nakamura, E. *Angew. Chem. Int. Ed.* **2009**, *48*, 2925.
25. Yoshikai, N.; Asako, S.; Yamakawa, T.; Ilies, L.; Nakamura, E. *Chem. Asian J.* **2011**, *6*, 3059.
26. (a) Hatakeyama, T.; Nakamura, M. *J. Am. Chem. Soc.* **2007**, *129*, 9844. (b) Sapountzis, I.; Lin, W.; Kofink, C. C.; Despotopoulou, A.; Knochel, P. *Angew. Chem.* **2005**, *117*, 1682. *Angew. Chem. Int. Ed.* **2005**, *44*, 1654. (c) Fürstner, A.; Leitner, A.; Méndez, M.; Krause, H. *J. Am. Chem. Soc.* **2002**, *124*, 13856.
27. Sirios, J. J.; Davis, R.; DeBoef, B. *Org. Lett.* **2014,** *16*, 868.
28. Shang, R.; Ilies, L.; Matsumoto, A; Nakamura, A. *J. Am. Chem. Soc.* **2013**, *135*, 6030.
29. (a) Vallée, F.; Mousseau, J. J.; Charette, A. B., *J. Am. Chem. Soc.* **2010**, *132*, 1514. (b) Knochel, P.; Gavryushin, A. *Synfacts* **2010**, *5*, 579.
30. Liu, W.; Cao, H.; Lei, A. *Angew. Chem. Int. Ed.* **2010**, *49*, 2004.
31. de Paolini, L.; Ribet, G. *Gazz. Chim. Ital.* **1932**, *62*, 1041–1048.
32. Horner, L.; Schwenk, E. *Ann. Chem. Justus Liebig* **1950**, *566*, 69–84.
33. Murahashi, S.; Naota, I.; Yonemura, K. *J. Am. Chem. Soc.* **1988**, *110*, 8256.
34. (a) Murata, S.; Miura, M.; Nomura, M. *J. Org. Chem.* **1989**, *54*, 4700–4702. (b) Murata, S.; Miura, M.; Nomura, M. *J. Chem. Soc. Chem. Commun.* **1989**, 116.
35. Volla, C. M. R.; Vogel, P. *Org. Lett.* **2009**, *11*, 1701.
36. (a) Richter, H.; Mancheño, O. G. *Eur. J. Org. Chem.* **2010**, 4460. (b) Yu. M. Shafran, V. A. Bakulev, V. S. Mokrushin, *Russ. Chem. Rev.* **1989**, *58*, 148–162 (*Usp. Khim.* **1989**, *58*, 250). (c) North, M. *Angew. Chem. Int. Ed.* **2004**, *43*, 4126.

37. (a) Surendra, K.; Krishnaveni, N. S; Mahesh, A.; Rama Rao, K. *J. Org. Chem.* 2006, 71, 2532. (b) Jarusiewicz, J.; Choe, Y.; Soo. Yoo, K.; Park, C. P.; Jung, K. W. *J. Org. Chem.* 2009, 74, 2873.
38. Enders, D.; Shilvock, J. P. *Chem. Soc. Rev.* 2000, 29, 359.
39. Singhal, S.; Jain, S. L.; Sain, B. *Adv. Synth. Catal.* 2010, 352, 1338.
40. Han, W.; Ofial, A. R. *Chem. Commun.* 2009, 5024.
41. Wagner, A.; Han, W.; Mayer, P.; Ofial, A. R. *Adv. Synth. Cat.* 2013, 355(14-15), 3058.
42. Han, W.; Ofial, A. R.; Mayer, P. *Act. Cryst.* 2010, E66, 379.
43. Wright, S. W.; Ammirati, M. J.; Andrews, K. M.; Brodeur, A. M.; Danley, D. E.; Doran, S. D.; Lillquist, J. S. et al. *J. Med. Chem.* 2006, 49, 3068.
44. Han, W.; Ofial, A. R. *Chem. Commun.* 2009, 6023–6025. Han, W.; Mayer, P.; Ofial, A. R. *Adv. Synth. Catal.* 2010, 352, 1667–1676.
45. Yoshikai, N.; Mieczkowski, A.; Matsumoto, A.; Ilies, L.; Nakamura, E. *J. Am. Chem. Soc.* 2010, 132, 5568.
46. Ratnikov, M. O.; Xu, X.; Doyle, M. P. *J. Am. Chem. Soc.* 2013, 135, 9475–9479.
47. (a) Ratnikov, M. O.; Doyle, M. P. *J. Am. Chem. Soc.* 2013, 135, 1549. (b) Catino, A. J.; Nichols, J. M.; Nettles, B. J.; Doyle, M. P. *J. Am. Chem. Soc.* 2006, 128, 5648.
48. (a) Casiraghi, G.; Battistini, L.; Curti, C.; Rassu, G.; Zanardi, F. *Chem. Rev.* 2011, 111, 3076. (b) Toure, B. B.; Hall, D. G. *Chem. Rev.* 2009, 109, 4439. (c) Wang, M.-Z.; Zhou, C.-Y.; Wong, M.-K.; Che, C.-M. *Chem. Eur. J.* 2010, 16, 5723. (d) Murahashi, S.-I.; Nakae, T.; Terai, H.; Komiya, N. *J. Am. Chem. Soc.* 2008, 130, 11005.
49. (a) Li, Z.; Li, C.-J. *J. Am. Chem. Soc.* 2005, 127, 3672. (b) Condie, A. G.; Gonzalez-Gomez, J. C.; Stephenson, C. R. J. *J. Am.Chem. Soc.* 2010, 132, 1464.
50. (a) Brillas, E.; Sires, I.; Oturan, M. A. *Chem. Rev.* 2009, 109, 6570. (b) Wang, B.; Yin, J.-J.; Zhou, X.; Kurash, I.; Chai, Z.; Zhao, Y.; Feng, W. *J. Phys. Chem. C* 2013, 117, 383.
51. Alberico, D.; Scott, M. E.; Lautens, M. *Chem. Rev.* 2007, 107, 174. Some examples: (a) Giri, R.; Maugel, N.; Li, J.; Wang, D.; Breazzano, S. P.; Saunders, L. B.; Yu, J. *J. Am. Chem. Soc.* 2007, 129, 3510. (b) Shi, Z.; Li, B.; Wan, X.; Chen, J.; Fang, Z.; Cao, B.; Qing, C.; Wang, Y. *Angew. Chem. Int. Ed.* 2007, 46, 5554.
52. Wen, J.; Zhang, J.; Chen, S.-Y.; Li, J.; Yu, X.-Q. *Angew. Chem. Int. Ed.* 2008, 47, 8897.
53. Wen, J.; Qin, S.; Ma, L.-F.; Dong, L.; Zhang, J.; Liu, S.-S.; Duan, Y.-S.; Chen, S.-Y.; Hu, C.-W.; Yu, X.-Q. *Org. Lett.* 2010, 12, 2694.
54. (a) Liang, J.-L.; Huang, J.-S.; Yu, X.-Q.; Zhu, N.; Che, C.-M. *Chem. Eur. J.* 2002, 8, 1563. (b) Cui, Y.; He, C. *J. Am. Chem. Soc.* 2003, 125, 16202. (c) Li, Z.; Capretto, D. A.; Rahaman, R.; He, C. *Angew. Chem. Int. Ed.* 2007, 46, 5184. (d) Yamawaki, M.; Tsutsui, H.; Kitagaki, S.; Anada, M.; Hashimoto, S. *Tetrahedron Lett.* 2002, 43, 9561. (e) Nageli, I.; Baud, C.; Bernardinelli, G.; Jacquier, Y.; Moran, M.; Müllet, P. *Helv. Chim. Acta.* 1997, 80, 1087. (f) Au, S.-M.; Huang, J.-S.; Che, C.-M.; Yu, W.-Y. *J. Org. Chem.* 2000, 65, 7858.
55. Wang, Z.; Zhang, Y.; Fu, H.; Jiang, Y.; Zhao, Y. *Org. Lett.* 2008, 10, 1863.
56. Pan, S.; Liu, J.; Li, H.; Wang, Z.; Guo, X.; Li, Z. *Org. Lett.* 2010, 12, 1932.
57. Matsubara, T.; Asako, S.; Ilieas, L.; Nakamura, E. *J. Am. Chem. Soc.* 2014, 136, 646.
58. Asako, S.; Ilies, L.; Nakamura, E. *J. Am. Chem. Soc.* 2013, 135, 17755.
59. Hennessy, E. T.; Betley, T. A. *Science* 2013, 340, 591.
60. Ortiz de Montellano, P. R., Ed., *Cytochrome P450: Structure, Mechanism, and Biochemistry,* Kluwer Academic/Plenum: New York, ed. 4, 2005.
61. Groves, J. T.; McClusky G. A.; White, R. E.; Coon, M. J.; *Biochem. Biophys. Res. Commun.* 1978, 81, 154.
62. (a) Fiori, K. W.; Espino, C. G; Brodsky, B. H.; Du Bois, J. *Tetrahedron* 2009, 65, 3042. (b) Liang, C.; Collet, F.; Robert-Peillard, F.; Müller, P.; Dodd, R. H.; Dauban, P. *J. Am. Chem. Soc.* 2008, 130, 343.
63. Nguyen, Q.; Sun, K.; Driver, T. G. *J. Am. Chem. Soc.* 2012, 134, 7262.
64. King, E. R.; Hennessy, E. T.; Betley, T. A. *J. Am. Chem. Soc.* 2011, 133, 4917.

65. This is evident from numerous papers that have been published recently. For example: (a) Michaudel, Q.; Thevenet, D.; Baran, P. S. *J. Am. Chem. Soc.* **2012**, *134*, 2547. (b) Du, B.; Jin, B.; Sun, P. *Org. Lett.* **2014**, *16*, 3032. (c) Li, Q.; Zhang, S.-Y.; He, G.; Ai, Z.; Nack, W. A.; Chen, G. *Org. Lett.* **2014**, *16*, 1764. (d) Pitts, C. R.; Bloom, S.; Woltornist, R.; Auvenshine, D. J.; Ryzhkov, L. R.; Siegler, M. A.; Lectka, T. J. *J. Am. Chem. Soc.*, **2014**, *136*, 9780.
66. Kim, C.; Chen, K.; Kim, J.; Que, L., Jr. *J. Am. Chem. Soc.* **1997**, *119*, 5964.
67. Chen, K.; Que, L., Jr. *J. Am. Chem. Soc.* **2001**, *123*, 6327.
68. (a) The first example of a non-heme iron catalyst demonstrated to effect stereospecific alkane hydroxylation: Okuno, T.; Ito, S.; Ohba, S.; Nishida, Y. *J. Chem. Soc., Dalton Trans.* **1997**, 3547. (b) Chen, K.; Que, L., Jr. *Chem. Commun.* **1999**, 1375.
69. White, M. C.; Doyle, A. G.; Jacobsen, E. N. *J. Am. Chem. Soc.* **2001**, *123*, 7194.
70. Chen, M. S.; White, M. C. *Science* **2007**, *318*, 783.
71. Fe(S,S-PDP) was a crystallographically characterized complex.
72. (a) Kolb, H. C.; VanNieuwenhze, M. S.; Sharpless, K. B. *Chem. Rev.* **1994**, *94*, 2483. (b) Sharpless, K. B. *Tetrahedron* **1994**, *50*, 4235.
73. Zhan, J.; Guo, H.; Dai, J.; Zhang, Y.; Guo, D. *Tetrahedron Lett.* **2002**, *43*, 4519.
74. Gómez, L.; Garcia-Bosch, I.; Company, A.; Benet-Buchholz, J.; Polo, A.; Sala, X.; Ribas, X.; Costas, M. *Angew. Chem. Int. Ed.* **2009**, *48*, 5720.
75. Chen, M. S.; White, M. C. *Science* **2010**, *327*, 566.
76. Gif chemistry: (a) Barton, D. H. R.; Doller, D. *Acc. Chem. Res.* **1992**, *25*, 504. (b) Barton, D. H. R. *Chem. Soc. Rev.* **1996**, *25*, 237. (c) Barton, D. H. R. *Tetrahedron* **1998**, *54*, 5805. (d) Stavropoulos, P.; Çelenligil-Çetin, R.; Tapper, A. E. *Acc. Chem. Res.* **2001**, *34*, 745. (e) Perkins, M. J. *Chem. Soc. Rev.* **1996**, *25*, 229. (f) Knight, C.; Perkins, M. J. *Chem. Commun.* **1991**, 925. Fenton chemistry: (a) Sawyer, D. T.; Sobkowiak, A.; Matsushita, T. *Acc. Chem. Res.* **1996**, *29*, 409. (b) Walling, C. *Acc. Chem. Res.* **1998**, *31*, 155. (c) MacFaul, P. A.; Wayner, D. D. M.; Ingold, K. U. *Acc. Chem. Res.* **1998**, 31.
77. Chen, K.; Que, Jr., L. *J. Am. Chem. Soc.* **2001**, *123*, 6327.
78. Bigi, M. A.; Reed, S. A.; White, M. C. *J. Am. Chem. Soc.* **2012**, *134*, 9721.

# 7 Copper-Mediated C—H Activation

*Nadia M. Ahmad*

## CONTENTS

## 7.1 INTRODUCTION

The prevalence of literature concerning C—H activations has increased dramatically over the last decade as more and more synthetically useful and efficient methodologies have become available. Several excellent reviews on C—H activation exist, including the use of different metal catalysts,[1] the use of C—H functionalizations in natural product synthesis,[2,3] and aryl–aryl bond formations,[4] and in catalytic C—H aminations.[5]

The use of copper in C—H activations affords many advantages. Copper is inexpensive and readily available, as well as being nontoxic. Generally, C—H activations employing copper do not require complex or bespoke ligands, or cocatalysts. Copper-catalyzed methods also tend to have good functional group tolerance, thus increasing their synthetic utility.

## 7.2 AROMATIC C—H ARYLATIONS, C(*sp²*)—H ARYL–ARYL BOND FORMATION

Copper was the first transition-metal shown to promote carbon–hydrogen bond arylation.[6–9] Daugulis et al. illustrate the use of copper in the arylation of acidic

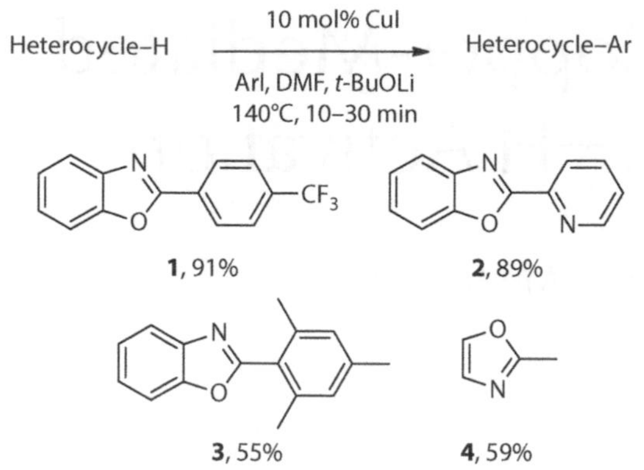

**SCHEME 7.1**   Copper-catalyzed arylation of acidic heterocycles.

heterocycles.[10] The use of a strong base such as *t*-BuOLi generates the organocopper species and the reaction proceeds in minutes at high temperatures (Scheme 7.1). However, for less acidic heteroaryls such as imidazole and triazole, a stronger base is required and the reaction is postulated to proceed via a benzyne mechanism.

Although good yields can be obtained, the benzyne mechanism results in formation of regioisomeric mixtures and the *tert*-butyl aryl ether side-product resulting from the reaction of the *tert*-butoxide base with the aryl halide. The authors found that addition of the widely used phenanthroline ligand allowed the use of *t*-BuOLi, a weaker lithium alkoxide base, in place of *t*-BuOK, and avoided the problems of a benzyne mechanism. In addition, the use of $K_3PO_4$ base in the arylation of the most acidic heterocycles, possessing DMSO $pK_a$'s < 27, resulted in good-to-excellent coupling yields. A representative example is shown (Scheme 7.2).

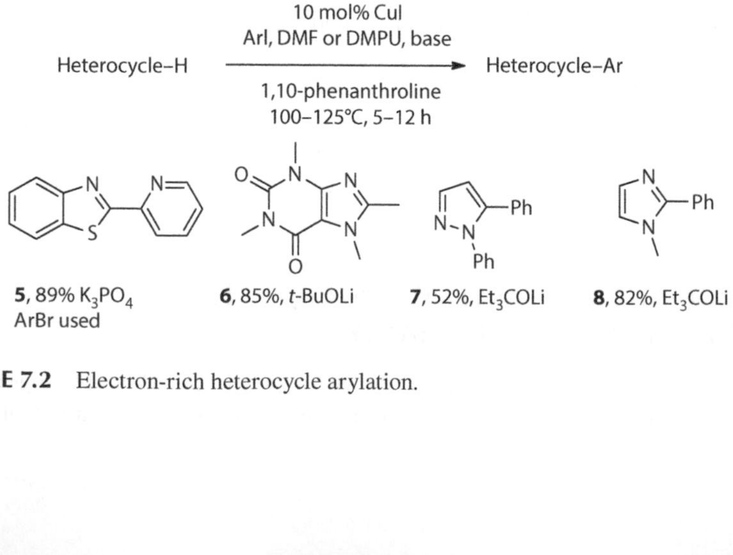

**SCHEME 7.2**   Electron-rich heterocycle arylation.

**TABLE 7.1**
**Formation of Bisoxazole**

| Entry | Substrate | Product | Yield (%) |
|-------|-----------|---------|-----------|
| 1 | | | 90 |
| 2 | | | 91 |
| 3 | | | 79 |
| 4 | | | 86 |

Bao et al. report an efficient and convenient air oxidative biaryl coupling reaction of azoles.[11] The authors did not find significant differences in the yields when comparing pure oxygen and air, and catalytic Cu(OAc)$_2$ in xylene provided the best conversions. The method was not only successfully applied to homo-coupling of various azoles (imidazoles, benzimidazoles, thiazoles, Table 7.1) but also to cross-coupling

**TABLE 7.2**
**Cross-Coupling of Two Different Benzimidazoles**

| Entry | X | Substrate | Yield % | | |
|---|---|---|---|---|---|
| | | | **12** | **13** | **14** |
| 1 | N—CH₃ | | 56 | 31 | 19 |
| 2 | | | 49 | 24 | 35 |

between two different azoles (Table 7.2). Although the cross-coupling yields were moderate, the cross-coupled yield was found to be the major product with minor yields of the homo-coupled side-products.

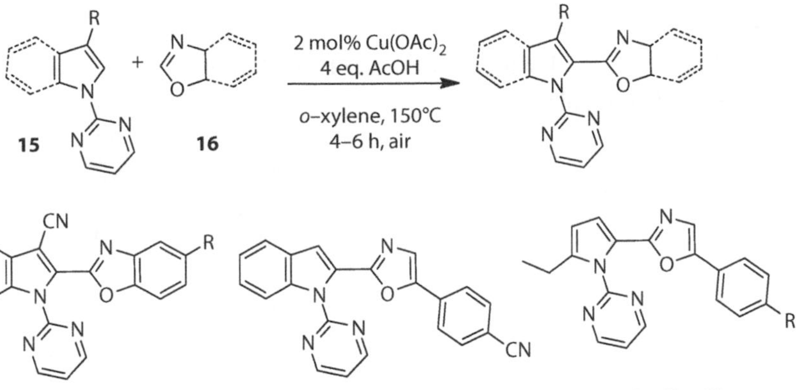


**17**, R = H, 81%
**18**, R = Me 70%

**19**, 70%

**20**, R = H, 63%
**21**, R = Cl 57%

**SCHEME 7.3** Copper-catalyzed cross-coupling of *N*-(2-pyrimidyl)indoles or pyrrole and 1,3-azoles (isolated yields).

Similarly, Miura and coworkers have developed a copper-mediated intermolecular cross-coupling of indoles and 1,3-azoles using a 2-pyrimidyl directing group (Scheme 7.3).[12] The process uses catalytic copper with oxygen as the sole oxidant and thereby is very clean, since water is the only side-product generated. The directing group was easily removed from the products with sodium methoxide in DMSO (Scheme 7.4).

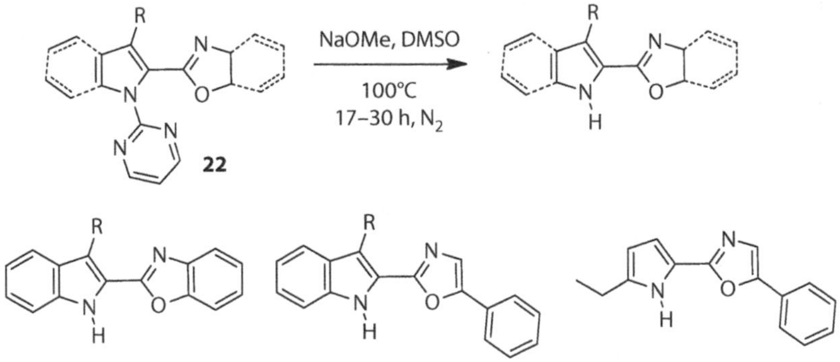


**23**, R = H 68%
**24**, R = CN 67%

**25**, R = H 88%
**26**, R = CN 76%

**27**, 56%

**SCHEME 7.4** Removal of 2-pyrimidyl protecting group.

**TABLE 7.3**
**Effect of Aryl Coupling Partners in the C—H Arylation**
**of 2-Phenyl-2H-Indazole**

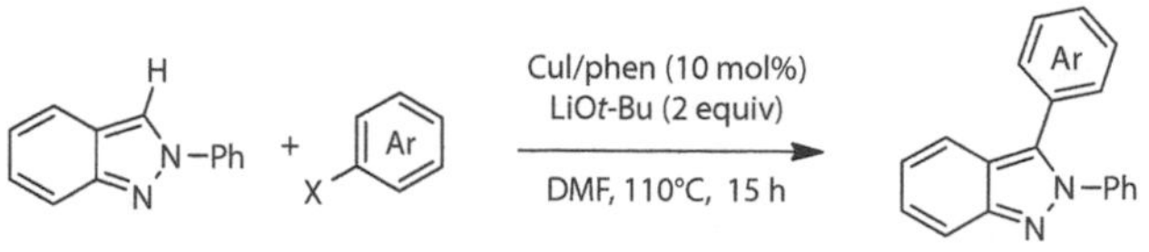

| Entry | Substrate | Product 29 | Yield (%) |
|-------|-----------|------------|-----------|
| 1 | 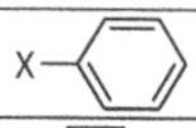 X—⬡ | X–I | 58 |
|   |           | X–Br | 4 |
| 2 | I—⬡—R | R = Cl | 46 |
|   |           | R = OMe | 56 |
|   |           | R = CF$_3$ | 52 |
| 3 | 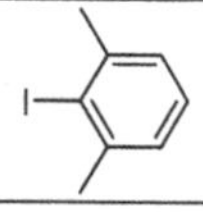 I—⬡ (dimethyl) |  | 40 |

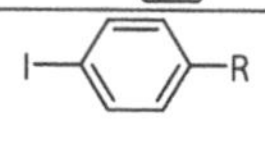

Itami and coworkers expand upon Daugulis' conditions (mentioned above) for the C—H arylation of 2H-indazoles, a motif inherently important in pharmaceuticals in themselves and as bioisosteres of indoles.[13] Aryl iodides were found to give significantly better yields than aryl bromides, with aryl iodides bearing either electron-withdrawing or electron-donating groups giving good yields. Steric hindrance only resulted in a slightly lower yield but, importantly, gave the desired product (Table 7.3).

Intriguingly, the copper-catalyzed C—H arylation of 1H-indazoles did not yield the desired product and, instead, gave rise to the triphenylamine derivative **29**. The authors found that repeating the reaction using palladium and a weaker base (potassium phosphate) gave the desired transformation.[13] Neither the copper nor palladium reactions proceeded with the indazole bearing a free NH (Scheme 7.5).

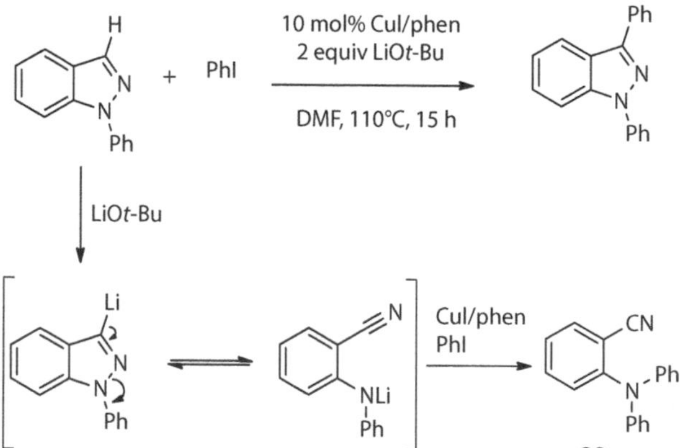

**SCHEME 7.5**   Unexpected reaction of 1-phenyl-1H-indazole when using CuI/phen/LiOt-Bu.

## TABLE 7.4
## Scope of Ligandless CuI–Catalyzed Intramolecular C—H Functionalization of Different Azoles

| Entry | Substrate | Product | Yield % |
|-------|-----------|---------|---------|
| 1 | | | 88 |
| 2 | | | 58 |
| 3 | | | 58 |
| 4 | | | 89 |

A highly practical ligand-free copper-catalyzed system for the intra-molecular direct C—H arylation of electron-rich azaheterocycles has been reported by Dominquez et al.[14] Complex hetero-fused compounds were synthesized under mild conditions, which did not have to be anhydrous, thus simplifying the protocol. A highlight was the tolerance of the reaction conditions to additional halogens; this, therefore, allows for further synthetic transformations. These are summarized in Table 7.4. The authors cite entries 3 and 4 to be the first direct copper-catalyzed C—H functionalizations of both 9H-purine and 4-azabenzimidazoles. These compounds are eminently important for their cardiovascular properties and use in the treatment of viral diseases such as AIDS and HIV.

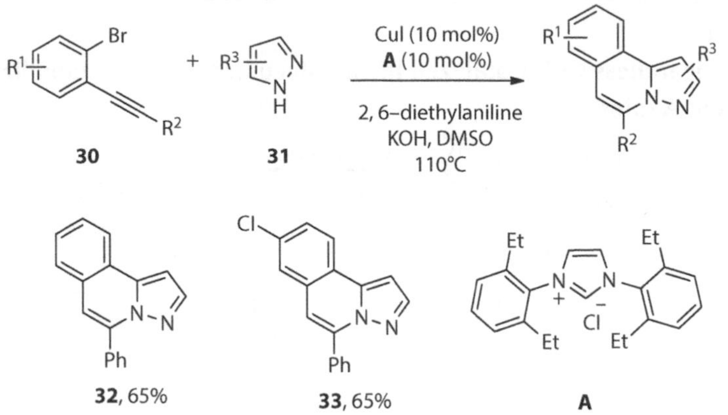

**SCHEME 7.6**  Copper-catalyzed tandem reaction of 2-alkynylbromobenzenes with pyrazoles.

The copper-catalyzed C—H activation reactions of 2-alkynylbromobenzene with pyrazoles, to afford pyrazolo[5,1-*a*]isoquinolines, have been reported by Wu and coworkers as part of their efforts to synthesize drug-like small molecules.[15] Several ligands were screened under different base/solvent combinations. The reaction was found to proceed well with 10 mol% of ligand **A**, in the presence of 10 mol% 2,6-diethylaniline, whose role in the catalytic cycle remains unclear (Scheme 7.6). The reader is referred to the reference for a discussion of the mechanistic cycle for this tandem copper-catalyzed hydroamination and C—H activation reaction.

Products were obtained in moderate-to-good yields with a range of different substituents present on both the aromatic ring of the 2-alkynylbromobenzene or attached on the pyrazole, under the optimized conditions. Although the generated products are quite specific, the report provides a good example of the use of C—H activation chemistry, utilizing catalytic copper in a tandem reaction.

## 7.3 AROMATIC C—H ALKENYLATIONS, C($sp^2$)—H ARYL–ALKYNE BOND FORMATION

The direct C—H alkynylation of 1,3,4-oxadiazoles with 1,2-dibromo-1-alkenes has been described in an eco-friendly manner by the use of CuBr and LiO*t*-Bu in PEG-400, a green solvent.[16] High yields were obtained at relatively low temperatures (80°C) in 2 h (Scheme 7.7).

As Table 7.5 shows aromatic, heteroaromatic, and aliphatic alkenes all resulted in excellent yields.

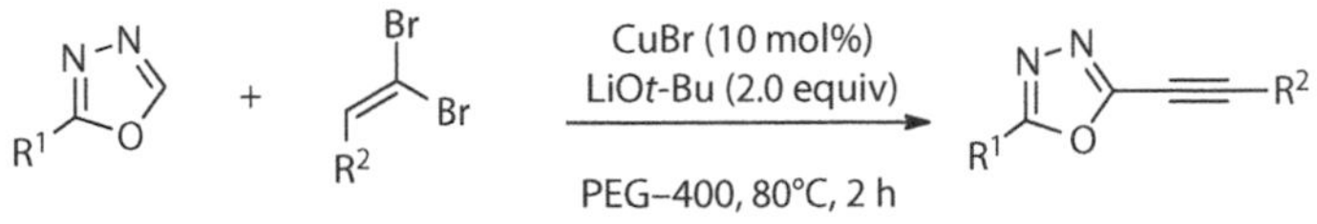

R$^1$ = aromatic, heteroaromatic
R$^2$ = aromatic, heteroaromatic, aliphatic, cyclic

**SCHEME 7.7**   Copper-catalyzed C—C cross-coupling.

**TABLE 7.5**

**Copper-Mediated Direct Cross-Coupling of 2-Phenyl-1,3,4-Ozadiazole with Different 1,1-Dibromo-1-Alkenes by Using PEG-400 as Solvent**

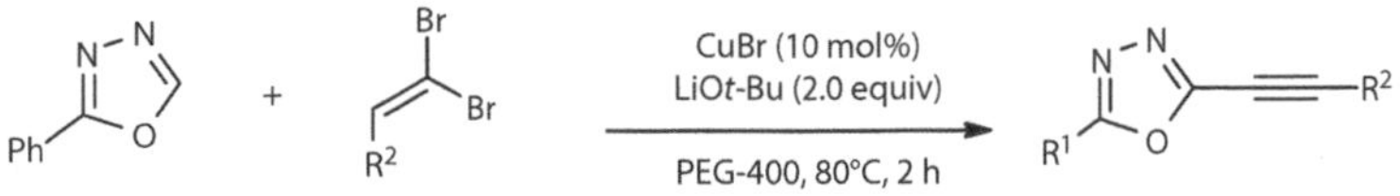

| Entry | R$^2$ | Product | Yield |
|-------|-------|---------|-------|
| 1 | 4-MeC$_6$H$_4$ | 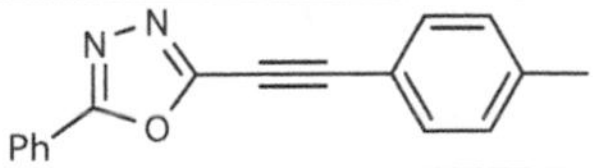 | 82 |
| 2 | 4-FC$_6$H$_4$ | 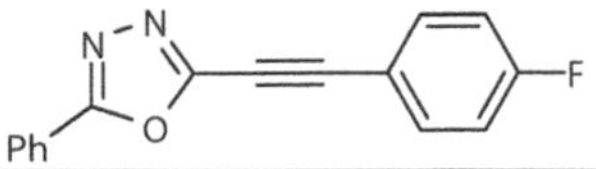 | 78 |
| 3 | 2-Thienyl | 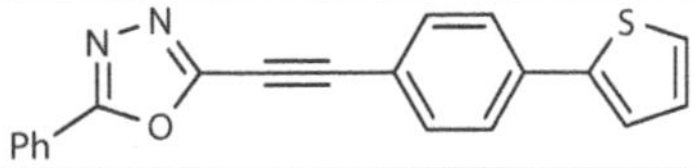 | 76 |
| 4 | C$_7$H$_{15}$ | 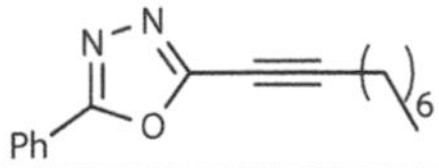 | 73 |

**TABLE 7.6**
**Copper-Mediated Direct Cross-Coupling of Various 1,3,4-Oxadiazole with Different 1,1-Dibromo-1-Alkenes by Using PEG-400 as Solvent**

| Entry | R¹ | R² | Product | Yield % |
|-------|-----|-----|---------|---------|
| 1 | 4-MeOC₆H₄ | 2-Thienyl | | 77 |
| 2 | 4-ClC₆H₄ | 2-Thienyl | | 78 |
| 3 | 2-Furyl | 4-FC₆H₄ | | 77 |
| 4 | 3-Nicotinyl | 4-MeC₆H₄ | | 82 |

The versatile scope of this reaction was further illustrated by the variations in the azole utilized. For example, as entries 1 and 2 (Table 7.6) show, electron-donating and electron-withdrawing groups on the aryl group of the oxadiazoles gave good yields, as well as the presence of *O*- and *N*-containing heterocycles.

## 7.4 AROMATIC C—H ALKYLATIONS, C($sp^2$)—H ARYL–ALKYL BOND FORMATIONS

The introduction of trifluoromethyl groups via copper-catalyzed oxidation under efficient conditions has been reported by Chu and Qing.[17] Moderate-to-excellent yields were obtained and common functionalities such as ester, nitrile, and bromine, were well-tolerated. The reaction works well on heteroarenes and electron-deficient arenes such as 1,3-oxadiazoles and benzo[$d$]imidazoles. Selected examples are shown here (Scheme 7.8).

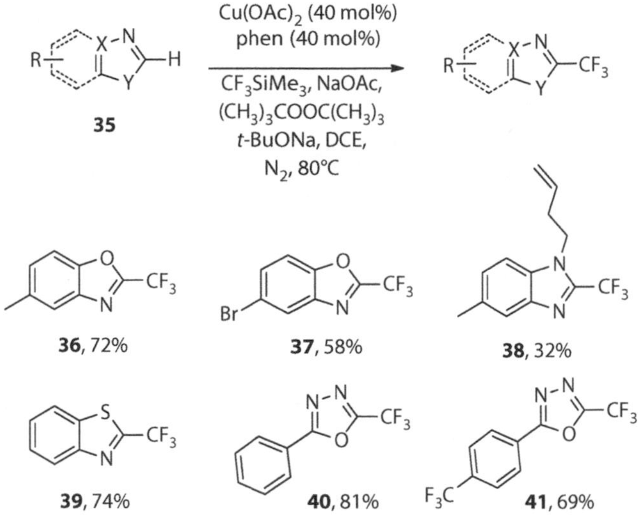

**SCHEME 7.8** Copper-catalyzed oxidative trifluoromethylation of heteroarenes and arenes with di-*tert*-butyl peroxide.

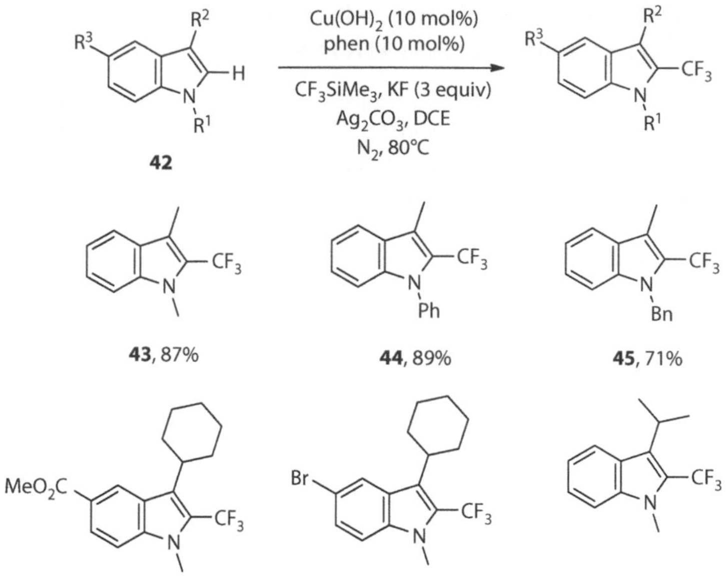

**SCHEME 7.9**  Copper-catalyzed oxidative trifluoromethylation of indoles.

Conditions to effect a similar reaction on indoles had to be optimized as the above set of conditions resulted in poor yields and variable regioselectivities. Consequently, 3-substituted indoles were found to undergo C—H activations to give products as shown, using silver carbonate as the oxidant (Scheme 7.9). Indoles substituted with an N-protected alkyl or aryl groups were compatible with the reaction conditions, and functional groups such as halogens and esters were tolerated. The oxidative trifluoromethylation was, however, significantly affected by the electron density of the pyrrole ring; N-tosyl indole gave only trace product, and indole bearing a methyl ester on the C3 was unreactive.

The use of nonactivated secondary alkyl halides as electrophiles in a copper-catalyzed direct alkylation of benzoxazoles has been described.[18] Although methods exist for the direct alkylation of aromatic heterocycles, such as radical alkylations[19–21] and coupling of heterocycles with alkyl electrophiles,[22–27] such methods are restricted to the incorporation of primary alkyl groups only. Secondary alkyl groups can be introduced by metal-catalyzed hydroarylation of olefins but in these cases examples are usually confined to activated alkyls such as benzyl or allyl.

**TABLE 7.7**
**Scope of Copper-Catalyzed Alkylation of Benzoxazoles**

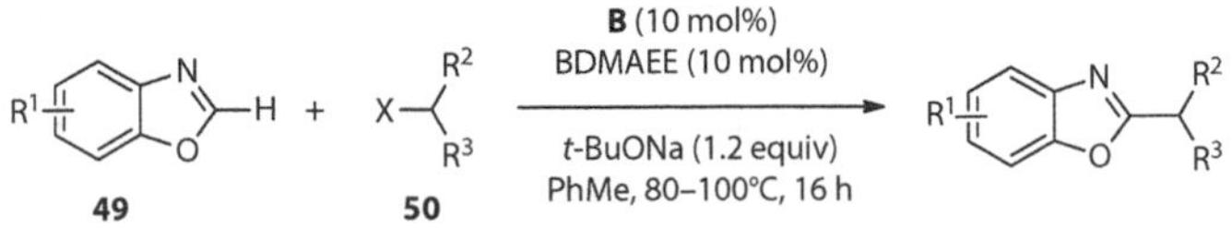


**49** **50**

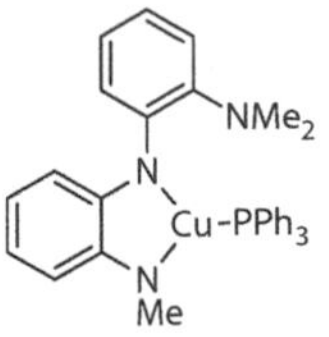

**B**

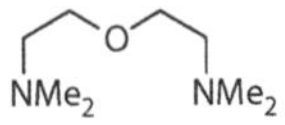

BDMAEE
additive

| Entry | Azole | Alkyl-X | Product | Yield % |
|---|---|---|---|---|
| 1 | 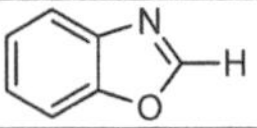 | 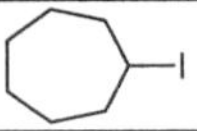 | 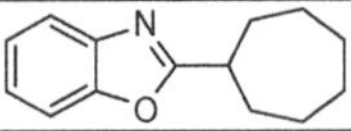 | 73 |
| 2 | 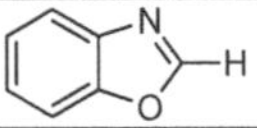 | 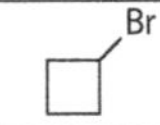 | 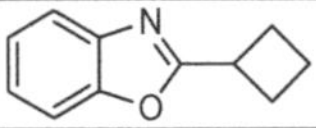 | 57 |
| 3 | 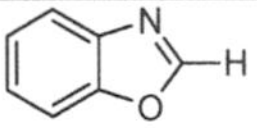 | 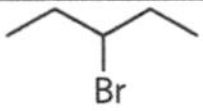 | 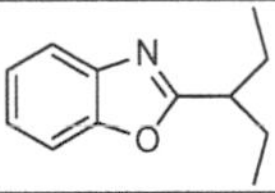 | 60 |
| 4 | 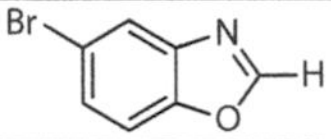 | 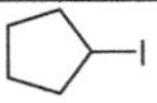 | 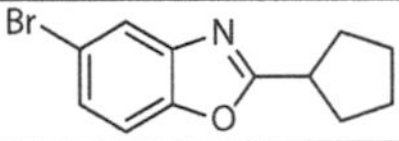 | 69 |
| 5 | 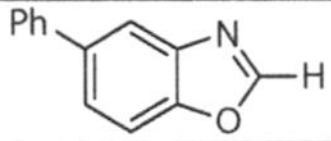 | 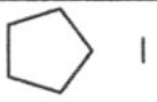 | 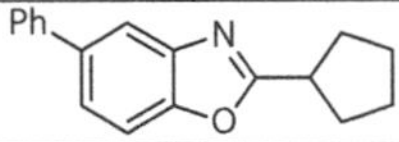 | 64 |

Hu and coworkers utilize the copper complex **B** shown, along with an ether additive, *bis*[2-(*N,N*-dimethylamino)ethyl]ether (BDMAEE).[18] Examples are shown here (Table 7.7); reactions using alkyl iodides proceeded at 80°C, whilst alkyl bromides required higher temperatures of 100°C.

The coupling of primary alkyl halides did not yield product, and oxazoles in place of benzoxazoles were also inefficient coupling partners. These results limit the scope of the reaction, nonetheless the method remains a useful one for the coupling of unactivated secondary alkyl halides.

## 7.5  ALKENE C—H ALKYLATIONS, C(*sp²*)—H ALKENE–ALKYL BOND FORMATIONS

Liu and Xu report a previously unprecedented reaction for Cu-catalyzed trifluoro-methylation of terminal alkenes via allylic C(*sp³*)—H bond activation.[28] Advantages of this method over others, which involve the use of Pd-catalyzed stannane cou-plings[29] or Cu-mediated nucleophilic trifluoromethylation of allyl halides, is the starting point of much simpler substrates and better functional group tolerance. The reaction can be conducted under mild conditions and exhibits good tolerance to moisture. Some examples are illustrated (Scheme 7.10).

The reaction mechanism is postulated to go through a Heck-like four-membered ring transition state as determined through experimental and theoretical analysis. The addition of water to the reaction did not lower the yield, and the use of palladium acetate in place of the copper catalyst gave no product. Importantly, the presence of the alkene group provides a useful handle for further functionalized trifluoromethyl-ated compounds.

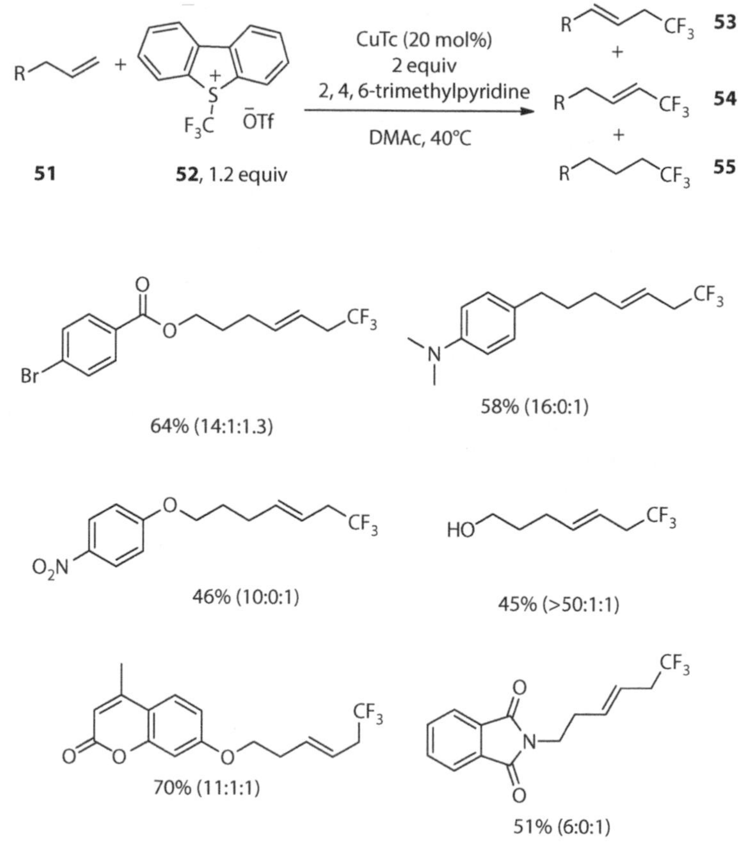

**SCHEME 7.10**  Scope of copper-catalyzed trifluoromethylation.

**TABLE 7.8**

**Synthesis of Various Pyrrole[3,2-d]Pyrimidines**

| Entry | Substrate | Product | Yield % |
|-------|-----------|---------|---------|
| 1 | | | 76 |
| 2 | | | 68 |
| 3 | | | 72 |
| 4 | | | 64 |

Majumder et al. report a specific uracil $C_6$—H functionalization to yield pyrrolo[3,2-d]pymidine motifs, which can be potentially useful bioactive structures.[30] Optimized conditions were found to be catalytic $Cu(OTf)_2$ (20 mol%) and potassium carbonate as the base (Table 7.8).

These types of pyrimidines and uracil-based molecules are prevalent in the literature as anticancer and antiviral compounds, for example, 3'-azido-3'-deoxythimidine (AZT) and (E)-5-(2-bromovinyl)-2'-deoxyuridine (BVDU).

## 7.6 FORMYL C—H ALKYLATIONS, C($sp^2$)—H ACYL BOND FORMATIONS

Wang and coworkers report a practical method for the copper-catalyzed amidation of aldehydes in the presence of NBS.[31] The method is operationally simple, requiring no additional ligands or catalysts, and is found to proceed in the presence of a range of functional groups. Examples are shown here (Table 7.9).

Imides were formed in good to excellent yields, including a coupling between a secondary amide and an aldehyde (entries 5 and 6). The method provides a viable alternative to the classic amide bond formation to form such imides via a functionalized carboxylic acid and an amine.

## TABLE 7.9
## Copper-Catalyzed Amidation of Aldehydes in the Presence of NBS

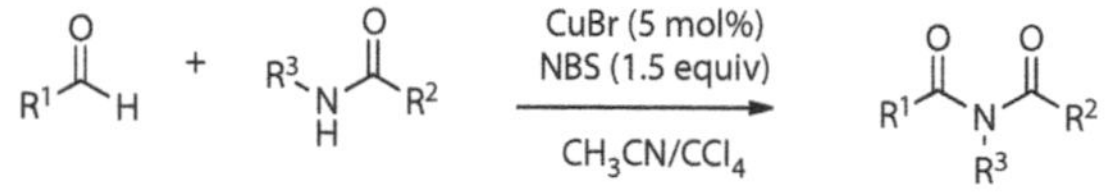

| Entry | Aldehyde | Amide | T (°C)/t (h) | Product | Yield % |
|---|---|---|---|---|---|
| 1 | 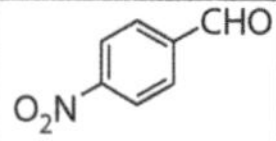 | 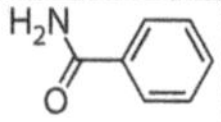 | 90/15 | 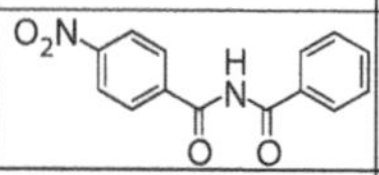 | 90 |
| 2 | 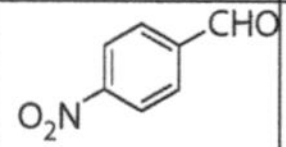 | 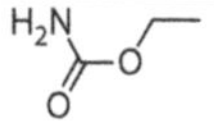 | 90/15 | 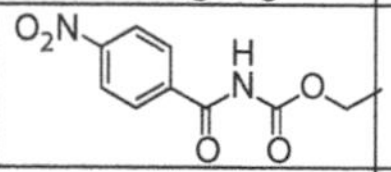 | 86 |
| 3 | 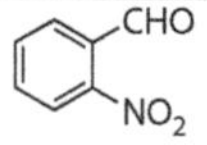 | 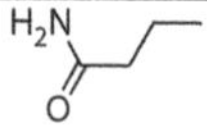 | 90/15 | 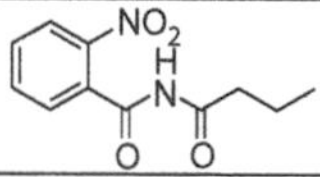 | 86 |
| 4 | 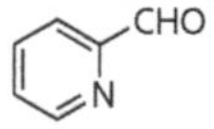 | 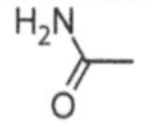 | RT/15 | 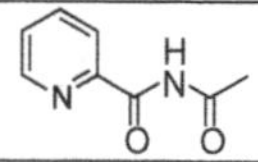 | 73 |
| 5 | 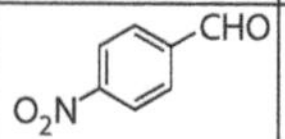 | 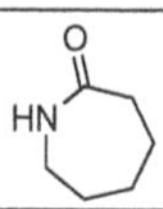 | 90/15 | 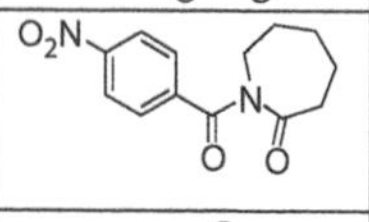 | 86 |
| 6 | 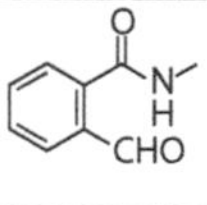 | – | 75/12 | 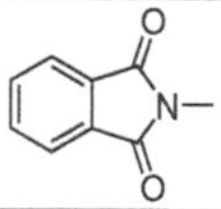 | 85 |

**TABLE 7.10**

**Reaction of Morpholine-4-Carbaldehyde with Various Thiols**

| Entry | Product | Yield % | Entry | Product | Yield % |
|-------|---------|---------|-------|---------|---------|
| 1 | | 87<br>71[α] | 5 | | 75 |
| 2 | | 89<br>77[α] | 6 | | 78 |
| 3 | | 82<br>72[α] | 7 | | 0 |
| 4 | | 82<br>61[α] | 8 | | 0 |

[α] Isolated yields with disulfides

The formation of dialkyl thiocarbamate compounds through a copper-catalyzed oxidative coupling of formamides with thiols has been demonstrated in high yields under solvent-free conditions by Yuan et al.[32] Aryl thiols bearing electron-donating or electron-withdrawing groups were well tolerated. However, aryl thiols bearing strongly electron-donating groups such as nitro- or trifluoromethyl groups were unreactive (Table 7.10). Formamides such as morpholine-4-carbaldehyde, piperidine-1-carbaldehyde and *N,N*-dimethylformamide were all found to undergo these conversions in moderate-to-good yields.

## 7.7   ALKYNYL C—H ALKYLATIONS, C(sp)—H ALKYNYL–ALKYL BOND FORMATIONS

Tu et al. report a facile, economic, and green method for the construction of propargylamines using copper-catalyzed C—H activation.[33] This so-called "A³" reaction involves the coupling of an acetylene, aldehyde, and amine (Table 7.11). The reaction proceeded in the microwave within 30 min; however, under conventional heating, the reaction took up to five days. A mechanism is shown here (Scheme 7.11). In addition, considering the prevalence of chiral propargyl amines in many important bioactive compounds, a substrate-controlled asymmetric coupling was carried out using

**TABLE 7.11**

**Coupling of Aldehyde, Alkyne, and Amine Catalyzed by CuI in Water**

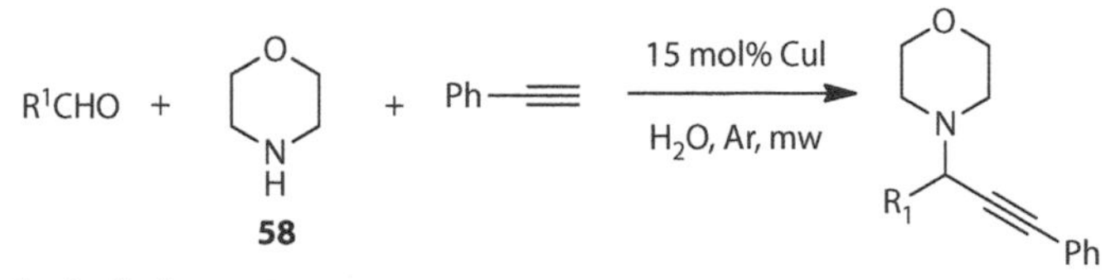

$$R^1CHO + \text{(58)} + Ph\text{—}\!\!\equiv\!\!\text{—} \xrightarrow[H_2O, Ar, mw]{15 \text{ mol\% CuI}} \text{product}$$

$R^1, R^2, R^3, R^4 = H, \text{alkyl, aryl}$

| Entry | R¹ | Time (min) | Yield % |
|-------|-----|-----------|---------|
| 1 | Ph | 20 | 90 |
| 2 | 2-FC$_6$H$_4$ | 20 | 89 |
| 3 | 4-MeOC$_6$H$_4$ | 30 | 82 |
| 4 | 4-MeOC$_6$H$_4$ | 30 | 85 |
| 5 | 4-NO$_2$C$_6$H$_4$ | 5 | 41 |
| 6 | 2-Furyl | 20 | 86 |

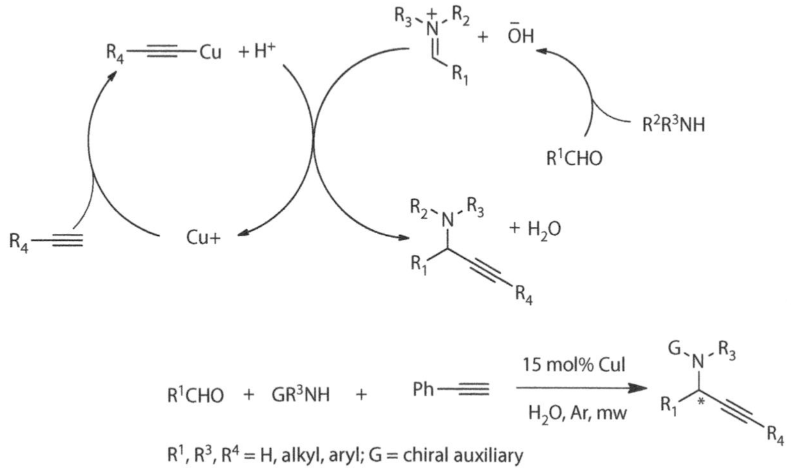

$$R^1CHO + GR^3NH + Ph\text{—}\!\!\equiv\!\!\text{—} \xrightarrow[H_2O, Ar, mw]{15 \text{ mol\% CuI}} \text{product}$$

$R^1, R^3, R^4 = H, \text{alkyl, aryl}; G = \text{chiral auxiliary}$

**SCHEME 7.11**   Mechanism of action in microwave reaction.

**TABLE 7.12**

**Diastereoselective A3 Coupling Induced with Chiral Amine Substrate**

| Entry | R¹ | Amine(G, R³) | dr | Yield % |
|-------|-----|--------------|-----|---------|
| 1 | Ph | (S)-Proline methyl ester | 95:5 | 88 |
| 2 | Ph | G = (S)-Ph(CH₃)CH, R³ = H | 67:33 | 83 |
| 3 | Ph | G = (S)-Ph(CH₃)CH, R³ = Bn | 67:33 | 81 |

the same conditions. The results showed that a highly diastereoselective coupling can be achieved using this method (Table 7.12).

In a similar fashion, the same coupling of an aldehyde, amine, and alkyne to generate propargylamines in a high yield can be carried out using a Cu(I) catalyst and an ionic liquid (Table 7.13).[34] The authors note that the yields are higher in ionic liquid such as bmim[PF₆] when using a catalyst such as CuBr. In addition, as the [bmim] PF₆ is a hydrophobic room-temperature ionic liquid, it is possible to exclude water, which is the only by-product in these reactions, and the catalyst is easily recyclable with only a small drop in activity after 5–10 reuses. Some representative examples are shown. A range of substrates were investigated; aryl and alkyl aldehydes both gave excellent yields and although the reaction was found to be impartial to electronic effects, steric effects resulted in the formation of only trace product. Similarly, sterically hindered amines gave no product (entry 8). Both alkyl and aryl acetylenes were well tolerated.

**TABLE 7.13**

**A³ Coupling Reaction with Copper Catalyst in [bmim][PF₆]**

$$R^1CHO \ + \ R^2R^3NH \ + \ R^2 \!\!-\!\!\!\equiv \xrightarrow[\text{[bmim]PF}_6, \ 120°C]{\text{CuCN 2 mol\%}}$$

| Entry | R¹ | Amine (R², R³) | R⁴ | Yield % |
|-------|-----|----------------|-----|---------|
| 1 | C₆H₅ | Piperidine | C₆H₅ | 85 |
| 2 | 4-(MeO)C₆H₄ | Piperidine | C₆H₅ | 79 |
| 3 | 3-(MeO)C₆H₄ | Piperidine | C₆H₅ | 84 |
| 4 | 4-BrC₆H₄ | Piperidine | C₆H₅ | 98 |
| 5 | 1-Naphthyl | Piperidine | C₆H₅ | 95 |
| 6 | 2-Thiophene | Piperidine | C₆H₅ | 83 |
| 7 | C₆H₅ | Morpholine | C₆H₅ | 95 |
| 8 | C₆H₅ | cis-2,6-Dimethyl piperidine | C₆H₅ | 0 |
| 9 | C₆H₅ | R² = Ph, R³ = H | C₆H₅ | 90 |
| 10 | C₆H₅ | Morpholine | (EtO)₂CH | 79 |

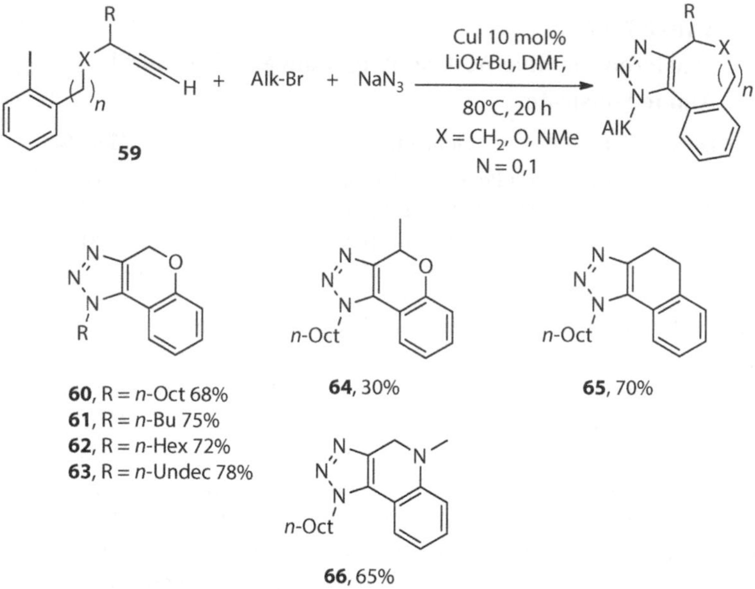

**SCHEME 7.12** Nonsequential cascade synthesis of fully substituted triazoles.

Ackerman and coworkers report a three-component one-pot cascade reaction to form annulated 1,2,3-triazoles.[35] The reaction proceeds with a copper-catalyzed 1,3-dipolar cycloaddition followed by an intra-molecular C—H bond arylation, and utilises inexpensive CuI for the formation of the one C—C and three C—N bonds in a site-selective manner (Scheme 7.12).

The reaction could be carried out in a non-sequential manner by employing equimolar amounts of the three substrates. The same conditions were also found to be effective for twofold C—H/N—H bond arylation on various azoles (Scheme 7.13).

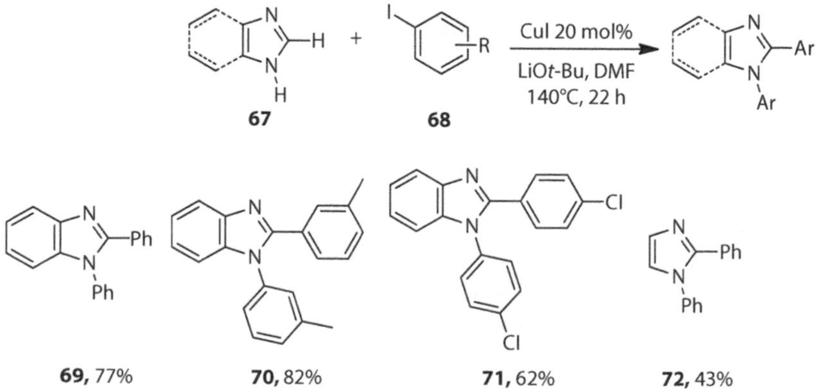

**SCHEME 7.13** Copper-catalyzed one-pot twofold C—H/N—H arylation with azoles.

## 7.8 ALKYL C—H ALKYLATIONS, C($sp^3$)—H ALKYL BOND FORMATIONS

Functionalization of the 1-position of tetrahydroisoquinolines (THIQs) has been neatly demonstrated using copper catalysis.[36] THIQs are common motifs in numerous pharmacological compounds and, thus, demand for their efficient synthesis is always high. The indolation, pyrrolation, and methoxyphenolation of THIQs is described by Schnürch et al. and can proceed in the presence of a free THIQ NH, although the yields were found to be higher with BOC-protected substrates. These reactions are summarized in Schemes 7.14 and 7.15. Representative examples of this reaction are shown and comparisons made with yields obtained using an iron nitrate catalyst instead (Table 7.14).

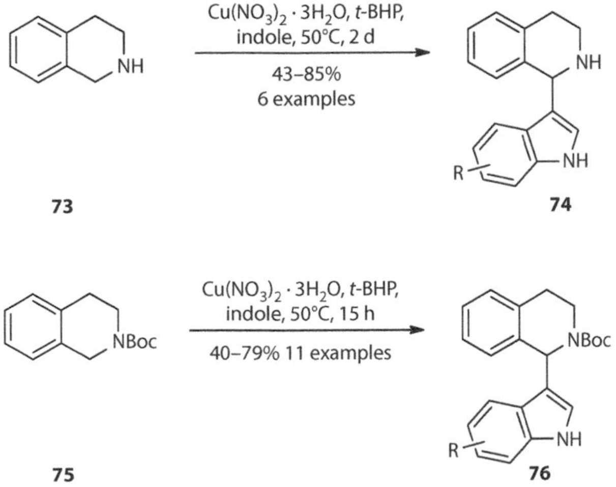

**SCHEME 7.14**   Indolylation of unprotected THIQ versus protection–deprotection pathway.

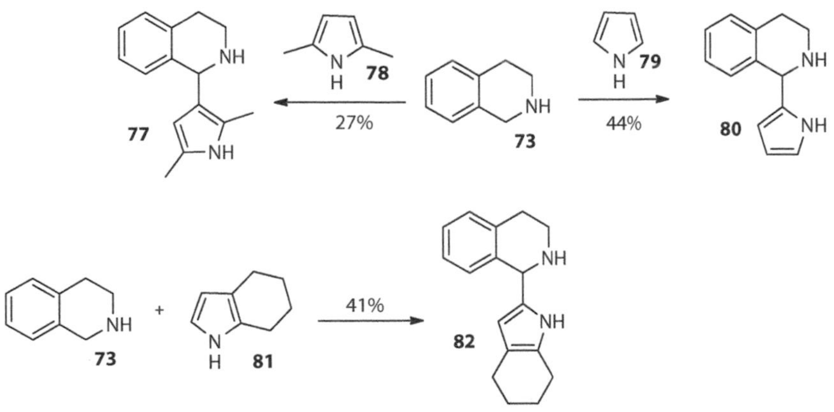

Conditions: Cu(NO$_3$)$_2$ · 3H$_2$O (5 mol%), t-BHP (1.3 equiv), 50°C, 15 h

**SCHEME 7.15**   Copper-catalyzed trifluoromethylation of C—H bond.

**TABLE 7.14**

**Scope of Methoxyphenylation on N-PG THIQ and Isochroman**

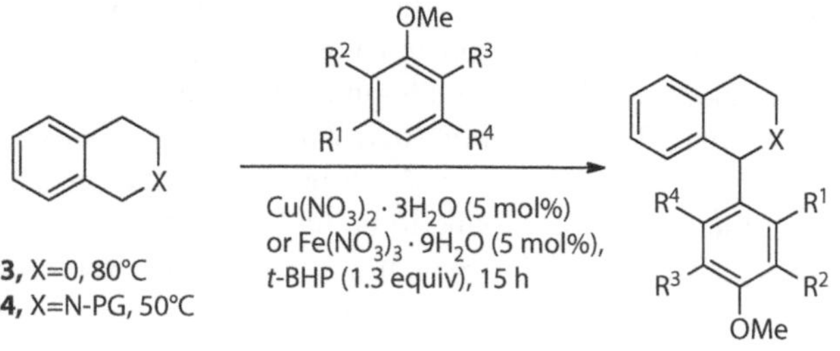

**83,** X=0, 80°C
**84,** X=N-PG, 50°C

Cu(NO₃)₂ · 3H₂O (5 mol%)
or Fe(NO₃)₃ · 9H₂O (5 mol%),
t-BHP (1.3 equiv), 15 h

| Entry | R¹ | R² | R³ | R⁴ | X | Yield % Cu | Yield % Fe |
|-------|-----|-----|------|------|------|------|------|
| 1 | OMe | H | H | OMe | NBoc | 76 | 81 |
| 2 | OMe | H | H | OMe | NBz | 46[a] | 54[a] |
| 3 | OMe | H | H | OMe | NCBz | 51[a] | 58[a] |
| 4 | OMe | H | OMe | H | NBoc | 23 | 47 |
| 5 | OMe | H | H | OMe | O | 51 | 55 |
| 6 | H | H | H | OMe | O | 23 | 12[a] |
| 7 | OMe | H | OMe | H | O | 32 | 15 |

[a] 36 h reaction time

**TABLE 7.15**

**Synthesis of Optically Active Cl-Substituted THIQ Derivatives**

| Entry | R¹ | R² | Yield (%) | ee (%) |
|-------|------|-------|-----------|--------|
| 1 | H | Ph | 67 | 63 |
| 2 | 4-MeO | Ph | 59 | 60 |
| 3 | 2-MeO | Ph | 54 | 73 |
| 4 | H | 4-BrPh | 72 | 64 |
| 5 | H | Hex | 65 | 26 |
| 6 | H | TMS | 11 | 30 |
| 7 | 2-MeO | 4-BrPh | 61 | 74 |
| 8 | 2-MeO | Pyr | 57 | 36 |

In a similar fashion, the cross-dehydrogenative coupling (CDC) of THIQs—to yield acetylated derivatives has been shown to occur stereoselectively by Zhang and Zhang.[37] A wide range of chiral ligands were screened. Some examples using PyBox are shown in Table 7.15.

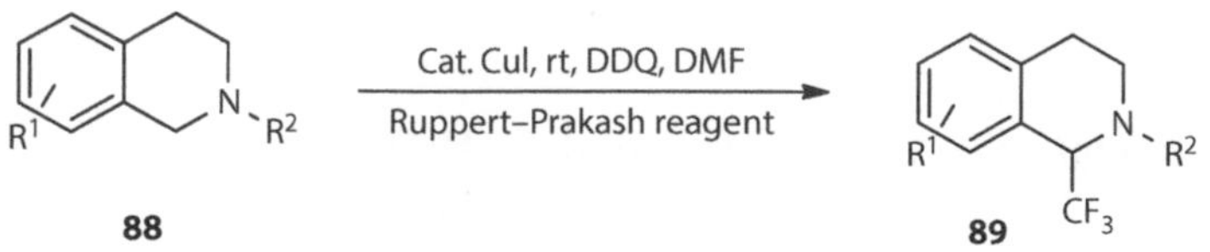

**SCHEME 7.16**    Copper-catalyzed pyrrolation of THIQ.

These examples illustrate several trends. Aliphatic substituted alkynes gave fair or low enantiomeric excesses, while aromatic substituted alkynes provided both good yields and ee (entries 4 and 7). Substitutions on the *N*-phenyl also had an effect. For example, a *para*-substituted methoxy group did not influence the enantioselectivity of the reaction whilst an *ortho*-methoxy group improved it.

Li and Mitsudera also describe the $sp^3$ C—H activation of THIQs but, in this case, introduce a trifluoromethyl group using DDQ and the Ruppert–Prakash reagent under mild conditions (Scheme 7.16).[38] The Ruppert–Prakash reagent has been successfully used to provide alpha-trifluoromethylated amine derivatives;[39–47] this protocol provides a more direct and simpler synthetic pathway.

R groups bearing electron-withdrawing groups such as $CF_3$ resulted in lower yields in the reaction; this was postulated to be due to the increased oxidative potential of the corresponding amine derivative (Table 7.16). Consequently, substrates containing stronger electron-withdrawing groups such as 2-acyl and an alkoxycarbonyl group did not undergo the reaction at all.

---

**TABLE 7.16**

**Copper-Catalyzed Trifluoromethylation of Amines**

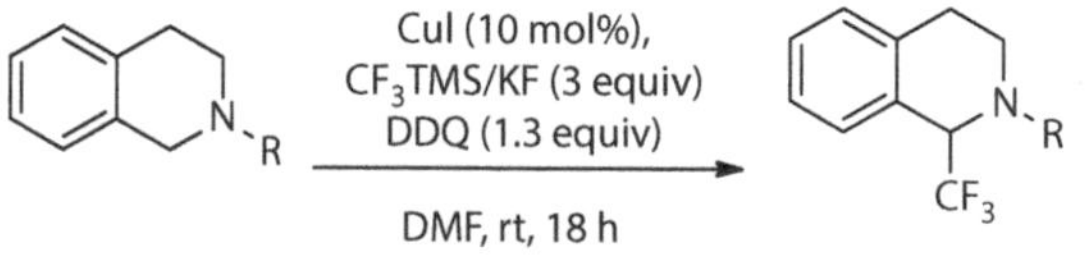

| Entry | R | Yield % |
|-------|-----|---------|
| 1 | Ph | 81 |
| 2 | 2-MeOC$_6$H$_4$ | 65 |
| 3 | 3-MeOC$_6$H$_4$ | 57 |
| 4 | 4-MeOC$_6$H$_4$ | 73 |
| 5 | 4-BrC$_6$H$_4$ | 69 |
| 6 | 4-CF$_3$C$_6$H$_4$ | 41 |
| 7 | 1-Naphthyl | 46 |
| 8 | 2-Pyridyl | 11 |
| 9 | 2-Pyridyl | 48 |

## 7.9   COPPER-CATALYZED DIRECT C—H AMINATION

Primary amines have been shown to act as directing groups for the C—H activation of anilines to give *ortho* azidation products.[48] This novel reaction occurs regioselectively under mild conditions. Subsequent conversion of the amino group to Cl, I, Br, CN, OH, or H, illustrates the versatility of the azido products. Non-*ortho*-substituted electron-deficient and electron-rich anilines resulted in both mono- and di-substituted products, which could be separated by flash chromatography. Several functional groups when substituted at the *ortho* position of the aniline were well tolerated under these reaction conditions; these include methyl, chloro, acetylene, and thiophenes. Examples are tabulated in Table 7.17.

**TABLE 7.17**

**Copper-Catalyzed Azidation of Anilines**

| Entry | Product | Yield % | Entry | Product | Yield % |
|-------|---------|---------|-------|---------|---------|
| 1 | | 68 | 6 | | 57 |
| 2 | | 68 | 7 | | 63 |
| 3 | | 52 | 8 | | 67 |
| 4 | | 68 | 9 | | 57 |
| 5 | | 54 | 10 | | 49 |

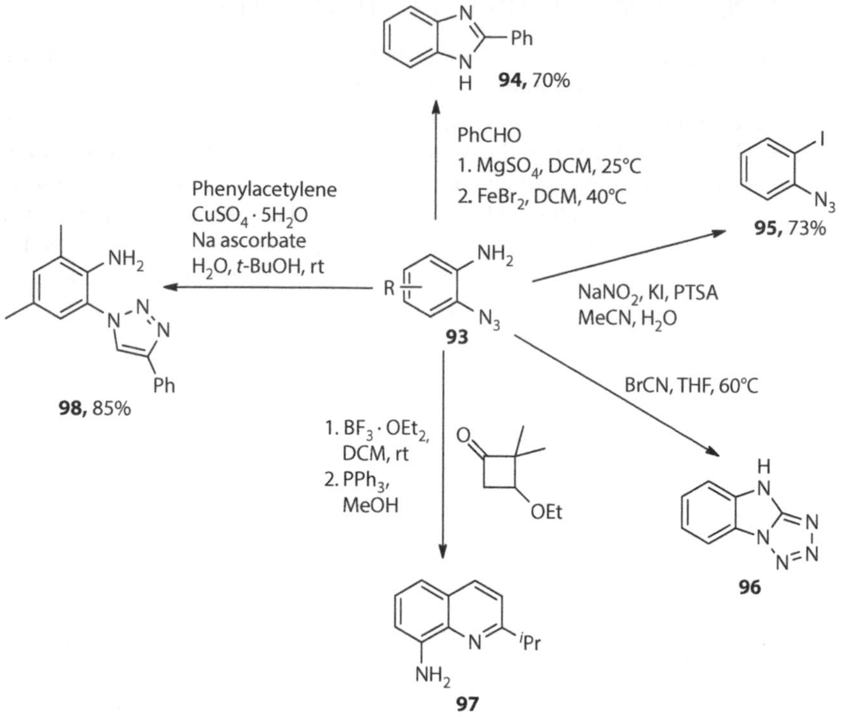

**SCHEME 7.17** Transformations of 2-azidoanilines.

The authors illustrate the versatility of these products by conducting further transformations leading to useful building blocks (Scheme 7.17).

Zhu et al. report a domino copper-catalyzed synthesis of benzimidazoles from boronic acids and amidines.[49] This involved a Chan–Lam–Evans N-arylation, C—H activation, and C—N bond formation (Scheme 7.18). These steps could be carried out in a one-pot reaction to give both primary and secondary benzimidazoles in good yields under mild reaction conditions (Schemes 7.19 and 7.20).

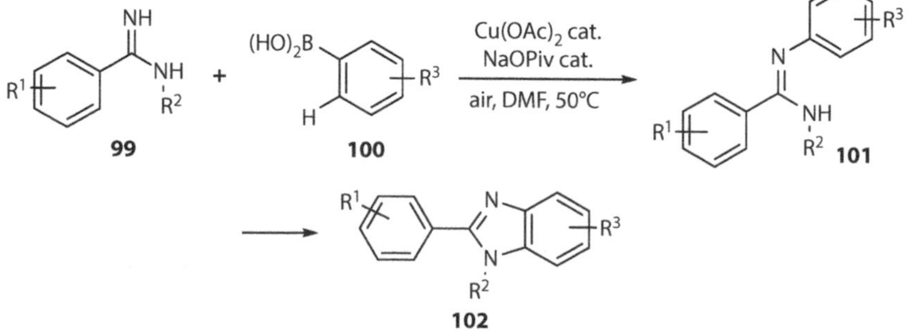

**SCHEME 7.18** Synthesis of benzimidazoles.

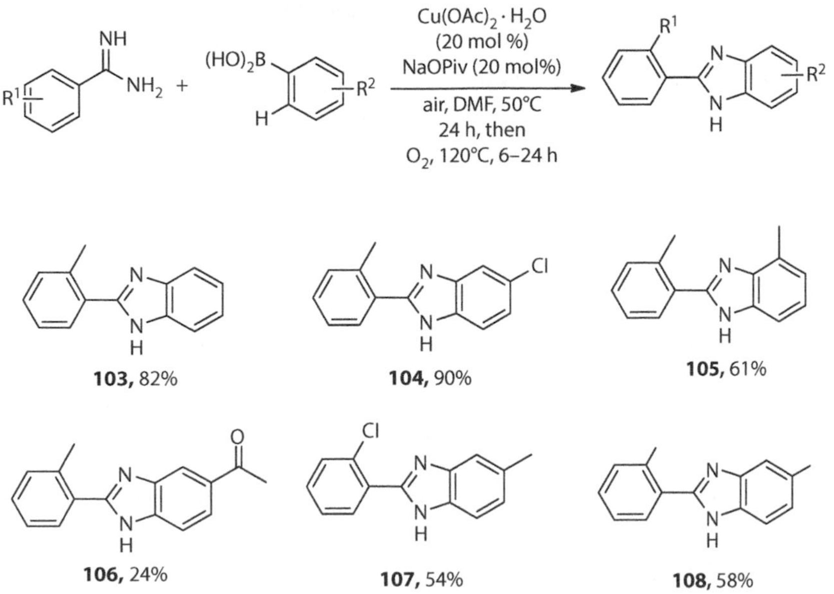

**SCHEME 7.19**   Scope of the copper-catalyzed synthesis of benzimidazoles from primary benzamidines.

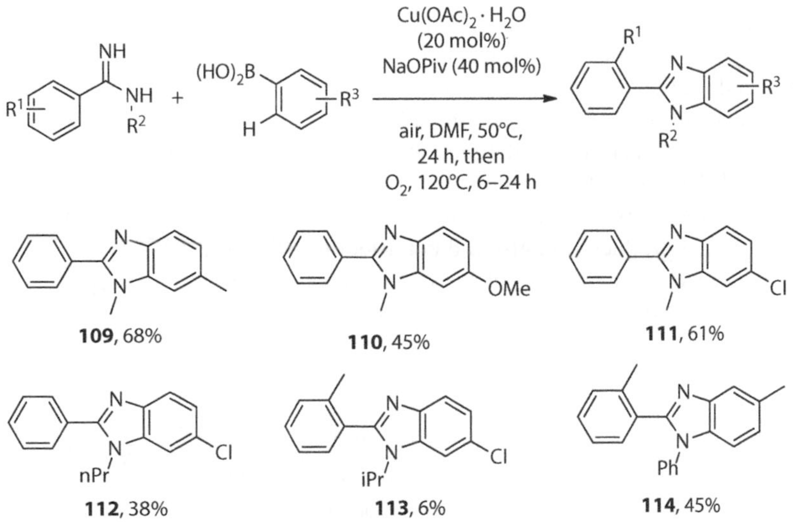

**SCHEME 7.20**   Scope of the copper-catalyzed synthesis of benimidazoles from secondary benzamidines.

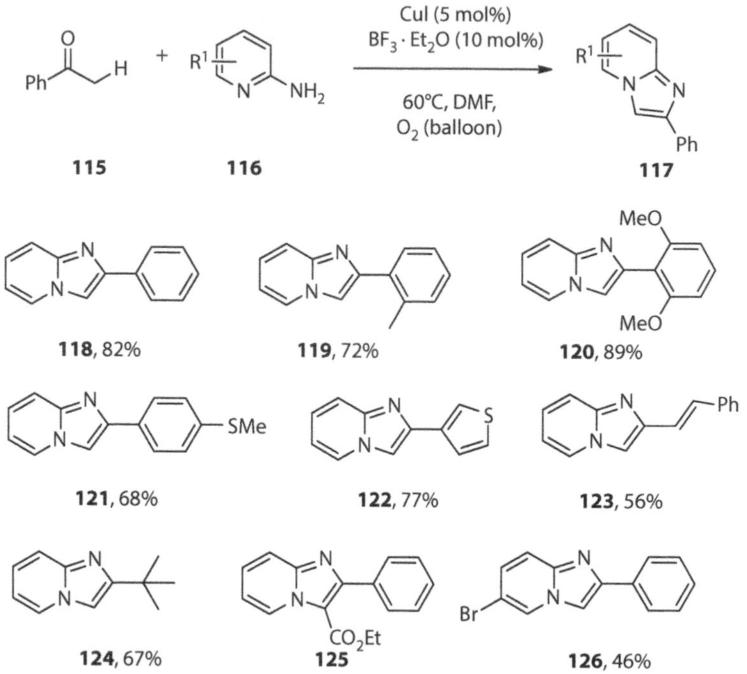
CuI (5 mol%)
BF3 · Et2O (10 mol%)
60°C, DMF,
O2 (balloon)
115
116
117
118, 82%
119, 72%
120, 89%
121, 68%
122, 77%
123, 56%
124, 67%
125
126, 46%

**SCHEME 7.21** Reaction of 2-aminopyridine with various ketones.

The copper-catalyzed synthesis of imidazo[1,2-*a*]pyridines via C—H activation using oxygen as the sole oxidant has been reported by Adimurthy et al.[50] The reaction proceeds in the presence of the radical-scavenger TEMPO, thus confirming that it does not occur via a radical reaction pathway.

Good yields, selectivity, and functional group compatibility are afforded with this method. Notably, electron-donating and electron-withdrawing substituents on the ketone did not prove detrimental to the process; steric factors, as exemplified by the formation of product **120**, also did not affect the (Scheme 7.21).

The authors demonstrate the utility of this reaction with the first report of a single-step synthesis of zolimidine, an antiulcer drug (Scheme 7.22).

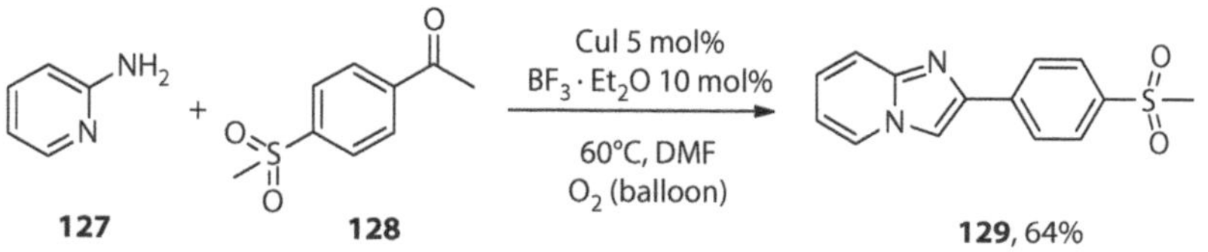
CuI 5 mol%
BF3 · Et2O 10 mol%
60°C, DMF
O2 (balloon)
127
128
129, 64%

**SCHEME 7.22** Synthesis of zomlidene.

## TABLE 7.18
### CuBr/NBS-Mediated Amidations of Benzylic $sp^3$ C—H Bonds

| Entry | 131 | Temp/Time | Product | Yield % |
|-------|-----|-----------|---------|---------|
| 1 | H₂NOC—⟨ ⟩—NO₂ | 40°C/12 h | | 78 |
| 2 | H₂N–S(O)(O)—⟨ ⟩— | 40°C/12 h | | 73 |
| 3 | H₂NOC—⟨ ⟩ | 50°C/12 h | | 68 |
| 4 | H₂NOC—⟨ ⟩ | 60°C/6 h | | 39 |

The copper-catalyzed amidation of saturated C—H bonds has been described by Zhao and coworkers, using CuBr and *N*-bromosuccinimide, or *N*-chlorosuccinimide, as the oxidant.[51] In particular, benzylic $sp^3$ C—H bonds and C—H bonds adjacent to a nitrogen atom underwent amidation. The reactions are insensitive to atmospheric moisture and oxygen, allowing for easy manipulation (Table 7.18). Benzamides with electron-withdrawing groups on the aromatic ring gave higher yields in this reaction than those containing electron-donating groups (entry 1 vs. entry 2). The reaction occurred selectively on the benzylic C—H adjacent to the oxygen atom when iso-chroman was the substrate.

**TABLE 7.19**

**Copper/NCS-Mediated Amidation of *N,N*-Dimethylaniline Derivatives**

| Entry | Anilines | Amide | Product | Yield % |
|-------|----------|-------|---------|---------|
| 1 | | | | 58 |
| 2 | | | | 43 |
| 3 | | | | 45 |
| 4 | | | | 64 |

Amidations next to nitrogen atoms were carried out in moderate yields as shown in Table 7.19. Aromatic amides showed better activity than aliphatic amides and primary amides resulted in higher yields than secondary amides. Succinimide was a good substrate. These reactions provide a viable alternative for the formation of amides of such substrates.

## 7.10   COPPER-CATALYZED DIRECT C—H OXYGENATION

Lei and coworkers report the oxidative functionalization of C—H bonds under the very mild conditions of using air as the oxidant at room temperature.[52] Excellent yields were obtained in reasonable reaction times (Table 7.20).

The reaction was determined not to be proceeding through a SET mechanism as the addition of either TEMPO or 1,1-diphenylethylene in radical-trapping experiments only slowed down the reaction slightly but did not stop it.

Gallardo-Donaire and Martin described a formal copper-catalyzed $C(sp^2)$—H hydroxylation assisted by benzoic acids.[53] Hydroxylated arenes are prepared under mild conditions with excellent chemoselectivity. The scope of the reaction thus far is limited to biaryl systems although both electron-donating and electron-withdrawing substituents on either aryl ring gave good-to-moderate yields; examples shown (Scheme 7.23).

**TABLE 7.20**

**Copper-Catalyzed Oxidation of Arenes and Heteroarenes**

$$Ar-H \quad + \quad O_2 \text{ (air)} \quad \xrightarrow[\text{$t$-BuONa, DMF, 25°C}]{\text{CuCl}_2 \text{ (5 mol\%),}} \quad Ar-OH$$

| Entry | Ar—H | Ar—OH | t (h) | Yield % |
|:---:|:---:|:---:|:---:|:---:|
| 1 | | | 1 | 70 |
| | | | | |
| 2 | R = H | | 4 | 84 |
| 3 | R = Me | | 3 | 84 |
| 4 | R = OMe | | 3 | 79 |
| 5 | R = Cl | | 6 | 37 |
| 6 | R = Br | | 6 | 76 |
| | | | | |
| 7 | R = H | | 5 | 73 |
| 8 | R = Me | | 5 | 67 |
| 9 | R = tBu | | 5 | 73 |

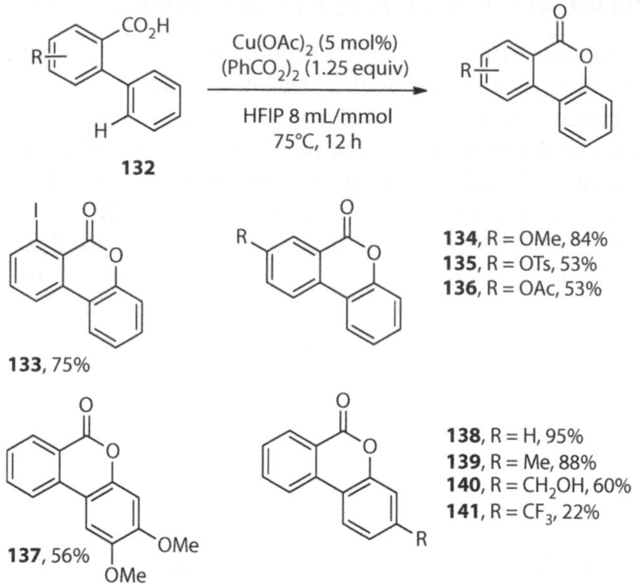

**SCHEME 7.23** Substituent effects.

This is an operationally simple protocol for $C(sp^2)$—H hydroxylations that required no pre-functionalization and with wide substrate scope.

Biaryl ethers can also be formed through copper-catalyzed C—H activation, this time using $Cu_2(OH)_2CO_3$, with air as the oxidant, and phenol or alcohol as the coupling partner.[54] Good functional group tolerance is observed, with ester, amine, nitrile, nitro, and halogen functionalities, all compatible with the reaction conditions. A pair of examples is shown for reaction scope with phenols, aliphatic alcohols, and amides, respectively (Tables 7.21 through 7.23).

## TABLE 7.21
## Reaction Scope with Respect to Phenols

| Entry | Phenol | Product | Yield % |
|-------|--------|---------|---------|
| 1 | 3-EtO$_2$C-Phenol | | 84 |
| 2 | 3-Aminophenol | | 57 |

## TABLE 7.22
## Reaction Scope with Respect to Aliphatic Alcohols

TMG - tetramethylguanidine

| Entry | Alcohol | Product | Yield % |
|-------|---------|---------|---------|
| 1 | Cyclopropylmethanol | | 75 |
| 2 | Cinchonine | | 85 |

**TABLE 7.23**

**Reaction Scope with Respect to Amides**

| Entry | Substrate | Product | Yield % |
|-------|-----------|---------|---------|
| 1 | 4-NO$_2$C$_6$H$_4$ | | 57 |
| 2 | 3,4-(OMe)$_2$C$_6$H$_4$ | | 54 |
| 3 | 4-C$_5$H$_4$N | | 47 |

Buchwald and Cheung have developed an efficient room-temperature catalytic oxidative cyclization of enamides to generate 2,5-disubstituted oxazoles via a vinylic C—H bond functionalization.[55] Utilization of catalytic CuBr$_2$ with K$_2$S$_2$O$_8$ as the oxidant yielded oxazoles bearing aryl, vinyl, and alkyl groups. Heteroaryl substituents were also tolerated with either electron-withdrawing or electron-donating motifs (Schemes 7.24 and 7.25).

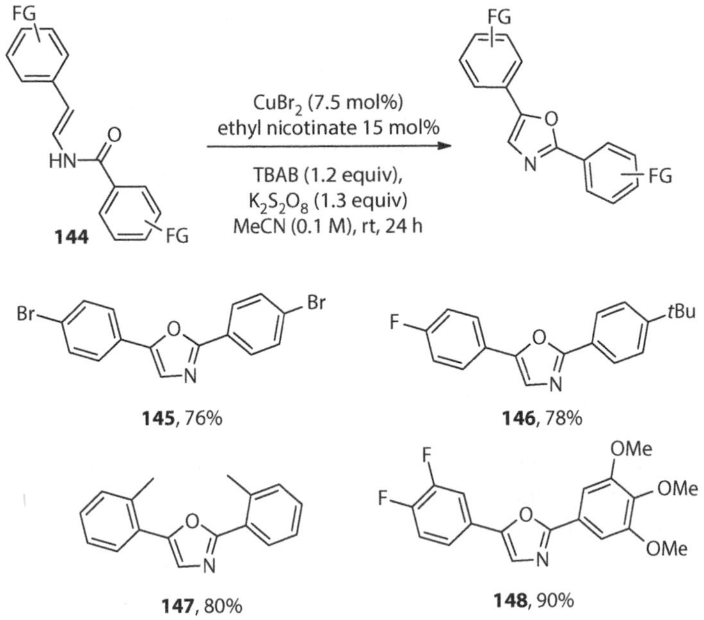

**SCHEME 7.24** Scope of enamides bearing aryl groups.

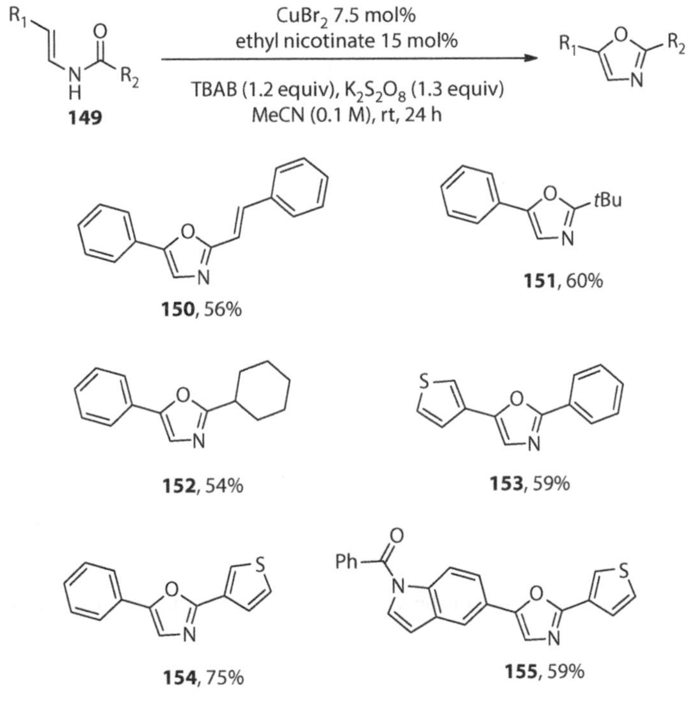

**SCHEME 7.25** Scope of enamides bearing vinyl, alkyl, heteroaryl groups.

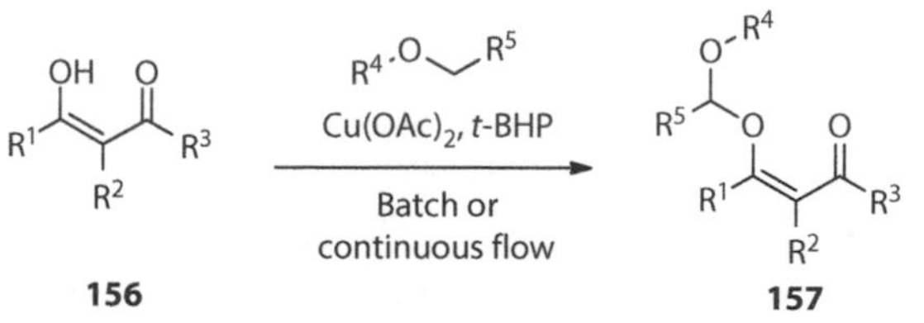

**SCHEME 7.26** Copper-catalyzed oxidative cross-dehydrogenative-coupling of β-ketoesters or 2-carbonyl-substituted phenols with ethers.

Kappe et al. have reported a novel copper-catalyzed oxidative C—O bond formation to yield unsymmetric acetal scaffolds.[56]

The synthesis requires TBHP as the oxidant at high temperatures; thus, combination of the oxidant with ether would result in obvious safety concerns. The group overcame this issue by successfully translating to a continuous-flow/microreactor protocol, thereby representing the first application of continuous-flow processing to C—H activation chemistry (Scheme 7.26).

Scheme 7.27 illustrates the substrate scope after optimisation of the reaction conditions. This is a particularly facile reaction to implement as it utilizes a cheap copper catalyst and a commercially available oxidant, TBHP. The authors compared continuous flow with batch protocols and found good correlations, thus allowing such reactions to be carried out without the specific need for a flow-reactor (Scheme 7.28).

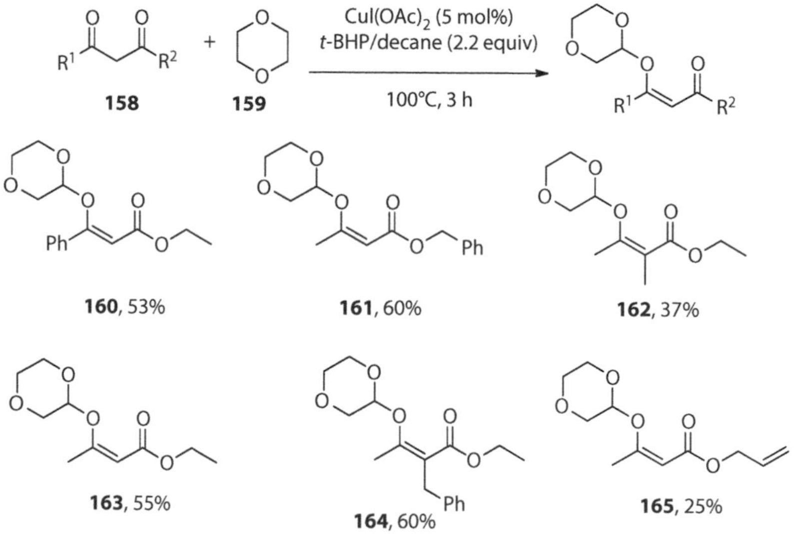

**SCHEME 7.27** Substrate scope in the Cu(OAc)₂-catalyzed coupling of β-ketoesters with dioxane.

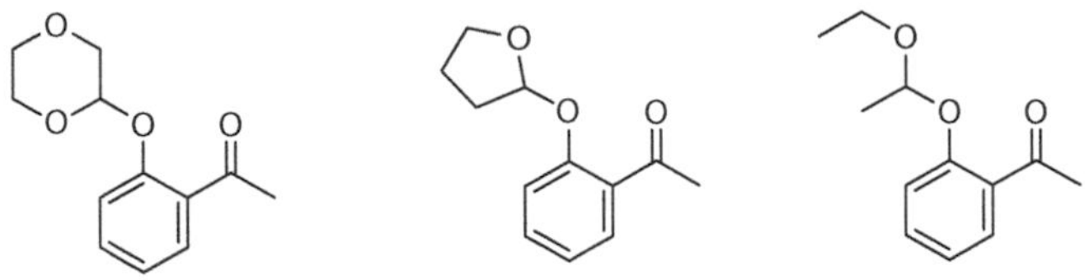

**166**
Flow 82%
MW 82%

**167**
Flow 84%
MW 96%

**168**
Flow 36%
MW 54%

**SCHEME 7.28**  Examples of comparisons between flow and microwave reaction conditions.

## 7.11  COPPER CATALYZED DIRECT C—H SULFINATION

Ranjit et al. report a route to 2-thio-substituted-1,3-benzothiazoles by direct C—H bond functionalization with alkyl or aryl thiols in the presence of copper[57] (Scheme 7.29).

Several examples are shown here (Table 7.24). The reaction uses stoichiometric copper at a 1:1 ratio with the ligand Bipy. A range of bases and solvents were screened with the highest yields resulting from the use of sodium carbonate and DMF.

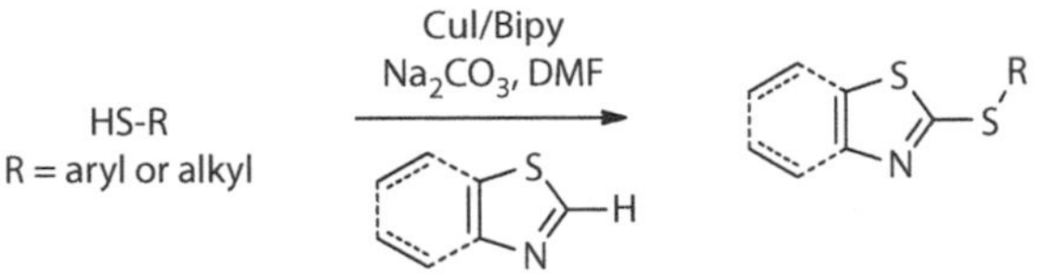

**SCHEME 7.29**  Formation of benzthiazoles.

The reaction was shown to work on substrates other than 1,3-benzothiazoles as shown in Scheme 7.30.

---

**TABLE 7.24**

**CuI/Bipy-Mediated Direct Sulfurization of Benzthiazoles with Thiols**

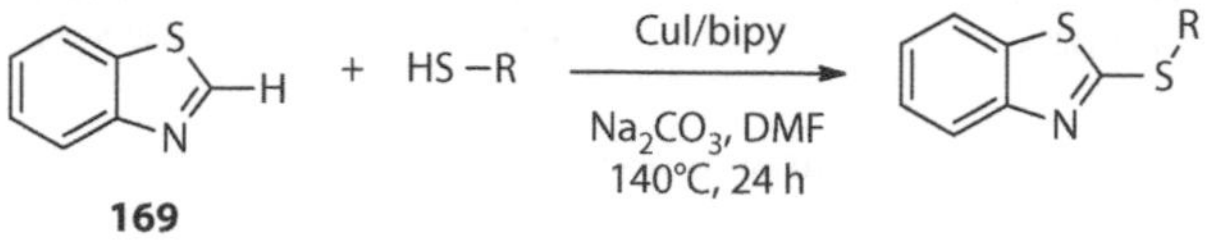

**169**

| Entry | Thiols | Product | Yield % |
|-------|--------|---------|---------|
| 1 | 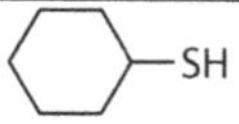 | 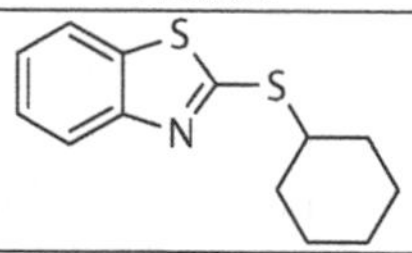 | 93 |
| 2 | 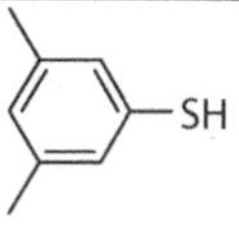 | 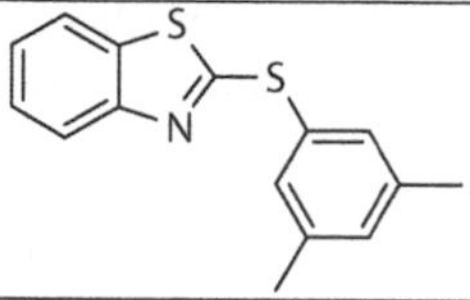 | 95 |
| 3 | 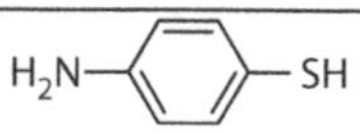 | 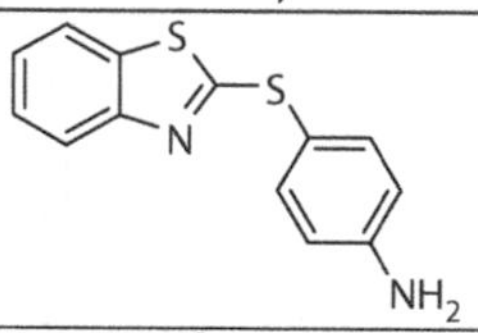 | 74 |
| 4 | 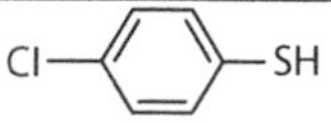 | 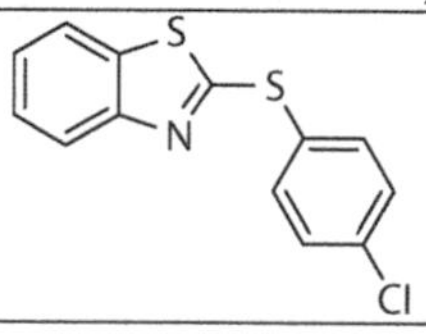 | 66 |

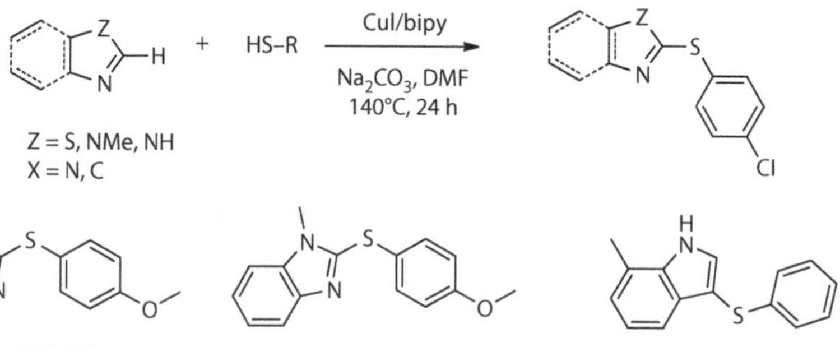

Z = S, NMe, NH
X = N, C

**170**, 79%          **171**, 96%          **172**, 85%

**SCHEME 7.30**   Scope of heteroarene coupling partners.

## 7.12 COPPER CATALYZED DIRECT C—H HALOGENATION

Shen et al. report the use of lithium halides under copper-catalyzed aerobic conditions to effect aromatic C—H bond activation resulting in bromo- and chloroarenes.[58] The catalyst employed $Cu(NO_3)_2 \cdot 3H_2O$ is cheap and readily available and, importantly, the reaction is compatible with both electron-donating and electron-withdrawing substituents on the aryl rings; examples are shown with the optimized reaction conditions, using LiCl as the halogen source. LiBr and NaCl also resulted in halogenation albeit in modest yields (Scheme 7.31).

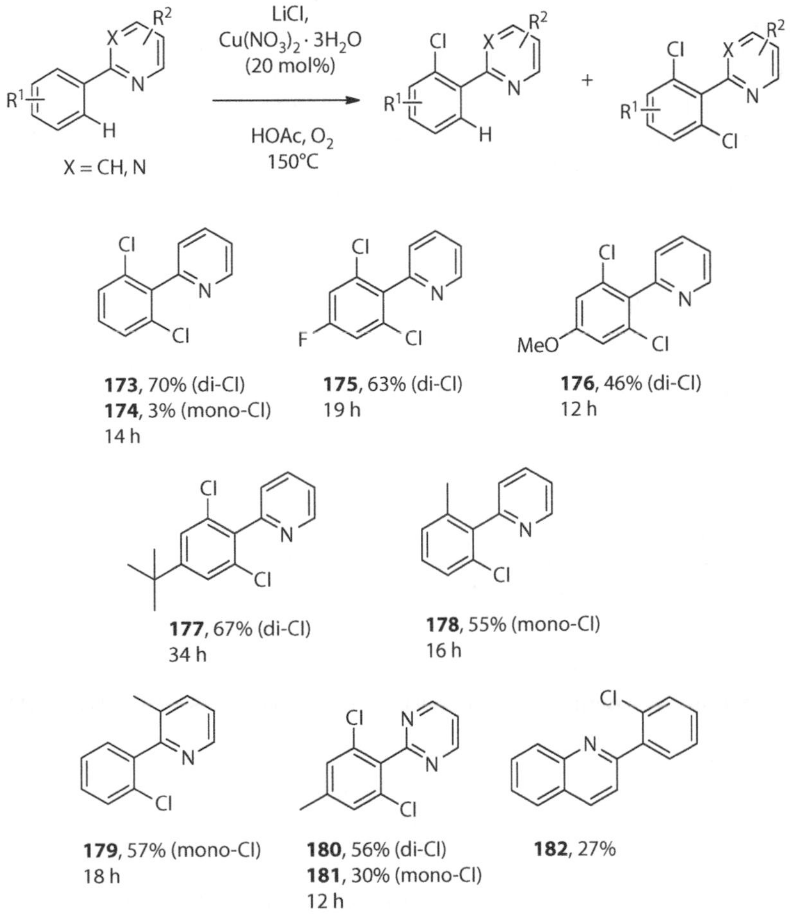
LiCl,
Cu(NO3)2 · 3H2O
(20 mol%)
HOAc, O2
150°C
X = CH, N
R1
R2
Cl
H
N
X
+
173, 70% (di-Cl)
174, 3% (mono-Cl)
14 h
F
MeO
175, 63% (di-Cl)
19 h
176, 46% (di-Cl)
12 h
177, 67% (di-Cl)
34 h
178, 55% (mono-Cl)
16 h
179, 57% (mono-Cl)
18 h
180, 56% (di-Cl)
181, 30% (mono-Cl)
12 h
182, 27%

**SCHEME 7.31** Copper-catalyzed chlorination of aromatic C—H bonds with oxygen as the terminal oxidant.

The reaction results in the dichlorinated species as the major product unless sterics prevail. This protocol avoids the use of potentially dangerous halogenating agents in electrophilic aromatic substitution reactions, or *ortho*-lithiation followed by halogen quench procedures. While examples of copper-catalyzed *ortho*-selective halogenation reactions have been reported by other groups, the above method overcomes some of their limitations, which include being restricted by the scope of electron-rich substrates, or the use of toxic halogen sources such as tetrachloroethane.[59-65]

## REFERENCES

1. Mousseau, J. J.; Charette, A. B. *Acc. Chem. Res.* **2013**, *46*, 412–424.
2. Yamaguchi, J.; Yamaguchi, A. D.; Itami, K. *Angew. Chem. Int. Ed.* **2012**, *51*, 8960–9009.
3. McMurray, L.; O'Hara, F.; Gaunt, M. J. *Chem. Soc. Rev.* **2011**, *40*, 1885–1898.
4. Alberico, D.; Scott, M. E.; Lautens, M. *Chem. Rev.* **2007**, *107*, 174.
5. Collet, F.; Lescot, C.; Dauban, P. *Chem. Soc. Rev.* **2011**, *40*, 1926–1936.
6. Steinkopf, W.; Leitsmann, R.; Hofmann, K. H. *Liebigs Ann. Chem.* **1941**, *546*, 180–199.
7. Bjorklund, C.; Nilsson, M. *Acta Chem. Scand.* **1968**, *22*, 2338–2346.
8. Ljusberg, H.; Wahren, R. *Acta Chem. Scand.* **1973**, *27*, 2717–2721.
9. Nilsson, M. *Tetrahedron Lett.* **1966**, *7*, 679–682.
10. Daugulis, O.; Do, H.-Q.; Shabashov, D. *Acc. Chem. Res.* **2009**, *42*, 1074–1086.
11. Li, Y.; Ji, J.; Qian, Q.; Bao, W. *Org. Biomol. Chem.* **2010**, *8*, 326–330.
12. Nishino, M.; Hirano, K.; Satoh, T.; Miura, M. *Angew. Chem. Int. Ed.* **2012**, *51*, 6993–6997.
13. Hattori, K.; Yamaguchi, K.; Yamaguchi, J.; Itami, K. *Tetrahedron*, **2012**, *68*, 7605–7612.
14. Barbero, N.; SanMartin, R.; Dominguez, E. *Org. Biomol. Chem.* **2010**, *8*, 841–845.
15. Pan, X.; Luo, Y.; Wu, J. *J. Org. Chem.* **2013**, *78*, 5756–5760.
16. Reddy, G. C.; Balasubramanyam, P.; Salvanna, N.; Das, B. *Eur. J. Org. Chem.* **2012**, *2012*, 471–474.
17. Chu, L.; Qing, F.-L. *J. Am. Chem. Soc.* **2012**, *134*, 1298–1304.
18. Ren, P.; Salihu, I.; Scopelliti, R.; Hu, X. *Org. Lett.* **2012**, *14*, 1748–1751.
19. Minisci, F.; Vismara, E.; Fontana, F. *Heterocycles* **1989**, *28*, 489–519.
20. Molander, G. A.; Colombel, V.; Braz, V. A. *Org. Lett.* **2011**, *13*, 1852–1855.
21. Bowman, W. R.; Storey, J. M. D. *Chem. Soc. Rev.* **2007**, *36*, 1803–1822.
22. Verrier, C.; Hoarau, C.; Marsais, F. *Org. Biomol. Chem.* **2009**, *7*, 647–650.
23. He, T.; Yu, L.; Zhang, L.; Wang, L.; Wang, M. *Org. Lett.* **2011**, *13*, 5016–5019.
24. Ackermann, L.; Barfusser, S.; Kornhaass, C.; Kapdi, A. R. *Org. Lett.* **2011**, *13*, 3082–3085.
25. Vechorkin, O.; Proust, V.; Hu, X. L. *Angew. Chem., Int. Ed.* **2010**, *49*, 3061–3064.
26. Yao, T.; Hirano, K.; Satoh, T.; Miura, M. *Chem. Eur. J.* **2010**, *16*, 12307–12311.
27. Ackermann, L.; Punji, B.; Song, W. F. *Adv. Synth. Catal.* **2011**, *353*, 3325–3329.
28. Xu, J.; Fu, Y.; Luo, D.-F.; Jiang, Y.-Y.; Xiao, B.; Liu, Z.-J.; Gong, T.-J.; Liu, L. *J. Am. Chem. Soc.* **2011**, *133*, 15300–15303.
29. Matsubara, S.; Mitani, M.; Utimoto, K. *Tetrahedron Lett.* **1987**, *28*, 5857–5860.
30. Roy, B.; Hazra, S.; Mondal, B.; Majumdar, K. C. *Eur. J. Org. Chem.* **2013**, *2013*, 4570–4577.
31. Wang, L.; Fu, H.; Jiang, Y.; Zhao, Y. *Chem. Eur. J.* **2008**, *14*, 10722–10726.
32. Yuan, Y.-q.; Guo, S.-r.; Xiang, J.-n. *Synlett* **2013**, *24*, 443–448.
33. Shi, L.; Tu, Y.-Q.; Wang, M.; Zhang, F.-M.; Fan, A.-A. *Org. Lett.* **2004**, *6*, 1001–1003.
34. Park, S. B.; Alper, H. *Chem. Commun.* **2005**, *2005*, 1315–1317.

35. Jeyachandran, R.; Potukuchi, H. K.; Ackermann, L. *Beilstein J. Org. Chem.* **2012**, *8*, 1771–1777.
36. Ghobrial, M.; Schnurch, M.; Mihovilovic, M. D. *J. Org. Chem.* **2011**, *76*, 8781–8793.
37. Zhang, M.; Zhang, A. *J. Heterocyclic Chem.* **2012**, *49*, 721–725.
38. Li, C.-J.; Mitsudera, H. *Tetrahedron Lett.* **2011**, *52*, 1898–1900.
39. Felix, C. P.; Khatimi, N. A.; Laurent, J. *Tetrahedron Lett.* **1994**, *35*, 3303–3306.
40. Nelson, D. W.; Easley, R. A.; Pintea, B. N. V. *Tetrahedron Lett.* **1999**, *40*, 25–28.
41. Nelson, D. W.; Owen, J.; Hiraldo, D. *J. Org. Chem.* **2001**, *66*, 2572.
42. Petrov, V. A. *Tetrahedron Lett.* **2000**, *41*, 6959–6962.
43. Prakash, G. K. S.; Mandeal, M.; Olah, G. A. *Angew. Chem. Int. Ed.* **2001**, *40*, 589–590.
44. Prakash, G. K. S.; Mandeal, M.; Olah, G. A. *Org. Lett.* **2001**, *3*, 2847–2850.
45. Prakash, G. K. S.; Mandeal, M. *J. Am. Chem. Soc.* **2002**, *124*, 6538–6539.
46. Kawano, Y.; Fujisawa, H.; Mukaiyama, T. *Chem. Lett.* **2005**, *34*, 422–423.
47. Mizuta, S.; Shibata, N.; Sato, T.; Fujimoto, H.; Nakamura, S.; Toru, T. *Synlett* **2006**, 267–270.
48. Tang, C.; Jiao, N. *J. Am. Chem. Soc.* **2012**, *134*, 18924–18927.
49. Li, J.; Benard, S.; Neuville, L.; Zhu, J. *Org. Lett.* **2012**, *14*, 5980–5983.
50. Mohan, D. C.; Donthiri, R.; Rao, S. N.; Adimurthy, S. *Adv. Synth. Catal.* **2013**, *355*, 2217–2221.
51. Liu, X.; Zhang, Y.; Wang, L.; Fu, H.; Jiang, Y.; Zhao, Y. *J. Org. Chem.* **2008**, *73*, 6207–6212.
52. Liu, Q.; Wu, P.; Yang, Y.; Zeng, Z.; Liu, J.; Yi, H.; Lei, A. *Angew. Chem. Int. Ed.* **2012**, *51*, 4666–4670.
53. Gallardo-Donaire, J.; Martin, R. *J. Am. Chem. Soc.* **2013**, *135*, 9350–9353.
54. Roane, J.; Daugulis, O. *Org. Lett.* **2012**, *15*, 5842–5845.
55. Cheung, C. W.; Buchwald, S. L. *J. Org. Chem.* **2012**, *77*, 7526–7537.
56. Kumar, G. S.; Piebar, B.; Reddy, K. R.; Kappe, C. O. *Chem. Eur. J.* **2012**, *18*, 6124–6128.
57. Ranjit, S.; Lee, R.; Heryadi, D.; Shen, C.; Wu, J.; Zhang, P.; Huang, K.-W.; Liu, X. *J. Org. Chem.* **2011**, *76*, 8999–9007.
58. Mo, S.; Zhu, Y.; Shen, Z. *Org. Biomol. Chem.* **2013**, *11*, 2713–2918.
59. Chen, X.; Hao, X. S.; Goodhue, C. E.; Yu, J.-Q. *J. Am. Chem. Soc.* **2006**, *128*, 6790–6791.
60. Menini, L.; Gusevskaya, E. V. *Chem. Commun.* **2006**, 209–211.
61. Menini, L.; Gusevskaya, E. V. *Appl. Catal. A,* **2006**, *309*, 122–128.
62. Menini, L.; da Cruz Santos, J. C.; Gusevskayaa, E. V. *Adv. Synth. Catal.* **2008**, *350*, 2052–2058.
63. Yang, L. J.; Lu, Z.; Stahl, S. S. *Chem. Commun.* **2009**, 6460–6462.
64. Wang, W.; Pan, C.; Chen, F.; Cheng, J. *Chem. Commun.* **2011**, *47*, 3978–3980.
65. Menimi, L.; Parreira, L. A.; Gusevskaya, E. V. *Tetrahedron Lett.* **2007**, *48*, 6401–6404.

# 8 Cobalt-Catalyzed C—H Activation

*Nicole L. Snyder and Eric J. Medici*

## CONTENTS

## 8.1 INTRODUCTION

Historically, cobalt catalysts were the first catalysts to be used in chelation-assisted C—H bond functionalization. The first report by Murahashi[2] in 1955, involved the *ortho*-carbonylation of a Schiff base **1** using dicobalt octacarbonyl to give the isoindoline derivative **2**.

Between 1955 and 2007 only a few examples of cobalt-catalyzed C—H functionalization were reported.[3–6] These examples were rather limited in scope, but provided a foundation for the substantial amount of work that has been done in this area since 2007.

Until recently, C—H functionalization reactions have relied heavily on expensive noble transition-metal catalysts. In the past 10 years, there has been a rise in the interest of low-valent cobalt catalysts for C—H functionalization.[1] These catalysts are earth-abundant, green, and can generally be used under milder reaction conditions than their noble transition-metal catalyst counterparts.

Cobalt-catalyzed C—H functionalization reactions are currently classified into two main categories based on their hypothetical catalytic cycles: (i) hydroarylation of alkynes and olefins and (ii) C—H/electrophile coupling. A separate category involving arylzincation of alkynes has also been proposed. This chapter explores cobalt-catalyzed hydroarylation and C—H electrophile coupling since 2007.

## 8.2 C—H FUNCTIONALIZATION VIA HYDROARYLATION OF ALKYNES AND OLEFINS

### 8.2.1 Hydroarylation of Alkynes to Form C—C Bonds

Yoshikai and coworkers explored the cobalt-catalyzed hydroarylation of alkynes by directed chelation-assisted C—H bond activation.[7,8] Screening studies revealed the reaction proceeded with the highest yields when derivatives of 2-phenylpyridine (3–7) and 4-octyne were treated with a ternary catalytic system consisting of a cobalt (II) bromide (CoBr$_2$) pre-catalyst, methyl diphenylphosphine (PMePh$_2$) as a ligand source, and methylmagnesium chloride (MeMgCl) as a reducing agent in THF at elevated temperatures. CoBr$_2$ was shown to be more effect then other pre-catalysts such as cobalt (II) chloride (CoCl$_2$), cobalt (II) acetylacetonate (Co(acac)$_2$), and cobalt (III) acetylacetonate and (Co(acac)$_3$). In addition, mondentate phosphine ligands such as triphenylphosphine (PPh$_3$), dimethylphenylphosphine (PMe$_2$Ph) and tricyclohexylphosphine (PCy$_3$) were not as effective, and bidentate phosphines such as 1,2-bis(diphenyl-phosphino)ethane (dppe), 1,3-bis(diphenylphosphino)propane (dppp), and 1,1′-bis(diphenylphosphino)ferrocene (dppf) completely retarded the reaction. The use of other Grignard reagents such as n-butylmagnesium bromide (n-BuMgBr), iso-propylmagnesium bromide (i-PrMgBr), tert-butylmagnesium bromide (t-BuMgBr), and tert-butylmethylmagnesium bromide (t-BuCH$_2$MgBr) were also shown to be relatively ineffective.

3, R = H
4, R = OCH$_3$
5, R = N(CH$_3$)$_2$
6, R = F
7, R = CF$_3$

8, R = H, 74–83%
9, R = OCH$_3$, 80%
10, R = N(CH$_3$)$_2$, 86%
11, R = F, 83%
12, R = CF$_3$, 74%

13, 62%          14, 35%          15, 75%

**16**, 68%                    **17**, 42%

Under the optimized conditions, the *ortho*-dialkenylated *syn* products **8–12** were obtained as the major products of the reaction when the R group was *para* to the pyridine ring. 2-Phenylpyridines with electron-donating and electron-withdrawing groups performed well equally in the reaction. In cases where the R group was *meta* (**13**) or *ortho* (**14**) to the pyridine ring, or the second *ortho* position was hindered as in the case of **15** where the pyridine ring is substituted at the 3-position with a methyl group, the *ortho* monoalkenylated *syn* products were obtained as the major products, albeit generally in lower yields, demonstrating the role of steric interactions on product distribution and yield. The authors explored a number of other interesting substrates including **16** and **17**, as well as different internal alkynes (not shown) to illustrate the scope and limitations of their system. The only limitation they found is the high temperatures required for conversion.

Lee, Fujita, and Yoshikai, used a similar cobalt-based ternary catalytic system to catalyze the room temperature addition of alkynes to aromatic imines.[9] Optimized conditions involving the reaction of aryl ketimines and internal alkynes with $CoBr_2$, tris(3-chlorophenyl)phosphane (P(3-$ClC_6H_5$)), and *t*-$BuCH_2MgBr$ in the presence of THF using pyridine as an additive, provided the corresponding *ortho* addition products **18–24** in high yields and with very good *E:Z* selectivity. Interestingly, the reaction was shown to prefer the more sterically hindered *ortho* position in cases where the starting material was substituted in the *meta* position. This is in contrast to the authors' previous work with 2-phenylpyridine derivatives. The reaction was also shown to tolerate a variety of functional groups including halogens, nitriles, and amides. A number of internal alkynes were shown to react smoothly under the reaction conditions reported, including unsymmetrical alkynes. In the latter case, unsymmetrical internal alkynes were shown to undergo C—C bond formation at the less sterically hindered acetylenic carbon. The authors rationalized this observation by noting that this would limit reduce steric interactions between the alkyne and the cobalt in the migratory insertion step of the catalytic cycle. The resulting products could be hydrolyzed to produce the corresponding benzofulvenes or ketone (not shown) depending on the nature of the substituents.

**18,** R = CH$_3$, 94% (89:11)     **20,** R = OCH$_3$, 87% (89:11)     **22,** R = H, 87%
**19,** R = CF$_3$, 85% (91:9)      **21,** R = F, 92% (88:12)          **23,** R = OCH$_3$, 87% (89:11)
                                                                       **24,** R = CF$_3$, 76% (87:13)

Recently, Yamakawa and Yoshikai attempted to extend this chemistry to aryl aldimines.[10] Limited success was achieved under the reaction conditions previously reported for aryl ketimines. However, careful screening and re-optimization of the reaction conditions eventually led to the desired addition products **28–33** in modest yields when tris(3-methylphenyl)phosphane (P(3-CH$_3$C$_6$H$_4$)$_3$) and *i*-PrMgBr were used instead of P(3-ClC$_6$H$_4$)$_3$ and *t*-BuCH$_2$MgBr.

**25,** OCH$_3$                                        **28,** OCH$_3$, 90%
**26,** PH                                             **29,** PH, 88%
**27,** CF$_3$                                         **30,** CF$_3$, 68%

**31,** 78%                    **32,** 66%                    **33,** 82% (63:37)

Yamakawa and Yoshikai also demonstrated that imine directing groups could be used to activate α,β-unsaturated imines.[11] Annulation of a series α,β-unsaturated imines and internal alkynes was accomplished using a simple and inexpensive cobalt-triarylphosphine catalyst. Reaction of a series of α,β-unsaturated imines bearing various electron-donating and electron-withdrawing groups (**34–40**) and diphenylacetylene with CoBr$_2$, P(3-ClC$_6$H$_5$) and *i*-PrMgCl in THF provided the corresponding dihydropyridine derivatives **41–47** in good yields under relatively mild conditions. Reduced yields were observed for α,β-unsaturated imines bearing bulkier substituents such as bromine and iodine, suggesting a steric component to the reaction. Careful screening of the reaction conditions revealed the reducing agent had the most significant influence on the reaction outcome. Common reducing agents for cobalt catalyzed C—H functionalization such as the primary Grignard reagents CH₃SiCH₂MgCl and CH₃MgCl, gave only minor yields of the desired products, while bulkier primary and secondary alkyl Grignard reagents such as CyMgBr and *i*-PrMgBr gave the best yields; however with minor amounts (<10%) of tertiary amine (not shown) formed from addition of the Grignard to the C=N. Suppression of the unwanted side product without loss of catalytic activity was achieved by reducing the loading of the reducing agent. From these observations, the authors hypothesized that the reaction proceeds first through alkenylation of the olelfinic C—H bond followed by a 6π electrocylization. They further demonstrated that a series of differentially functionalized dihydropyridines could be synthesized in good to excellent yields from different α,β-unsaturated imines and symmetrical and unsymmetrical acetylene derivatives (not shown).

| | |
|---|---|
| **34,** R =H | **41,** R = H, 91% |
| **35,** R = CH₃ | **42,** R = CH₃, 95% |
| **36,** R = OCH₃ | **43,** R = OCH₃, 95% |
| **37,** R =Cl | **44,** R =Cl, 86% |
| **38,** R = Br | **45,** R = Br, 70% |
| **39,** R =I | **46,** R = I, 23% |
| **40,** R = CN | **47,** R = CN, 84% |

Ding and Yoshikai also explored the addition of oxazoles,[12] thiazoles,[13] and indoles[14] to internal alkynes. In the case of the reaction between oxazoles **48–50** and **54–56** with 4-octyne, modest to high yields were achieved using a ternary catalytic system consisting of CoBr₂, bis[(2-diphenylphoshpino)phenyl] ether (DPEphos) and (trimethylsilyl)methylmagnesium chloride (Me₃SiCH₂MgCl) in THF at room temperature.[12] The corresponding C2 addition products were shown to occur in a *syn* fashion and preceded smoothly with differentially substituted benzoxazoles (**51–53**) and 5-aryl oxazoles (**57–59**), albeit with reduced yields for the latter. Electron-donating and electron-withdrawing substituents seemed to have little to no impact on reaction yields. However, the use of other bidentate phosphine ligands led to reduced yields, as did the use of other reducing agents. Modest to high yields were also observed for other internal alkynes including unsymmetric alkynes (not shown).

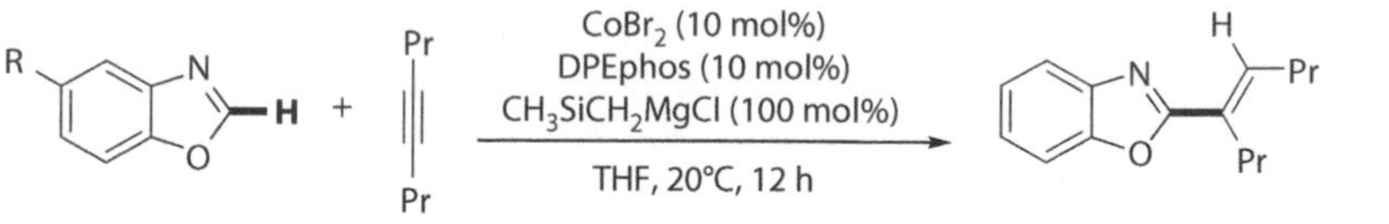

**48,** R = H          (1.2 equiv)
**49,** R = CH$_3$
**50,** R = Cl

**51,** R = H, 86%
**52,** R = CH$_3$, 77% (25:1)
**53,** R = Cl, 85% (12:1)

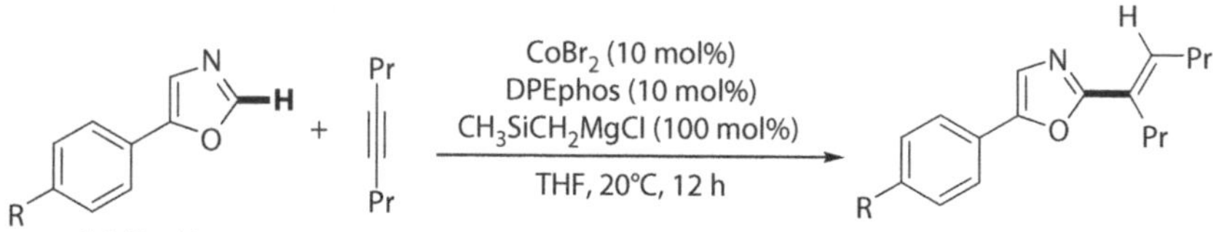

**54,** R = H          (1.2 equiv)
**55,** R = OCH$_3$
**56,** R = Cl

**57,** R = H, 66%
**58,** R = OCH$_3$, 55%
**59,** R = Cl, 34%

A similar substitution could be achieved for thiazoles.[13] However, a slight tuning of the reaction conditions was required to achieve similar yields. For example, reaction of benzothiazole **60** with 4-octyne using a ternary catalytic system consisting of a CoBr$_2$, 4,5-bis-(diphenylphosphino)-9,9-dimethylxanthene (Xantphos) and Me$_3$SiCH$_2$MgCl in THF/toluene gave high yields of the desired C2 substituted product **61**.

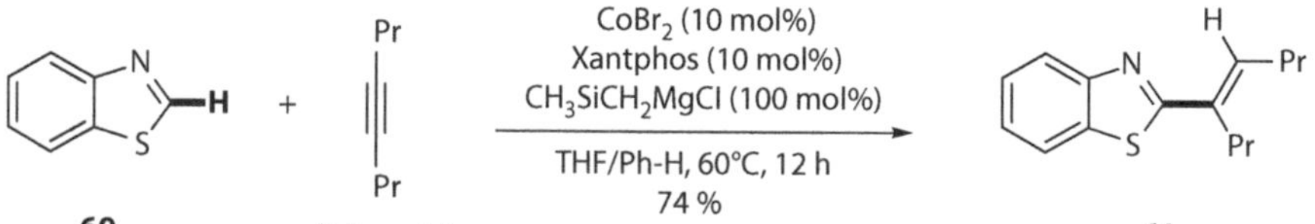

**60**          (1.2 equiv)

**61**

More recently, Ding and Yoshikai extended their methodology to N-pyrimidylindoles.[14] Once again they found that tuning of the reaction conditions was required to achieve high yields of the desired products. Reaction of 6-substituted indoles such as **62–66** with 4-octyne using a ternary catalytic system consisting of a CoBr$_2$, 2-(diphenyl-phosphinoethyl)-pyridine (pyphos) and t-BuCH$_2$MgCl in THF gave high yields of the desired C2 substituted products **67–71** with only a slight decrease in yields for indoles bearing electron-withdrawing groups. Notably, high yields were also achieved for 7-substituted indoles (**72–73**) as well as benzimidazole **74**.

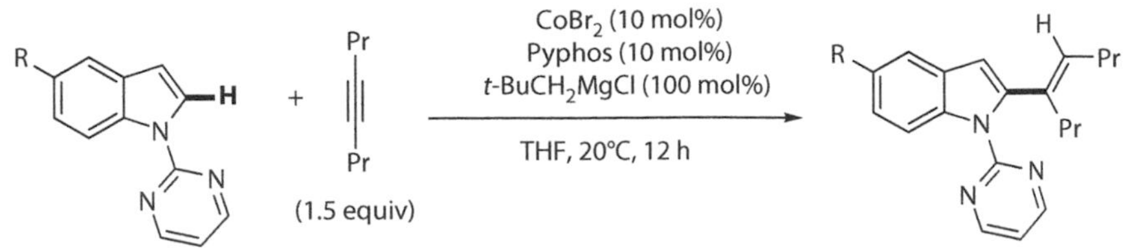

62, R = H
63, R = OCH₃
64, R = Cl
65, R = F
66, R = CN

67, R = H, 96%
68, R = OCH₃, 89%
69, R = Cl, 85%
70, R = F, 87%
71, R = CN, 75%

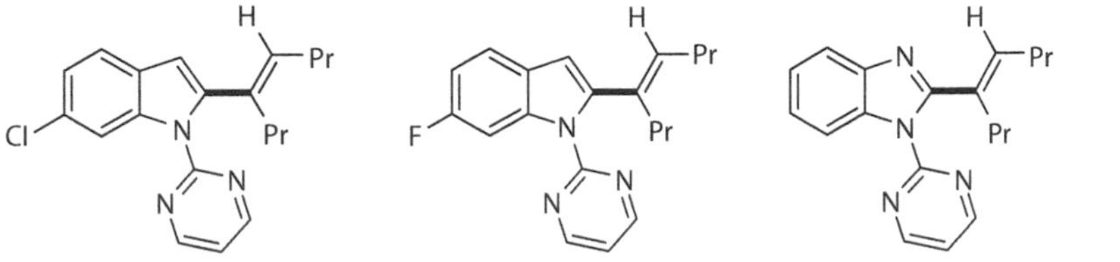

72, 89%          73, 92%          74, 82%

Combined, the work above demonstrates the sensitivity of these reactions to the choice of the ligand and reducing (Grignard) agent. However, in general, reducing reagents without β-hydrogens generally perform better than other reagents because they: (i) limit the formation of a cobalt hydride species thereby limiting the reduction of the alkyne, and (ii) are nonophilic thus reducing possible addition of the Grignard reagent.

## 8.2.2 HYDROARYLATION OF ALKENES TO FORM C—C BONDS

Gao and Yoshikai developed the first regioselective-switchable method for the hydroarylation of styrenes,[15] based on their previous work with the hydroarylation of alkynes. The reaction between a series of 2-phenylpyridine and styrene derivatives with a ternary cobalt-N-heterocyclic carbene N,N-bis-(mesityl)imidazolium chloride (Co-IMesHCl) gave the corresponding linear addition products 75–84A (A:B ratio), while treatment with a ternary cobalt-phosphine catalyst (Co-PCy₃) gave the corresponding branched addition products 75–84B (A:B ratio). In both cases, high yields and very good regioselectivities were observed regardless of the nature (electron-donating versus electron-withdrawing) and positions of the substituents on the 2-phenylpyridine and styrene derivatives reported. It is noteworthy that in some cases, for example, 84 where R=CF₃, the electronic nature of the R group had a larger impact than ligand control.

**Co-IMesHCl**

**75,** $R^1$ = H, $R^2$ = Ph, 84% (3:97)
**76,** $R^1$ = H, $R^2$ = 4-$CH_3C_6H_4$, 69% (1:99)
**77,** $R^1$ = H, $R^2$ = 2-$CH_3C_2H_4$, 78% (<1:99)
**78,** $R^1$ = H, $R^2$ = 4-$PhC_6H_4$, 72% (9:91)
**79,** $R^1$ = H, $R^2$ = 4-$OCH_3C_6H_4$, 74% (10:90)
**80,** $R^1$ = H, $R^2$ = 4-$FC_6H_4$, 69% (2:98)
**81,** $R^1$ = H, $R^2$ = 2-$FC_6H_4$, 84% (2:98)
**82,** $R^1$ = 4-$OCH_3$, $R^2$ = Ph, 74% (2:98)
**83,** $R^1$ = 4-F, $R^2$ = Ph, 82% (14:86)
**84,** $R^1$ = 4-$CF_3$, $R^2$ = Ph, 81% (85:15)

**Co-PCy₃**

**75,** $R^1$ = H, $R^2$ = Ph, 88% (96:4)
**76,** $R^1$ = H, $R^2$ = 4-$CH_3C_6H_4$, 93% (>99:1)
**77,** $R^1$ = H, $R^2$ = 2-$CH_3C_6H_4$, 34% (92:8)
**78,** $R^1$ = H, $R^2$ = 4-$PhC_6H_4$, 80% (99:1)
**79,** $R^1$ = H, $R^2$ = 4-$OCH_3C_6H_4$, 68% (99:1)
**80,** $R^1$ = H, $R^2$ = 4-$FC_6H_4$, 86% (96:4)
**81,** $R^1$ = H, $R^2$ = 2-$FC_6H_4$, NR
**82,** $R^1$ = 4-$OCH_3$, $R^2$ = Ph, 82% (96:4)
**83,** $R^1$ = 4-F, $R^2$ = Ph, 83% (85:15)
**84,** $R^1$ = 4-$CF_3$, $R^2$ = Ph, 31% (79:21)

The authors used mechanistic studies to rationalized the observed regioselectivities. Deuterium labeling studies revealed that the C—H bond cleavage and olefin insertions steps are reversible and are competing under the two different sets of reaction conditions. In the case of the Co-PCy₃ catalyst, the authors argued that the thermodynamic preference for a benzylcobalt species governs the selectivity. With the Co-IMES catalyst, the preference of the cobalt to avoid steric repulsion governs the selectivity.

The reaction was also shown to work well for the addition of aryl imines. However, reduced yields and regioselectivities were observed. Treatment of the *ortho*-alkylated products with acid readily gave the corresponding ketone in high yields (**85–88**, A:B).

**Co-IMesHCl**

**85,** $R^1$ = H, $R^2$ = Ph, 57% (8:92)
**86,** $R^1$ = H, $R^2$ = 2-Naphthyl, 23% (63:37)
**87,** $R^1$ = $OCH_3$, $R^2$ = Ph, 50% (12:88)
**88,** $R^1$ = Ph, $R^2$ = Ph, 46% (23:77)

**Co-PCy₃**

**85,** $R^1$ = H, $R^2$ = Ph, 64% (90:10)
**86,** $R^1$ = H, $R^2$ = 2-Naphthyl, 73% (99:1)
**87,** $R^1$ = $OCH_3$, $R^2$ = Ph, 53% (90:10)
**88,** $R^1$ = Ph, $R^2$ = Ph, 76% (96:4)

Lee and Yoshikai extended their *ortho*-alkylation methodology to a series of aldimines[16] and ketimines[17] with moderate to high yields. For aldimines, a ternary system consisting of $CoBr_2$, $P(4-CH_3C_6H_4)_3$, and $Me_3SiCH_2MgCl$ in warm THF gave desired products such as **89–97** in good to excellent yields. However, the authors revealed that one of the the *ortho* positions must be substituted in order to limit the formation of dialkylated addition products.

(1.2 eqiv)

**89,** R = CH$_3$, Ar = Ph, 80%
**90,** R = Ph, Ar = Ph, 70%
**91,** R = Morpholine, Ar = Ph, 81%

**92,** Ar = 4-OCH$_3$-C$_6$H$_4$, 53%
**93,** Ar = 4-Cl-C$_6$H$_4$, 93%
**94,** Ar = 4-F-C$_6$H$_4$, 76%

**95,** 89%              **96,** 62%              **97,** 87%

For ketimines, the reaction conditions optimized for aldimines gave only modest yields.[17] Therefore, a ternary system of $CoBr_2$, $P(4-FC_6H_4)_3$, and CyMgBr was developed to generate desired products such as **99** in high yield.

**98**         (1.2 eqiv)                                                           **99**

Ding and Yoshikai reported a unique cobalt-catalyzed intra-molecular olefin hydroarylation for the production of dihydropyrroloindoles and tetrahydropyridoindoles.[18] These compounds are important because they provide the scaffolding for a number of biologically relevant compounds including indole alkaloids such as strychnine, brucine, and vincamine. The process was shown to be switchable depending on the ligand used for the reaction. For example, treatment of indole **100** with the ternary cobalt-carbene catalyst SIMes gave the corresponding 6,5,5-tricycic compound **101**, while treatment with a ternary cobalt-carbene catalyst IPr gave the corresponding 6,5,6-tricycic compound **102**. Moderate to high yields and good regioselectivities were observed in both cases. Once again, the ligand-controlled regiodiverence observed was rationalized based on the olefin insertion step and the steric nature of the ligand. The author extended their methodology to generate a series of interesting heterocyclics such as compounds **103–105**.

**103,** 73% (9:1)          **104,** 40%          **105,** 63%

Gao, Yoshikai and Ding were the first to report on the *ortho*-alkylation of aromatic imines with vinyl silanes.[19] Addition of vinyl silane to aryl imines substituted with various functional groups proceeded with moderate to good yields to give the desired products **106–115** using a ternary catalyst system consisting of CoBr$_2$, phenanthroline (Phen), and *t*-BuCH$_2$MgBr in THF. In some examples, such as the imine derived from 2-fluorenyl methyl ketone, a 1:1 mixture of two regioisomers was formed from the reaction (**114** and **115**).

**106,** R = OCH₃, 81%     **109,** R = CH₃, 74%     **112,** 92%     **113,** 89%
**107,** R = Ph, 79%        **110,** R = Cl, 58%
**108,** R = Cl, 70%        **111,** R = CF₃, 41%

**114,** 37%                     **115,** 39%

The authors also demonstrated that a similar catalytic system could be used to promote the addition of aryl ketimines to aliphatic olefins at elevated temperatures with good success to produce the desired products **116–119** after acid-catalyzed hydrolysis.

**116,** 76%          **117,** 76%          **118,** 52%          **119,** 18%

Ding and Yoshikai also demonstrated the direct C2 alkylation of *N*-pyrimidylindole rings with vinyl silanes.[20] Their work built on previous research using internal alkynes to perform a similar transformation. Initial screenings revealed that the reaction between *N*-pyrimidylindoles **62–65** and **120** with vinyltrimethylsilane proceeded best in the presence of CoBr₂, 2,9-dimethyl-4,7-diphenylphenanthroline (Bathocuproine or Bathocup), and CyMgBr in refluxing THF. Interestingly, *N*-heterocyclic carbene and phosphine ligands did not improve the catalytic efficiency despite their utility in previous systems. The reaction was also shown to be sensitive to the type of Grignard (reducing) reagent used and the loading. Reduced loading (60 mol%) of primary and secondary alkyl Grignard reagents promoted the reaction while suppressing cross-coupling between the indole and the Grignard reagent. Overall, the reaction conditions proved tolerant of a variety of functional groups and moderate yields were generally obtained regardless of the type and placement of the substituent. Notably, the highest yield was obtained with pyridylindole 126. Compounds 127 and 128 demonstrate the effects of sterics on product yield. The authors also demonstrated that different vinyl silanes could be used with moderate success (not shown).

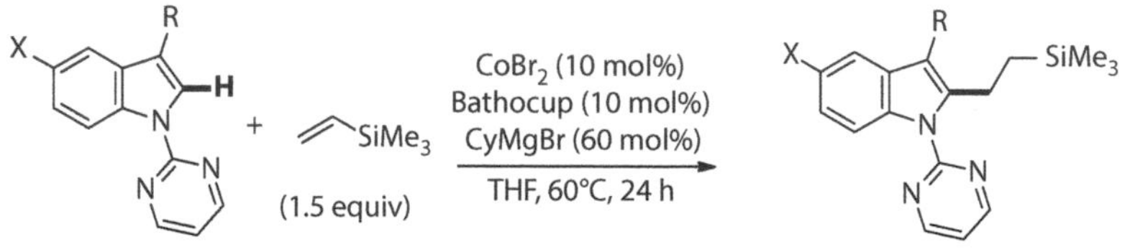

62, X = H, R = H
63, X = OCH₃, R = H
64, X = Cl, R = H
65, X = F, R= H
120, X = H, R = CH₃

121, X = H, R = H, 69%
122, X = OCH₃, R = H, 62%
123, X = Cl, R = H, 39%
124, X = F, R = H, 55%
125, X = H, R = CH₃, 42%

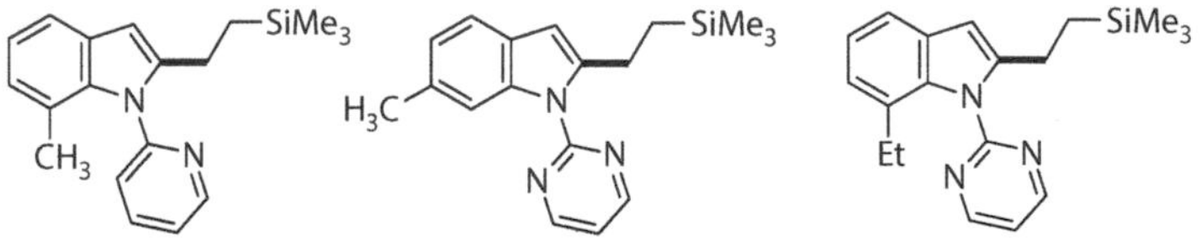

126, 80%          127, 71%          128, 58%

Reactions were also attempted with *N*-pyrimidylindole and simple alkenes such as norborene, 1-octene, and styrene. While modest yields were obtained with nor-borene 129, low yields were obtained with 1-octene 130 and styrene (not shown).

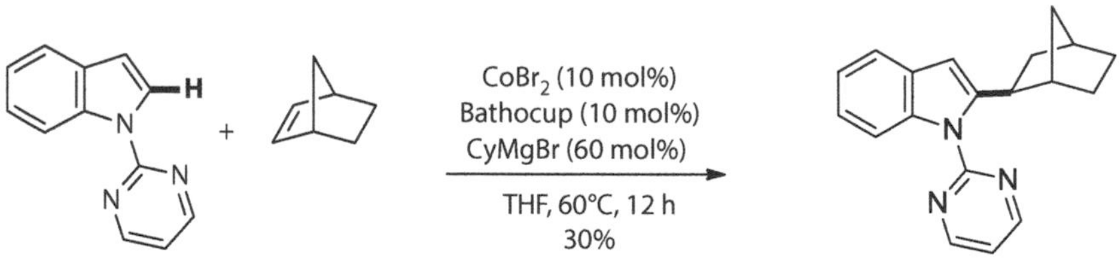

62                                                                    129

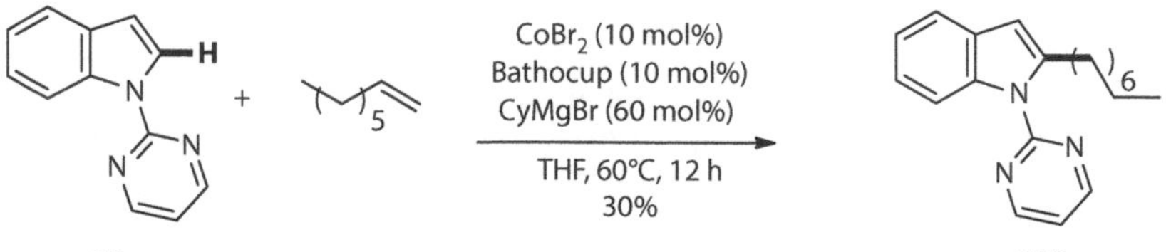

62                                                                    130

Li and Jones investigated the cobalt-catalyzed synthesis of quinolones from dial-lylanilines using C—H and C—N activations in an effort to overcome the environ-mentally unfriendly harsh conditions commonly used for the synthesis of these compounds.[21] Reaction of various diallylanilines 131–134 under optimized reaction conditions with $Co_2(CO)_8$ in the presence of 1 atm CO (to stabilize the Co complex) in THF at elevated temperatures gave the desired 2-ethyl-3-methyl quinolones 135–138 in modest to good yields. Several additional examples, such as 139–142, were also reported. The reaction was shown to favor diallylaniline derivatives with electron-donating substituents in the *meta* and/or *para* position(s); introduction of groups at the *ortho* positions (electron-donating or electron-withdrawing) tended to inhibit the reaction either through sterics or by chelation with the reactive cobalt species.

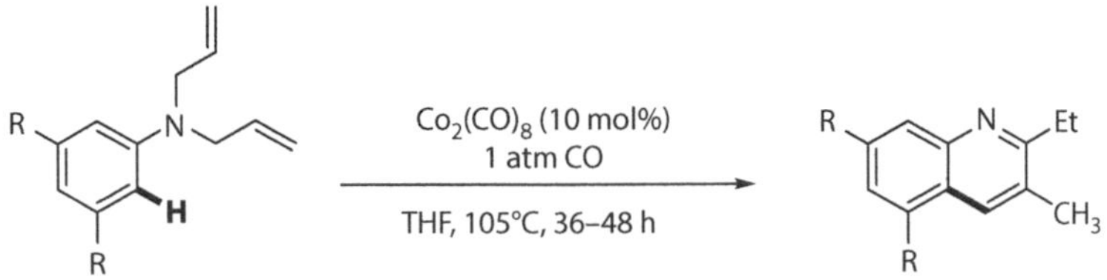

131, R = H
132, R = CH_3
133, R = OCH_3
134, R = CF_3

135, R = H, 65%
136, R = CH_3, 67%
137, R = OCH_3, 49%
138, R = CF_3, 0%

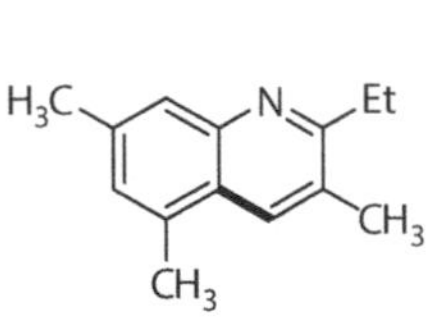

139, 67% 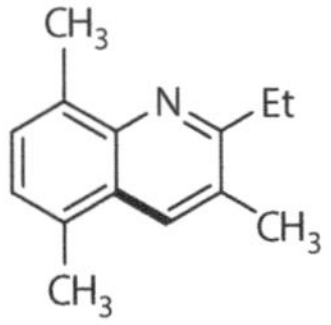          140, 40%

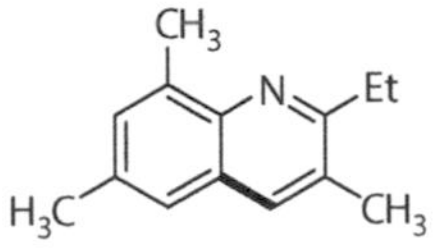

141, 32% 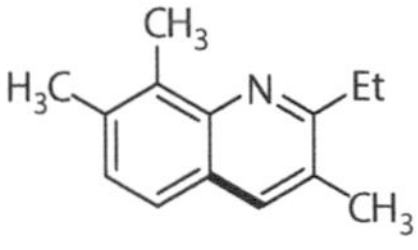          142, 28%

## 8.2.3 Hydrosilylation of Alkenes to Form C—Si Bonds

Deng and coworkers recently developed the first silyl-donor-functionalized *N*-heterocyclic carbene (NHC) cobalt complexes for use as catalysts in hydrosi-lylation reactions.[22] Of the catalysts screened, 144 proved superior in rapidly pro-ducing the desired silylated products 145 and 146 in reactions between 1-octene 143 and $PhSiH_3$ while minimizing isomerization side products 147 and 148. The properties of their new catalyst proved superior to two additional catalysts, $Co_2(CO)_8$ and $(IPr_2Me_2)CoPh_2$, which were also studied.

## 8.3  C—H/ELECTROPHILE COUPLING

### 8.3.1  C—C BOND FORMATION

Li and coworkers studied the ability of a several transition metals including niobium, molybdenum, and cobalt, to catalyze the direct C—H functionalization of aromatic C—H bonds.[23] Aryl bromides **149–155** and arenes such as benzene were reacted with each catalyst in the presence of an appropriate ligand for 48 h. In nearly all of the studies performed, the niobium ($NbCl_5$) and molybdenum ($Mo(OAc)_2$)-based catalysts outperformed cobalt ($Co(acac)_3$). In addition, a few general trends were observed. First, aryl bromides with electron-donating substituents were shown to promote the cross coupling reaction regardless of their position, while more electron-deficient derivatives showed significantly decreased yields. Steric hindrance was also shown to impact efficiency as demonstrated with **163–165**. Finally, the authors noted that the couplings between benzene and **154** demonstrated the compatibility of these catalysts for systems containing C—F bonds. Despite the inferior yields obtained with the cobalt catalyst employed in this study, these initial reactions demonstrated the utility of using cobalt for C—H functionalization via electrophilic coupling.

**149,** R = H
**150,** R = CH₃
**151,** R = OCH₃
**152,** R = C(CH₃)₃
**153,** R = Ph
**154,** R = F
**155,** R = CN

**156,** R = H, [Nb] = 62%, [Mo] = 69%, [Co] = 55%
**157,** R = CH₃, [Nb] = 62%, [Mo] = 65%, [Co] = 52%
**158,** R = OCH₃, [Nb] = 66%, [Mo] = 69%, [Co] = 66%
**159,** R = C(CH₃)₃, [Nb] = 78%, [Mo] = 73%, [Co] = 59%
**160,** R = Ph, [Nb] = 51%, [Mo] = 57%, [Co] = 54%
**161,** R = F, [Nb] = 20%, [Mo] = 69%, [Co] = 31%
**162,** R = CN, [Nb] = 36%, [Mo] = 52%, [Co] = trace

**163,** [Nb] = 59%, [Mo] = 59%, [Co] = 52%

**164,** [Nb] = 13%, [Mo] = 37%, [Co] = 36%

**165,** [Nb], [Mo], [Co] = NR

For: [Nb] = NbCl₅ (10 mol%), bathophen. (30 mol%), KOtBu(3.0 equiv)
         2 mL benzene, 80°C, 48 h
     [Mo] = Mo(Ac)₂ (10 mol %), bathophen. (30 mol%), KOtBu (3.0 equiv)
         2 mL benzene, 80°C, 48 h
     [Co] = Co(acac)₃ (15 mol%), DMEDA (60 mol%), KOtBu (3.0 equiv)
         4 mL of benzene 100°C, 48 h

C−H bond activation of aromatic compounds to form aryl–alkyl bonds using cobalt catalysts was extended by Nakamura and coworkers to the activation of carboxamides.[24] The inexpensive and benign cobalt salt Co(acac)₂ was successfully used as a catalyst in the presence of ethyl magnesium chloride (EtMgCl) and 1,2-dimethyl-3,4,5,6-tetrahdyro-2(1$H$)-pyrimidinone (DMPU) to convert benzamide **166** to the *ortho* and dialkylated products **167** and **168** with the major product being dialkylated **168** in 79% yield. A series of additional carboxamides were examined and the authors noted a few general trends. First, the reaction was shown to proceed smoothly regardless of the electronic nature of the substituent (e.g., electron-donating **169** and electron-withdrawing **170**). Second, *N*-methyl carboxamides such as **166** provided higher yields in comparison to *N*-phenyl **171** and *N*-isopropyl **172** derivatives. However, the later groups were also shown to be more effective at suppressing the formation of dialkylated product. Interestingly, a tertiary amide (not shown) did not provide any products mono- or dialkylated.

**166**

Co(acac)₂ (10 mol%)
EtMgCl (5.8 equiv)
⟶
DMPU (30 equiv)
THF, rt, 12 h

**167,** <2%     +     **168,** 79%

**169,** 61%         **170,** 69%         **171,** 52%         **172,** 52%

The authors extended their methodolgy-phenylpyridine derivatives to produce the corresponding products in high yield. Interestingly, for the parent compound **3**, the mono alkylated product **173** was produced in preference to the dialkylated product **174**. This is in contrast to the authors work with *N*-methyl carboxamides where the dialkylated addition products were generally produced in higher yield. In addition, high yields were achieved for 2-phenyl and aryl pyridine regardless of the substitution patterns as demonstrated with addition products **175–177**.

**3**          Co(acac)₂ (10 mol%) EtMgCl (5.8 equiv) / DMPU (30 equiv) THF, rt, 12 h          **173,** 80%          +          **174,** 17%

**175,** 93%                    **176,** 87%                    **177,** 83%

**166,** R = H
**178,** R = Ph          Co(acac)₂ (10 mol%) CyMgCl (3.0 equiv) / DMPU (12 equiv) Et₂O, rt, 12 h          **179,** R = H, 63%
**180,** R = Ph, 81%          +          **181,** R = H, 22%
**182,** R = Ph, >5%

**183,** 64%                    **184,** 5%                    **185,** 79%

Chen, Ilies, and Nakamura employed a similar C—H activation process to form aryl-alkyl bonds between aromatic carboxamides and alkyl halides, with the only difference being that the Grignard reagent was used as a reducing agent instead of an alkylating agent.[25] In the process, the authors also found that Co(acac)₂ could be used to catalyze *ortho* alkylations of aromatic carboxamides with alkyl chlorides that would normally undergo isomerization reactions under Friedel–Crafts conditions.

Carboxamides substituted at the *meta* position with phenyl groups such as **178** were found to be superior in comparison to unsubstituted **166** or para-substituted benzamides (not shown). In addition, monoalkylated products **179–180** were produced in higher yields than the corresponding dialkylated products **181–182**. The authors further demonstrated the utility of this reaction by using naphthyl and pyridine based carboxamides to demonstrate produce the desired addition products **183–185**. They also showed that primary, secondary, and tertiary alkyl halides, as well as differentially substituted alkyl halides could be used with success to produce the desired products **186–190**, albeit lower yields.

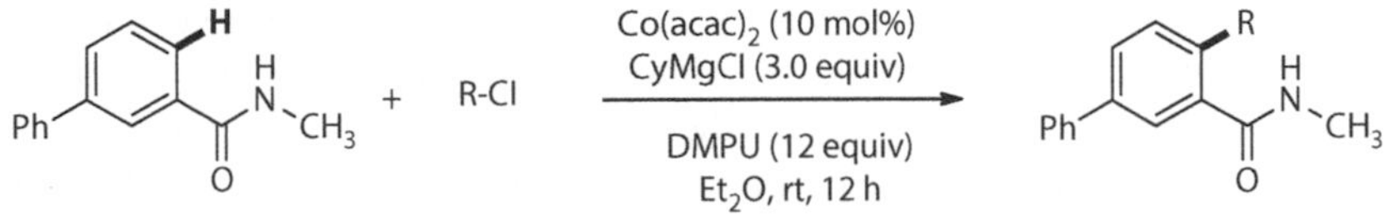

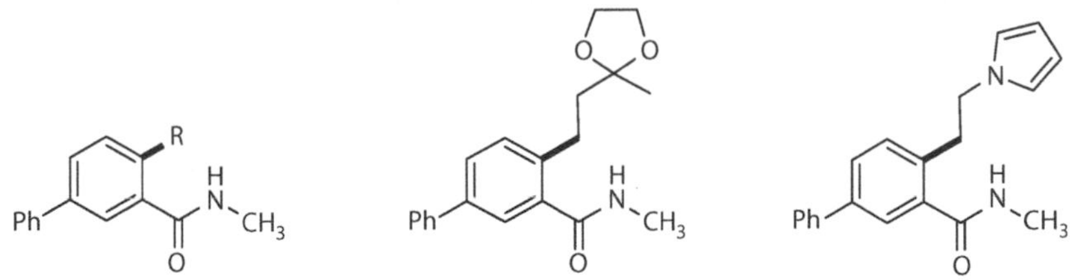

**186**, R = Cy, 15%
**187**, R = *t*Bu, 83%
**188**, R = CH$_2$(CH$_2$)$_8$CH$_3$, 68%

**189**, 64%

**190**, 59%

Yoshikai and coworkers used their regioselective hydroarylation to demonstrate that their *N*-heterocyclic carbene cobalt catalyst could be used to perform the *ortho*-arylation of aromatic imines with aryl chlorides under mild conditions.[26] Their work was inspired by previous studies by Li and coworkers with 2-arylpyridine derivatives.[27] Treatment of aryl imines **191–193** with CoBr$_2$, IMesHCl, *t*-BuCH$_2$MgBr in THF, followed by aqueous HCl gave the corresponding ketones **194–196** in modest to good yields. The authors were able to generate various imine derivatives such as **197–199** to show the utility of their catalytic system.

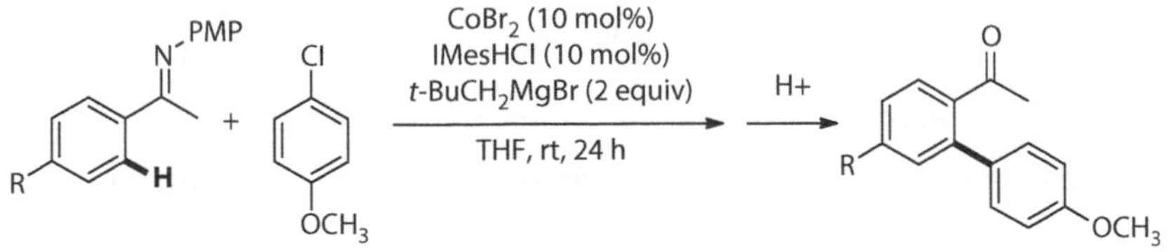

**191**, R = OCH$_3$
**192**, R = Ph
**193**, R = F

(2 equiv)

**194**, R = OCH$_3$, 79%
**195**, R = Ph, 61%
**196**, R = F, 66%

**197,** 68%          **198,** 36%          **199,** 87%

Recently, Ackermann and coworkers expanded on Nakamura's methodology to prepare 2-phenylpyridine analogs via C—H bond alkylation and arylation.[28] The authors demonstrated that high yields of the desired products could be obtained if IMesHCl was used as a ligand in the presence of Co(acac)$_2$, CyMgCl, and 1,3-dimethyl-3,4,5,6-tetrahydro-2-pyrimidinone (DMPU). Interestingly, the more sterically congested product was produced with 2-(3-methoxyphenyl) pyridine analog **200** and in higher yields than 2-(4-methoxyphenyl) pyridine **201**, demonstrating a strong electronic effect versus steric effect.

Ackermann also reported on the C—H bond functionalization of aryl pyridine and indole derivatives with alkyl chlorides. Again, Co(acac)$_2$ was used as a precatalyst, this time, in the presence of IPrHCl to generate a Co-IPr catalyst. The *ortho* alkylation products were achieved in high yields for a number of aryl pyridine (**203–206**) and indole derivatives such as **207** with little differences yields between derivatives bearing electron-donating groups (**204**) and electron-withdrawing groups (**205**). Additional alkyl halides were examined by the authors with similar results (not shown).

**200,** 89%          **201,** 66%          **202,** 94%

**203**, 90%     **204**, 80%     **205**, 91%     **206**, 79%

**207**     **208**, 97%

Song and Ackermann further expanded on their work with *N*-heterocyclic carbenes to develop economical cobalt-catalyzed direct arylation and benzylation processes via C—H/C—O bond cleavage with sulfamates, carbamates, and phosphates.[29] After carefully screening reaction conditions, the authors determined that the optimized catalytic system involved the reaction of Co(acac)$_2$, IMesHCl, and CyMgCl in DMPU. Using these conditions, direct arylation could be achieved using *N*-heterocyclic biaryls and *N*-substituted indoles with sulfamates and carbamates to produce the desired addition products **209–210**, **202** and **207**, and **211–212**. The later examples are the first examples of direct arylation using carbamates. Notably, arenes with electron-donating and electron-withdrawing substituents gave good yields of the desired products. However, competition experiments revealed that arenes bearing electron-withdrawing substituents reacted faster due to a more acidic C—H bond. In the same report, Song and Ackermann also demonstrated that their catalytic system could be used for direct benzylation of indoles such as **207** with phosphates under mild reaction conditions.

**209**, 55%                    **210**, 80%                    **202**, 95%

**209**, 63%                    **210**, 90%                    **202**, 96%

**207**, R = H                                              **212**, R = H, 91%
**211**, R = CH₃                                            **213**, R = CH₃, 93%

**207**                                    **214**

Yoshikai and Gao reported on the cobalt-catalyzed arylation of aldimines using a directed C—H bond functionalization.[30] A series of 2-arylpyridines were treated with *N*-phenylbenzaldimine in the presence of CoBr$_2$, IPrHCl, and *t*-BuCH$_2$MgBr followed by an acid workup to give the desired *ortho* addition products **215–217** in good yield.

**215, 84%**        **216, 76%**        **217, 70%**

Yoshikai and coworkers extended their methodology to the cobalt-catalzyed *ortho* C—H alkylation of 2-arylpyridines via the ring opening of functionalized aziridines under mild conditions.[31] The desired *ortho* addition products **218–223** were obtained in modest to good yields.

**218, 72%**        **219, 61%**        **220, 67%**

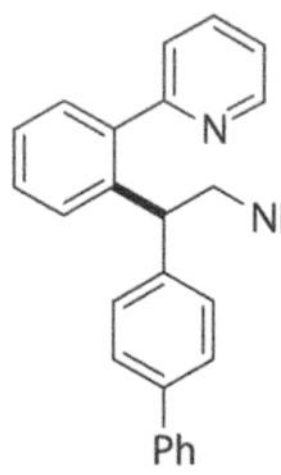
N
NHPh
Ph

**221,** 73%

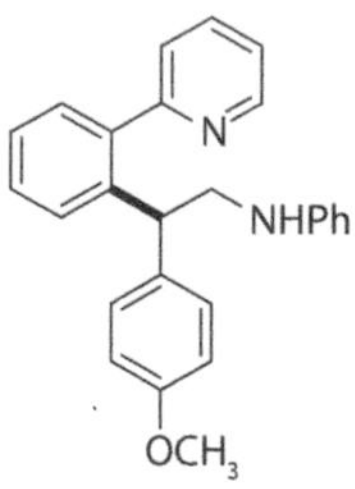
N
NHPh
OCH3

**222,** 63%

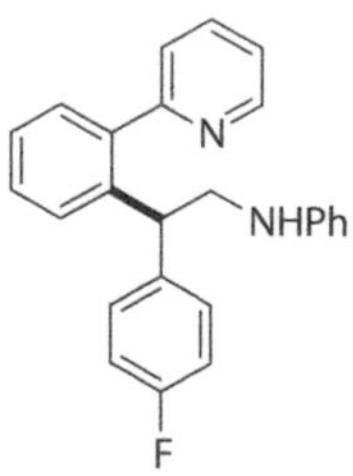
N
NHPh
F

**223,** 46%

Zhao, Toste and Bergman reported on a one-pot direct Michael addition of a series of alkenes using a cobalt-dinitrosyl mediated vinylic C—H functionalization reaction.[32] Treatment of various Michael acceptors **224–228** with cobalt catalyst **229** in Verkade's base at elevated temperatures led to the corresponding tetra-alkyl-substituted alkene products **230–234** in moderate to good yields. The authors demonstrated that tricyclic systems such as **235–237** could also be prepared using this methodology. The increased yield of **235** over **236** was rationalized by the rigidity of the corresponding five-membered ring cobalt complex, which is more stable then the more flexible six-membered ring complex. Inter-molecular couplings with norborene and five- and six-membered cyclic enones also provided the desired products **238** and **239**, respectively in good yield and with high diastereoselectivity. The authors noted that the only disadvantage with this system was the limited turnover due to base composition and dimerization of the catalyst.

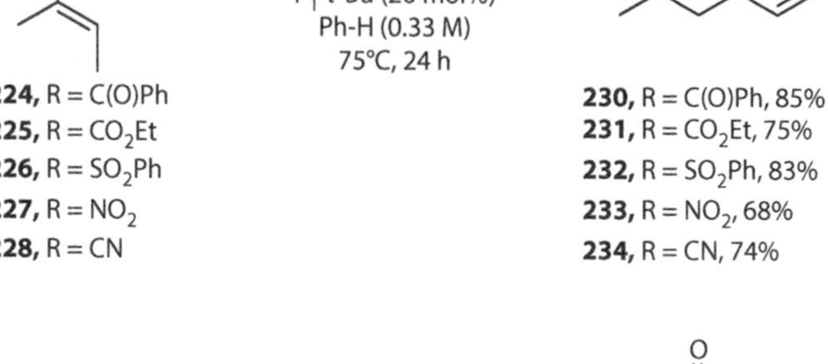
R
H
O
N
Co
N
O
229
(1.2 equiv)
P1-t-Bu (20 mol%)
Ph-H (0.33 M)
75°C, 24 h
R

**224,** R = C(O)Ph
**225,** R = $CO_2Et$
**226,** R = $SO_2Ph$
**227,** R = $NO_2$
**228,** R = CN

**230,** R = C(O)Ph, 85%
**231,** R = $CO_2Et$, 75%
**232,** R = $SO_2Ph$, 83%
**233,** R = $NO_2$, 68%
**234,** R = CN, 74%

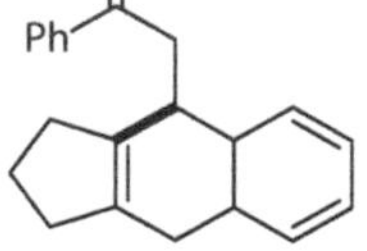
O
Ph

**235,** 91%

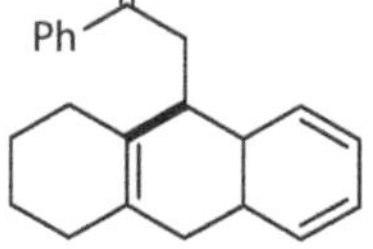
O
Ph

**236,** 19%

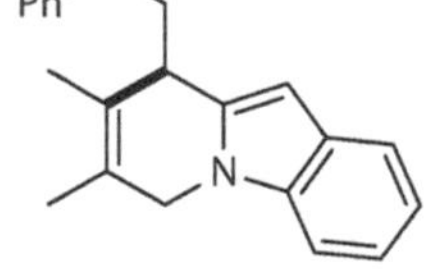
O
Ph
N

**237,** 77%

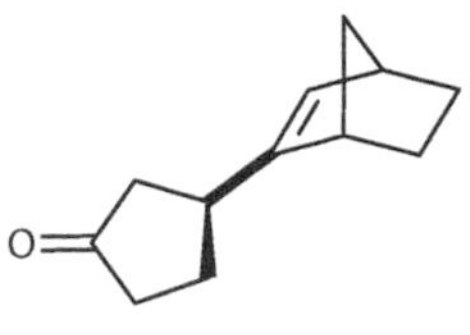

**238,** 74%, 92:8 *dr*

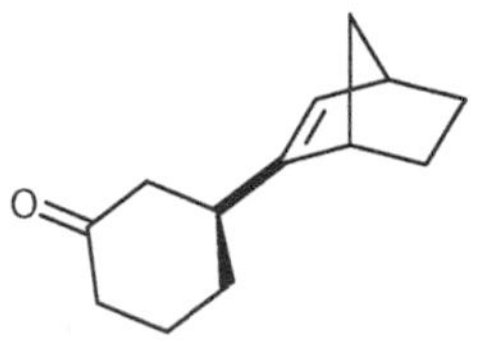

**239,** 79%, 93:7 *dr*

Yoshino and coworkers recently employed a cationic high-valent Cp * Co (III) complex for directed C—H bond activation.[33] Treatment of 2-phenylpyridine with sulfonyl imines in the presence of [Cp * CoIII(benzene)]PF$_6$)$_2$ catalyst in 1,2-dichloroethane (DCE) at elevated temperatures gave the desired products **246–253** in moderate to good yields for a variety of sulfonyl imines.

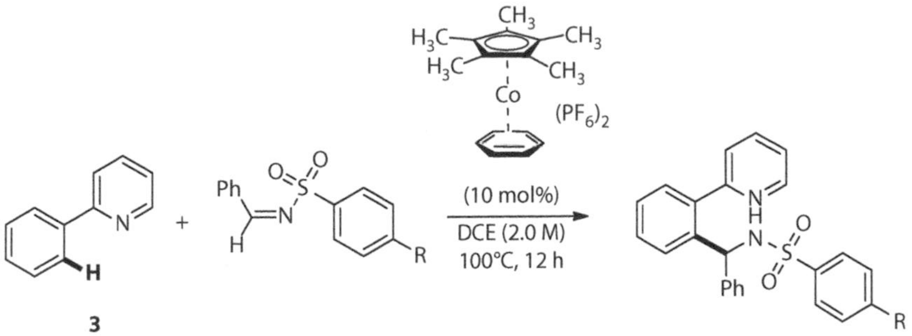

**240,** R = H
**241,** R = CH$_3$
**242,** R = OCH$_3$
**243,** R = Cl
**244,** R = Br
**245,** R = CF$_3$

**246,** R = H, 80%
**247,** R = CH$_3$, 76%
**248,** R = OCH$_3$, 57%
**249,** R = Cl, 83%
**250,** R = Br, 72%
**251,** R = CF$_3$, 64%

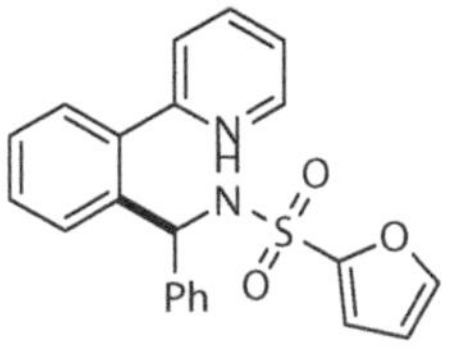

**252,** 66%

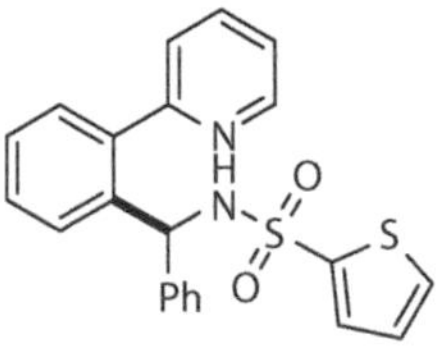

**253,** 69%

The authors expanded their work to include α,β-unsaturated ketones. Again a series of reactions were performed with 2-phenylpyridine and α,β-unsaturated ketones **254–257** this time using THF as a solvent with a longer reaction periods. The desired Michael addition products **258–261** were produced in high yield regardless of the substituent. Cyclic α,β-unsaturated ketones such as cyclopent-2-enone also gave high yields of the desired products **264–266**.

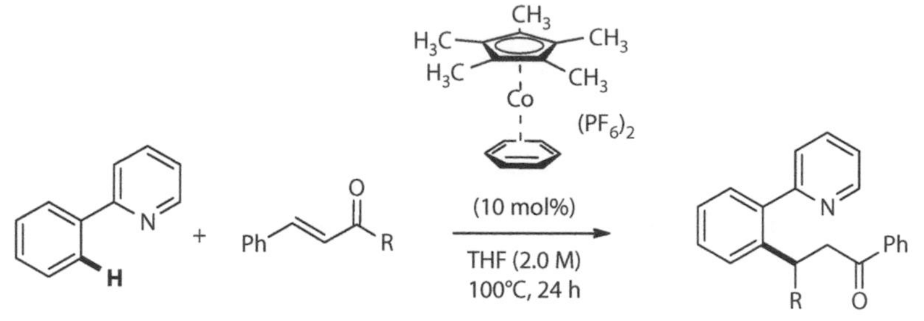

**3**

**254,** R = CH₃
**255,** R = Ph
**256,** R = p-OCH₃-C₆H₄
**257,** R = p-Br-C₆H₄

**258,** R = CH₃, 78%
**259,** R = Ph, 80%
**260,** R = p-OCH₃-C₆H₄, 76%
**261,** R = p-Br-C₆H₄, 75%

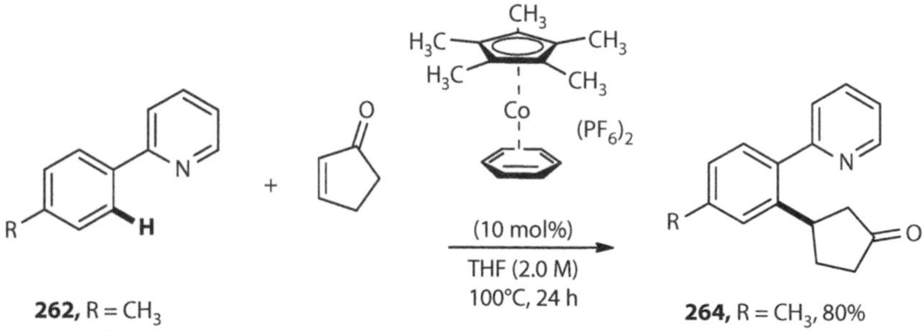

**262,** R = CH₃
**4,** R = OCH₃
**263,** R = Br

**264,** R = CH₃, 80%
**265,** R = OCH₃, 87%
**266,** R = Br, 74%

## 8.3.2   C—N Bond Formation

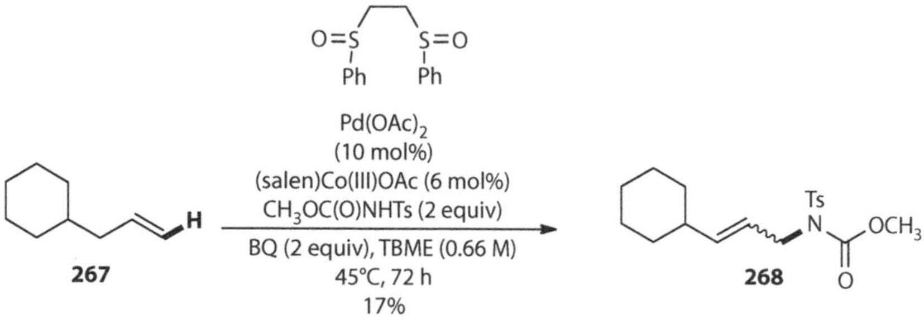

**267**

**268**

Reed and White provided a rare example of an inter-molecular linear allylic C—H amination (LAA) using a heterobimetallic catalysts in which one of the catalysts was cobalt-based.[34] Treatment of allylcyclohexane **267** with N-(methoxycarbonyl)-p-toluenesulfonamide in the presence of commercially available catalysts bis-sulfoxide/Pd(OAc)₂ and (salen)Co(III)OAc, benzoquinone (BQ) and tert-butyl methyl ether (TBME) gave the desired linear E-allylic amine **268** in 17% yield with high regio- (>100:1 linear:bicyclic) and stereo- (91:1 *E:Z*) selectivity. Although higher yields were obtained with the corresponding chromium chloride salen, this work constitutes the first example of its kind using a cobalt catalyst. The authors went on to demonstrate that the Pd(II)/bis-sulfoxide catalyst is responsible for promoting allylic C—H cleavage, while the salen catalyst and benzoquinone promote amination of the π-allylPd intermediate.

Ye and coworkers described an efficient benzylic C—H amination via dehydrogenative coupling with various sulfonamides, carboxamides, and carbamates using an inexpensive catalyst/oxidant system.[35] Reactions of substituted diphenylmethane derivatives **269–274** and ethylbenzene derivatives **281–282** with tosylamine in the presence of CoBr$_2$, di-*tert*-butyl peroxide (DTBP), and 1,1,2-trichloroethylene (TCE) using acetic acid as an additive, resulted in good to excellent yields of the corresponding sulfonamides for diphenylmethane (**275–277**) and ethylbenzene (**283**) derivatives bearing electron-donating groups. Moderate yields were observed for diphenylmethane derivatives bearing electron-withdrawing groups (**278–280**) and no reaction occurred for ethylbenzene bearing bromine in the *para* position (**284**). The authors hypothesized that the reduced yields for the latter derivatives are due to the destabilization of the intermediate benzyl cation species formed during the catalytic cycle. Additionally, sulfonamides bearing electron-withdrawing groups in the *para* position also resulted in reduced yields, presumably due to decreased electron density on the nitrogen atom. Interestingly, the authors found that other readily available cobalt salts including CoCl$_2$, Co(OAc)$_2$, and hydrated CoBr$_2$ were inactive. Similar trends were observed when carboxamides and carbamates (not shown) were used instead of the sulfonamides.

**269,** R = H
**270,** R = CH$_3$
**271,** R = OCH$_3$
**272,** R = Br
**273,** R = CN
**274,** R = CO$_2$CH$_3$

**275,** R = H, 93%
**276,** R = CH$_3$, 81%
**277,** R = OCH$_3$, 82%
**278,** R = Br, 65%
**279,** R = CN, 64%
**280,** R = CO$_2$CH$_3$, 55%

**281,** R = H
**282,** R = Br

**283,** R = H, 90%
**284,** R = Br, 0%

Sun and coworkers developed an air-stable carbonyl(pentamethyl-cyclopenta-dienyl)cobalt iodide complex Cp * Co(CO)I$_2$ which could be used in combination with silver hexafluoroantimonate (AgSbF$_6$) and potassium acetate (KOAc) for the directed C2 selective C—H amidation of indoles.[36] A series of indoles **62, 285–289** were shown to undergo amidation with tosyl azide in dichloroethylene (DCE) when catalyzed with cationic (pentamethyl-cyclopentadienyl)cobalt (III) Cp * Co(III) (C$_6$H$_6$)(PF$_6$)$_2$ generated *in situ* from the reaction between Cp * Co(CO)I$_2$ and AgSbF$_6$. The desired C2 functionalized indoles **290–295** were produced in high yields, and the reaction showed excellent tolerance of a variety of electron-donating and electron-withdrawing functional groups. A series of differentially 4-, 5-, and 6-substituted indoles and aryl and alkyl azides were also explored, and the cor-responding C2 functionalized indoles were also produced in excellent yields (not shown). The authors noted that both AgSbF$_6$ and KOAc were essential for the gen-eration of the active catalytic species, and that the reaction could be accomplished with as little as 1.25% catalyst loading of Cp * Co(CO)I$_2$ with only a slight reduc-tion in yield.

(1.5 equiv)
**62,** R = H
**285,** R = CH$_3$
**286,** R = OCH$_3$
**287,** R = F
**288,** R = Cl
**289,** R = Br

**290,** R = H, 92%
**291,** R = CH$_3$, 92%
**292,** R = OCH$_3$, 85%
**293,** R = F, 89%
**294,** R = Cl, 89%
**295,** R = Br, 90%

Zhang and coworkers recently reported on the use of a cobalt (II) tetraphenyl porphyrin complex, Co(TPP), as a catalyst for the intra-molecular amination of sulfonyl azides.[37] Their work builds on early studies where Cenini and cowork-ers used Co(TPP) for the amination of benzylic C—H bonds with aryl azides.[38] Reactions of primary, secondary, and tertiary sulfonyl azides with Co(TPP) in chlorobenzene in the presence of molecular sieves and under an inert atmosphere at 80°C gave the desired sultams **296–301** in high yield. The authors noted that tertiary C—H bonds reacted to give higher product yields then than secondary and primary C—H bonds, respectively, in the formation of five-membered heterocyclic ring structures.

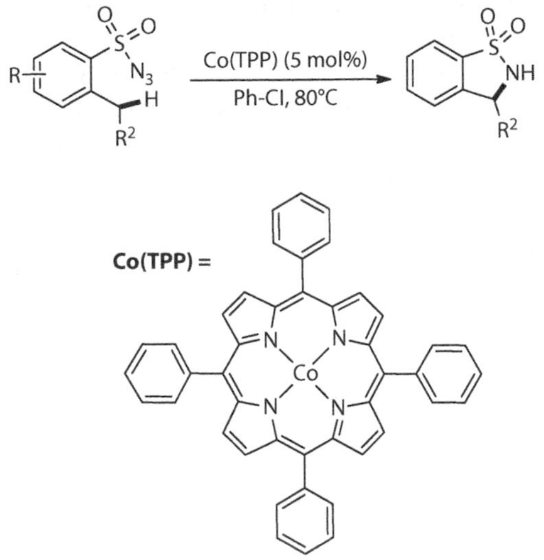

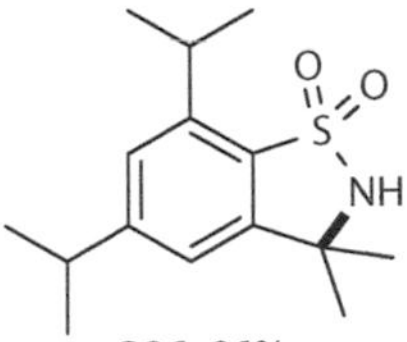

**296,** 96%

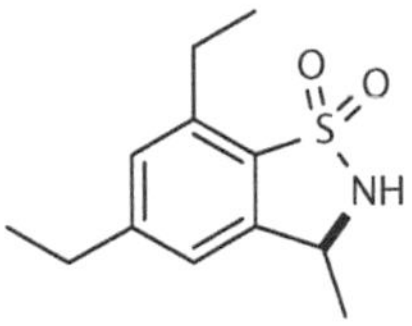

**297,** 90%

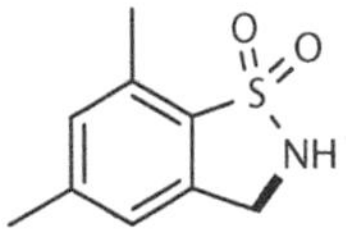

**298,** 96%

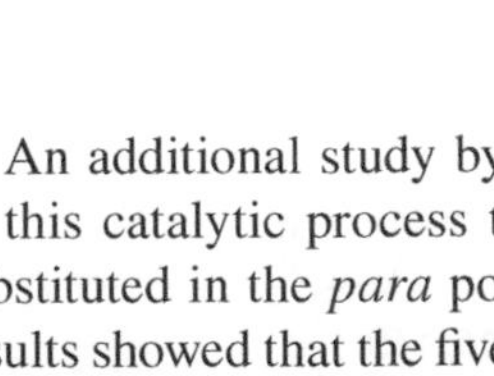

**299,** 87%

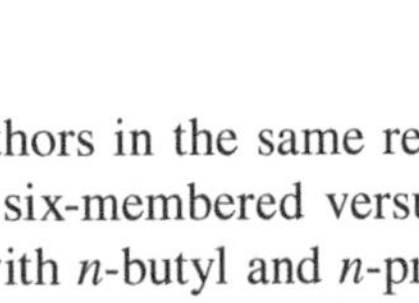

**300,** 93%

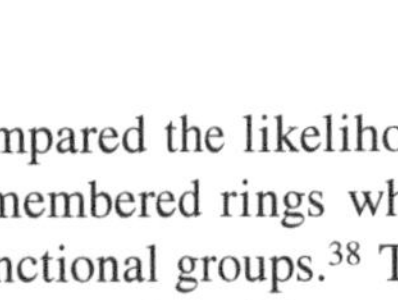

**301,** 99%

An additional study by the authors in the same report compared the likelihood of this catalytic process to form six-membered versus five-membered rings when substituted in the *para* position with *n*-butyl and *n*-propyl functional groups.[38] The results showed that the five-membered ring formation was always preferred due to its thermodynamic stability, but the *n*-propyl aryl-sulfonyl substrate yielded near-equal yields of both ring formations.

Nitrene insertion of C—H bonds using bromamine-T and a cobalt-based porphyrin catalysts, Co(TDCIPP), was also explored by Zhang and coworkers for use in constructing tosyl-protected amines.[39] Amination of the sp³ C—H bond of indan **302** using bromamine-T in acetonitrile produced the desired amine **303** in 73%

yield. Further studies were carried out using various substrates with moderate to good yields. The use of tetralin as a substrate yielded 66% of desired amine product **304**. Interestingly, phthalan produced imine **305** in 50% yield instead of the desired protected amine product. Fluorene, containing an activated benzylic C—H bond, produced only 14% yield of the desired product **306**. Finally, ethylnaphthalene derivatives containing both primary and secondary C—H bonds gave rise to the corresponding benzylic products **307** and **308** in yields of 33–34%, respectively. In a later report, Zhang and coworkers extended this approach to generate Troc-protected amines using TrocN$_3$ and Co(TPP) to catalyze the desired C—H aminations in good to excellent yields (not shown).[40]

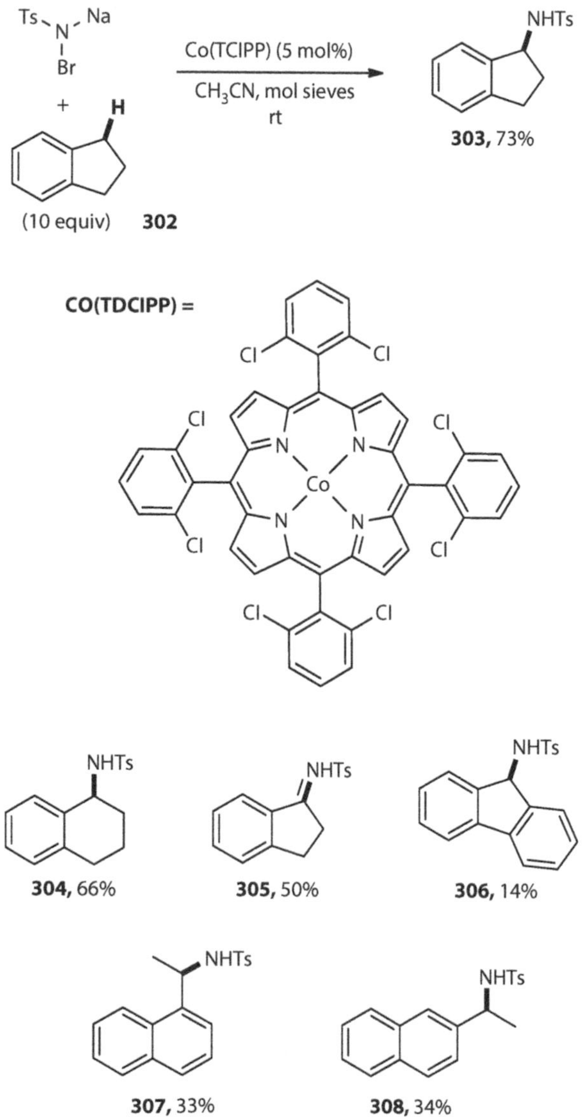

**303**, 73%

**CO(TDCIPP) =**

**304**, 66%

**305**, 50%

**306**, 14%

**307**, 33%

**308**, 34%

Zhang and coworkers later explored the possibility of using phosphoryl azides and cobalt-based porphyrin catalysts to produce six- and seven-membered heterocyclic cyclophosphoramidates via an intra-molecular amination.[41] These compounds are important intermediates of cyclophosphamides which have been shown to be potent anticancer drugs. The corresponding C—H aminations were shown to occur under mild conditions in trifluorotoluene at 80°C overnight, producing nitrogen gas as the only byproduct. A number of cobalt-based porphyrin catalysts were screened to determine the optimal catalyst. Co(P1) catalyzed the reaction of **309** to **310** in 79% yield, while the Co(P2) and Co(P3) catalyzed the same reaction to give 98% and 96% yields of the desired products, respectively. The more sterically hindered catalysts Co(P4) and Co(P5) produced only minor amounts of the desired product. Additional studies were undertaken using Co(P2) to determine the effectiveness of this cobalt porphyrin catalyst in catalyzing C—H amination with other phosphoryl azides. In most cases the desired products, such as **311–316** were produced in high yield.

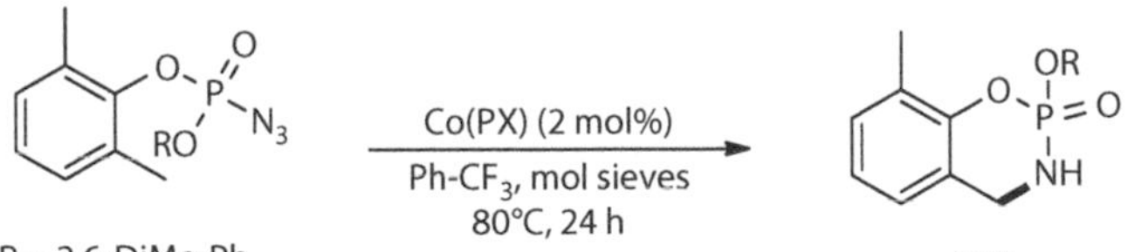

R = 2,6-DiMe-Ph
**309**

Co(PX) (2 mol%)
Ph-CF$_3$, mol sieves
80°C, 24 h

**310**

Co(P**X**) =

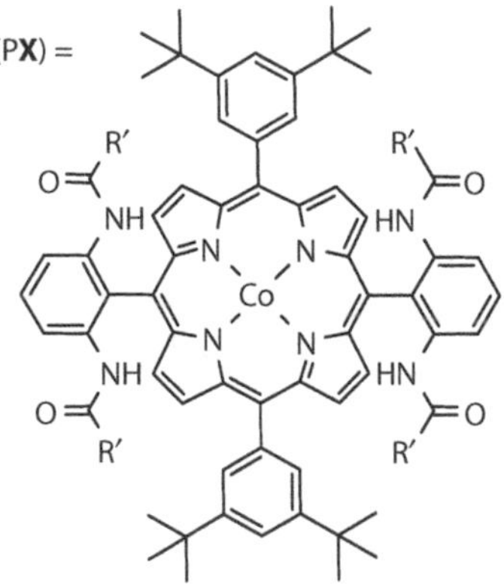

**X**
**1:** R′ = i-Pr, 79%
**2:** R′ = Me, 98%
**3:** R′ = Et, 96%
**4:** R′ = t-Bu, <5%
**5:** R′ = Ph, 0%

Co(P**1**):

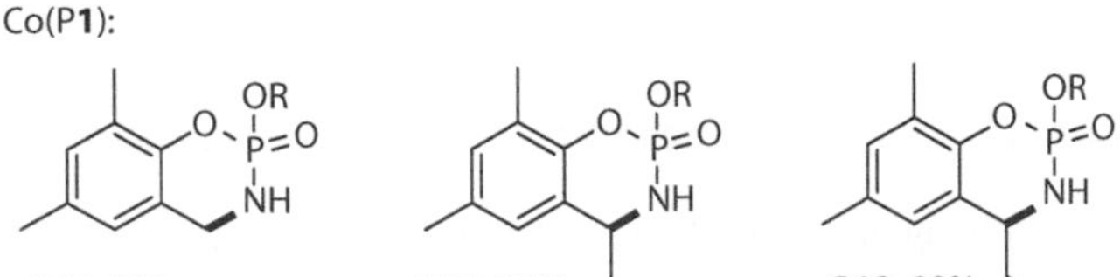

**311**, 99%

**312**, 94%

**313**, 90%

Co(P2):

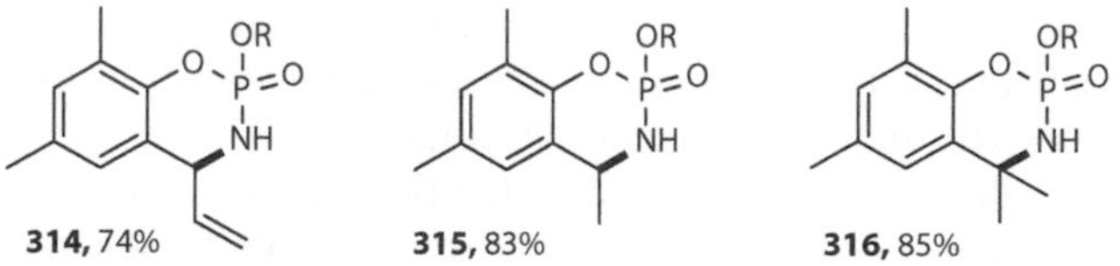

**314**, 74%          **315**, 83%          **316**, 85%

Interestingly, substrate **317** provided insight to the preference of the catalyzed amination to form six-membered rings such as **318** versus seven-membered rings such as **319**. The observed yields show preference for the formation of the six-membered ring, most likely due to the preference for benzylic primary C—H bonds. However, the amount of seven-membered ring product demonstrates the capability of the catalytic system to aminate nonbenzylic primary C—H bonds. In further studies it was shown that this system was capable of producing seven-membered ring amination products in high yields in the absence of benzylic C—H bonds (not shown).

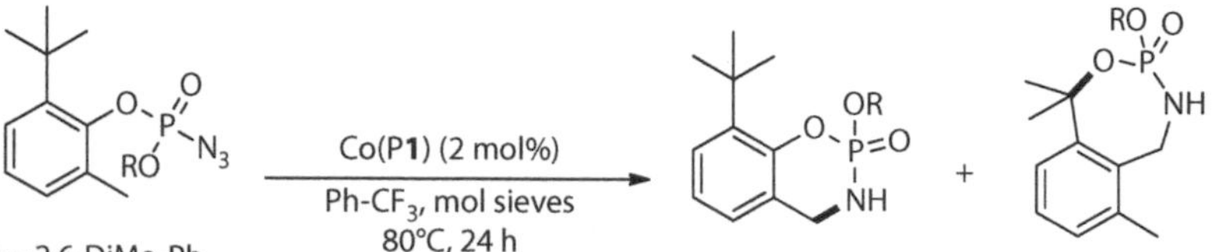

R = 2,6-DiMe-Ph
**317**

Co(P1) (2 mol%)
Ph-CF$_3$, mol sieves
80°C, 24 h

**318**, 76–84%     +     **319**, 16–24%

Co(P1) (2 mol%)
Ph-H, mol sieves
40°C, 20 h

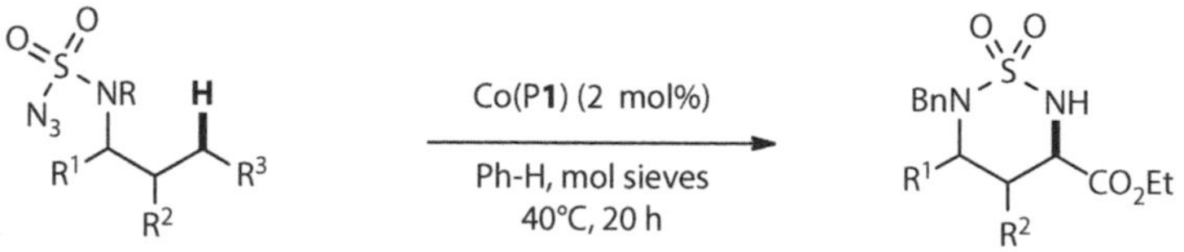

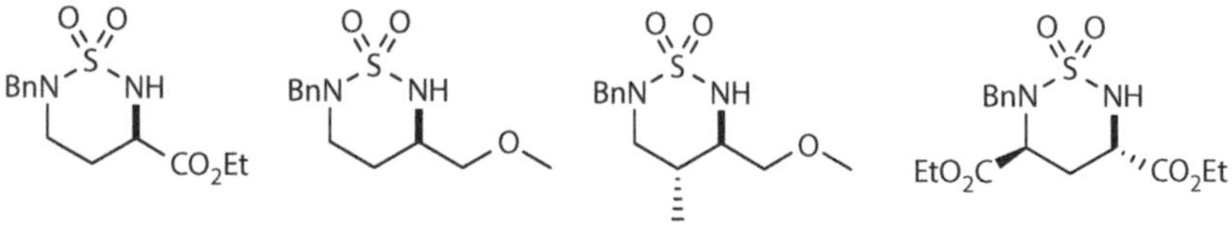

**320**, 98%          **321**, 99%          **322**, 91%          **323**, 99%

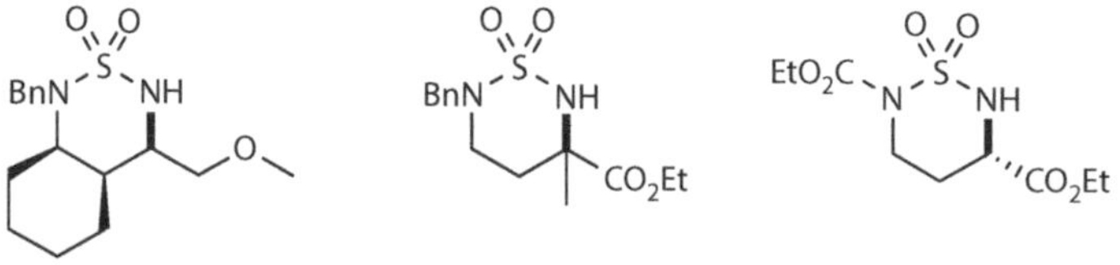

**324**, 95%          **325**, 91%          **326\***, 65%

More recently, Zhang and coworkers studied the ability of cobalt-based porphyrin systems to catalyze the intra-molecular amination of electron-deficient C—H bonds, allylic to electron-withdrawing groups.[42] The ability to efficiently produce these aminated products is important because they can serve as α-amino acid derivatives. However, they are often difficult to produce because the hydrogen from the electron-deficient C—H bond is difficult to abstract using an electrophilic metallonitrene radical intermediate. Initially, Co(TPP) was used as the catalyst of choice due to its successful use in catalyzing azide amination reactions. However, Co(P1) was shown to be more efficient. A variety of substrates bearing various electron-donating and electron-withdrawing groups, when treated with Co(P1) in warm benzene, gave high yields of the desired products 320–326. The higher yields observed with Co(P1) are attributed to hydrogen bonding between the catalyst and substrate.

Notably, the authors tested the stereoselectivity of the amination reaction by looking at azide substrates containing not only an α-, but also a β-substituted carbon, with respect to the electron-withdrawing group. Compound 322 demonstrated the preference for electron-deficient C—H bonds, yielding the 1,6-amination products even with electron-rich C—H bonds bonds present. The same can be observed when there is γ-substitution; the electron-poor C—H bond is the preferred amination site, as shown in 323. Further, an example exhibiting β,γ-disubstitution, with a cyclohexane branching from the β and γ carbons, produced 324 in 95% yield. Both *cis-* and *trans-*cyclohexane (not shown) examples were tested. Substrates containing α,α-disubstituted carbons allowed researchers Zhang and coworkers to look at how this reaction faired when provided with a tertiary proton. As product 325 shows, yields remained high. Finally, the authors analyzed the effects of the presence of a nitrogen-based electron-withdrawing group at the 3-position on the ability of the reaction to catalyze the 1,6-amination product. Reaction conditions for this example required 5 mol% of catalyst and temperatures of 80°C for 1 h. Results showed that the product 326 could be formed in moderate yields. However, the reaction was very sluggish.

### 8.3.3  C—B Bond Formation

Recently, Obligacion, Semproni and Chirik demonstrated the first example of a C—H borylation using pincer-ligated cobalt complexes.[43] Treatment of benzofuran 327 or N-methyl indole 330 with 1 mol% of cobalt pincer catalyst 328 in the presence of pinacol borane (HBPin) gave the corresponding borylation products 329 and 331 in near quantitative yields. Good to excellent yields were also obtained for various furan and thiophene derivatives (not shown).

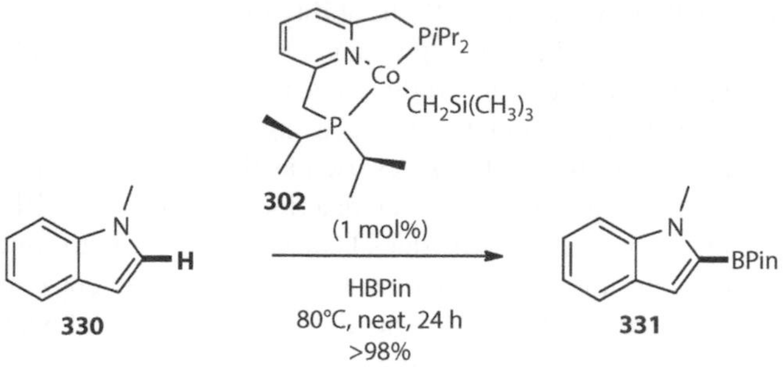

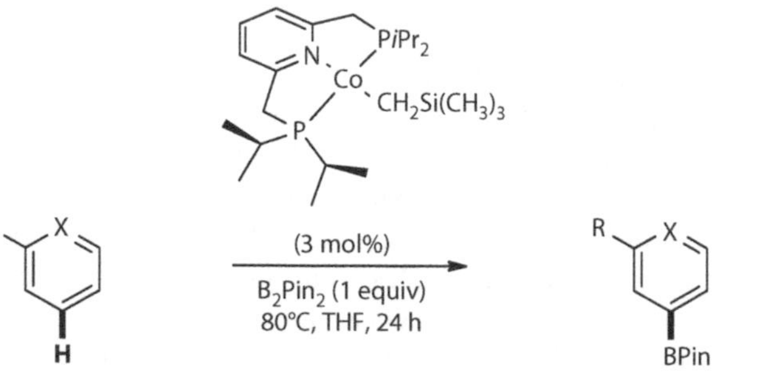

**302**
(1 mol%)

**330**

HBPin
80°C, neat, 24 h
>98%

**331**

(3 mol%)

B₂Pin₂ (1 equiv)
80°C, THF, 24 h

**332,** X = N, R = CH₃
**333,** X = N, R = OCH₃
**334,** X = C, R = CH₃
**335,** X = C, R = OCH₃
**336,** X = C, R = F

**337,** X = N, R = CH₃, 98% 79:12:9 (4:5:6)
**338,** X = N, R = OCH₃, 98% 82:18 (4:5)
**339,** X = C, R = CH₃, 98% 70:30 (m:p)
**340,** X = C, R = OCH₃, 98% 81:10:9 (m:p:o)
**341,** X = C, R = F, 98% 89:11 (o:m)

The authors also investigated the C—H borylation of a series of substituted pyridine **332–333** and arene **334–336** derivatives. Careful examination of the reaction conditions revealed that high yields could be obtained in the presence of 3 mol% catalyst and the diborane B₂Pin₂ (HBPin is formed *in situ*) in 0.55 M THF. In these cases, THF was favored over running the reaction neat since the pyridine and arene derivatives were often solids. Through their studies the authors determined that borylation occurred at the most sterically accessible C—H bonds, with the selectivity arising from the C—H oxidative addition step. In addition, electron-poor arenes borylated much more readily than the electron-rich examples due to the acidity of the C—H bond.

## REFERENCES

1. For a recent review please see: Gao, K.; Yoshikai, N. *Acc. Chem. Res.* **2014**, *47*, 1208–1219.
2. Murahashi, S. *J. Am. Chem. Soc.* **1955**, *77*, 6403–6404.
3. Halbirtter, G.; Knoch, F.; Wolski, A.; Kisch, H. *Angew Chem. Int. Ed.* **1994**, *33*, 1603–1605.
4. Lenges, C. P.; Brookhart, M. *J. Am. Chem. Soc.* **1998**, *120*, 6965–6979.

5. Lenges, C. P.; White, P. S.; Brookhart, M. *J. Am. Chem. Soc.* **1997**, *119*, 3165–3166.
6. Bolig, A. D.; Brookhart, M. *J. Am. Chem. Soc.* **2007**, *129*, 14544–14545.
7. Gao, K.; Lee, P.-S.; Fujita, T.; Yoshikai, N. *J. Am. Chem. Soc.* **2010**, *132*, 12249–12251.
8. Yoshikai, N. *Synlett* **2011**, *132*, 1047–1051.
9. Lee, P.-S.; Fujita, T.; Yoshikai, N. *J. Am. Chem. Soc.* **2011**, *133*, 17283–17295.
10. Yamakawa, T.; Yoshikai, N. *Tetrahedron* **2013**, *69*, 4459–4465.
11. Yamakawa, T.; Yoshikai, N. *Org. Lett.* **2013**, *15*, 196–199.
12. Ding, Z.; Yoshikai, N. *Org. Lett.* **2010**, *12*, 4180–4183.
13. Ding, Z.; Yoshikai, N. *Synthesis* **2011**, 2561–2566.
14. Ding, Z.; Yoshikai, N. *Angew. Chem. Int. Ed.* **2012**, *51*, 4698–4701.
15. Gao, K.; Yoshikai, N. *J. Am. Chem. Soc.* **2011**, *133*, 400–402.
16. Lee, P.-S.; Yoshikai, N. *Angew. Chem. Int. Ed.* **2013**, *53*, 1240–1244.
17. Dong, J.; Lee, P.-S.; Yoshikai, N. *Chem. Lett.* **2013**, *42*, 1140–1142.
18. Gao, K.; Yoshikai, N. *Angew. Chem. Int. Ed.* **2013**, *52*, 8574–8578.
19. Gao, K.; Yoshikai, N. *Angew. Chem. Int. Ed.* **2011**, *50*, 6888–6892.
20. Ding, Z.; Yoshikai, N. *Beilstein. J. Org. Chem.* **2012**, *8*, 1536–1542.
21. Li, L.; Jones, W. D. *J. Am. Chem. Soc.* **2007**, *129*, 10707–10713.
22. Mo, Z.; Liu, Y.; Deng, L. *Angew. Chem. Int. Ed.* **2013**, *52*, 10845–10849.
23. Li, H.; Sun, C-L.; Yu, M.; Yu, D.-G.; Li, B.-J.; Shi, Z.-J. *Chem. Eur. J.* **2011**, *17*, 3593–3597.
24. Chen, Q.; Ilies, L.; Yoshikai, N.; Nakmura, E. *Org. Lett.* **2011**, *13*, 3232–3234.
25. Chen, Q.; Ilies, L.; Nakamura, E. *J. Am. Chem. Soc.* **2011**, *133*, 428–429.
26. Gao, K.; Lee, P.-S.; Long, C.; Yoshikai, N. *Org. Lett.* **2012**, *14*, 4234–4237.
27. Li, B.; Wu, Z.-H.; Gu, Y.-F.; Sun, C. L.; Wang, B.-Q.; Shi, Z.-J. *Angew. Chem. Int. Ed.* **2011**, *50*, 1109–1113.
28. Punji, B.; Song, W.; Shevchenko, G. A.; Ackermann, L. *Chem. Eur. J.* **2013**, *19*, 10605–10610.
29. Song, W.; Ackermann, L. *Angew. Chem. Int. Ed.* **2012**, *51*, 8251–8254.
30. Gal, K.; Yoshikai, N. *Chem. Commun.* **2012**, *48*, 4305–4307.
31. Gao, K.; Paira, R.; Yoshikai, N. *Adv. Synth. Catal.* **2014**, *356*, 1486–1490.
32. Zhao, C.; Toste, F. D.; Bergman, R. G. *J. Am. Chem. Soc.* **2011**, *133*, 10787–10789.
33. Yohsino, T.; Ikemoto, H.; Matsunaga, S.; Kanai, M. *Angew. Chem. Int. Ed.* **2013**, *52*, 2207–2211.
34. Reed, S. A.; White, M. C. *J. Am. Chem. Soc.* **2008**, *130*, 3316–3318.
35. Ye, Y.-H.; Zhang, J.; Wang, G.; Chen, S.-Y.; Yu, X.-Q. *Tetrahedron* **2011**, *67*, 4649–4654.
36. Sun, B.; Yoshino, T.; Matsunaga, S.; Kanai, M. *Adv. Synth. Catal.* **2014**, *356*, 1491–1495.
37. Ruppel, J. V.; Kamble, R. J.; Zhang, X. P. *Org. Lett.* **2007**, *9*, 4889–4892.
38. Ragaini, F.; Penoni, A.; Gallo, E.; Tollari, S.; Gotti, C. L.; Lapadula, M.; Mangioni, E.; Cenini, S. *Chem. Eur. J.* **2003**, *9*, 249–259.
39. Harden, J. D.; Ruppel, J. V.; Guang-Yao, G.; Zhang, X. P. *Chem. Commun.* **2007**, 4644–4646.
40. Lu, H.; Subbarayan, V.; Tao, J.; Zhang, X. P. *Organomettalics* **2010**, *29*, 389–393.
41. Lu, H.; Tao, J.; Jones, J. E.; Wojtas, L.; Zhang, X. P. *Org. Lett.* **2010**, *12*, 1248–1251.
42. Lu, H.; Hu, Y.; Jiang, H.; Wojtas, L.; Zhang, X. P. *Org. Lett.* **2012**, *14*, 5158–5161.
43. Obligacion, J. V.; Semproni, S. P.; Chirik, P. J. *J. Am.Chem. Soc.* **2014**, *136*, 4133–4136.

# 9 Fluorination and Trifluoromethylation of Arenes and Heteroarenes via C—H Activation

*Ji Zhang and Timothy T. Curran*

## CONTENTS

## 9.1 INTRODUCTION

Fluorine-containing molecules are of particular interest in the fields of biomedicine, agriculture, and material sciences.[1] Over the past decade, with the progress of fluorine chemistry research and in-depth understanding of fluorine atoms and fluorine substituent groups,[2] drug discovery scientists are continuing to utilize fluorinated molecules as therapeutic agents (Scheme 9.1).[3] The latest statistics show that the current global annual sales of fluorine-containing drugs at about US$40 billion; among 200 of the world's top-selling drugs, fluorinated drugs accounted for 29, having total sales of $32 billion.[4] It is estimated that 25%–30% of new drug development programs are based on the fluorine chemical raw material. Fluorine remains an important element in blocking metabolic activity, altering chemical properties and has been used as a functional group. According to the literature, about 15%–20% of the new drug candidate compounds contain a fluorine atom or a trifluoromethyl group. Based on our statistics,[5] a total of 163 listed fluorinated drugs will have been approved by the FDA by the end of 2013, indicating that other than the chlorine atom, the fluorine atom is the second most utilized halogen atom selected by pharmaceutical chemists.[6] Drug research which utilizes structure–activity relationship (SAR) methodology in tuning molecules as potential drugs coupled with the ready availability of fluorinated intermediates has provided a huge boost in drug discovery.

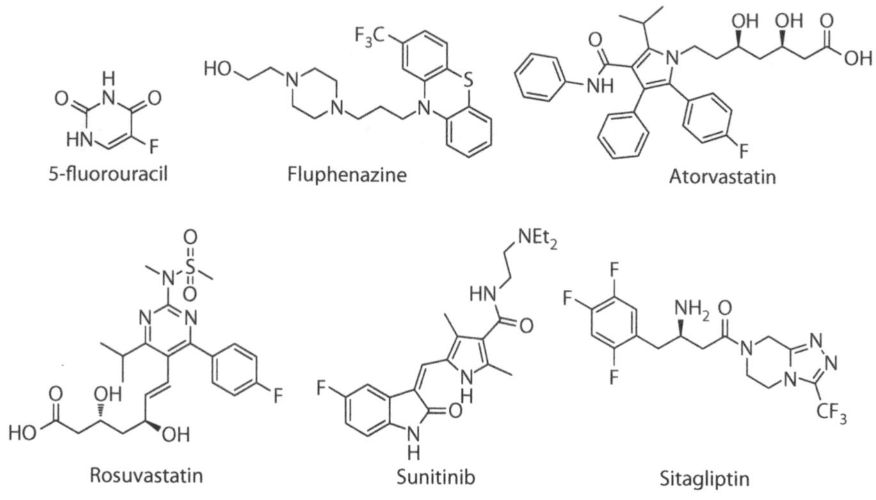

5-fluorouracil            Fluphenazine                              Atorvastatin

Rosuvastatin              Sunitinib                    Sitagliptin

**SCHEME 9.1**    Examples of fluoro- and trifluoromethyl-containing therapeutic agents.

No doubt, in recent years, the progress in organic fluorine chemistry plays an important role and is in part driven by improved methods to produce these molecules. In this chapter, we will discuss some novel methods for fluorination of arenes and heteroarenes via C—H activation.[7]

Unlike other halogens and due to its unique electronic structure, the fluorine atom is the most electronegative atom and mimics the size and bonding capabilities of the hydrogen atom. Thus, from an SAR perspective, a fluorine atom can be more easily substituted for a hydrogen atom and reasonably fine-tune the molecular structure of the drug. Such modifications serve to block sites of drug metabolism and reduce the metabolic rate, increasing the drug half-life in the body. The fluorine atom may also manifest its effect via inter-molecular hydrogen bonding, thereby increasing the bioavailability of drugs and selectivity.[8] Most noteworthy is that the trifluoromethyl group, due to its strong electron-withdrawing capability, its increase in lipophilicity (in comparison to the $CH_3$ group) and stability characteristics, exhibits ideal fat-solubility-enhancing biological permeability and target selectivity. Due to these observations, many drugs and drug candidates often contain the trifluoromethyl group.[9]

In 2011, the USFDA approved a total of 35 small molecule drugs, which contained seven new molecular entities having a fluorine atom (Scheme 9.2). In 2012 there were 33 small molecule drugs approved, of which six were fluorine-containing organic molecules (Scheme 9.3). In 2013, a total of eight fluorinated drugs have been approved (Scheme 9.4). According to incomplete statistics, dozens of fluorinated drugs entered clinical studies, some of which were fluorinated drugs listed in Scheme 9.5. For example, cholesterol transport protein (CEPT) inhibitors anacetrapib from Merck and Lilly's evacetrapib, both contain fluorinated aromatic and/or trifluoromethyl aromatic structural units. Both clinical candidates are expected to be blockbuster drugs! Daiichi reported the efficient squalene synthetase inhibitor,

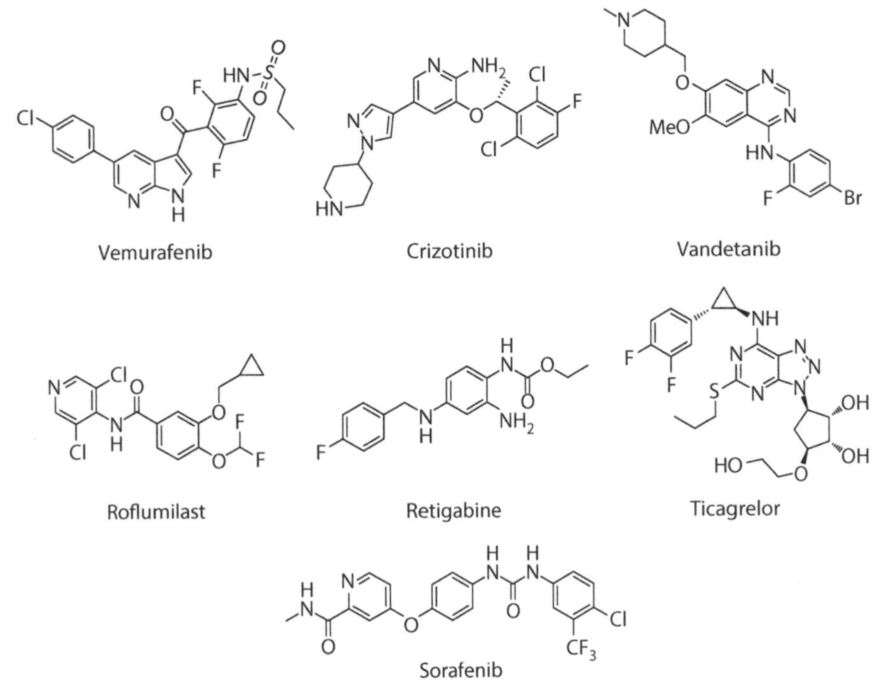

Vemurafenib

Crizotinib

Vandetanib

Roflumilast

Retigabine

Ticagrelor

Sorafenib

**SCHEME 9.2** Examples of fluoro- and trifluoromethyl-containing compounds in development.

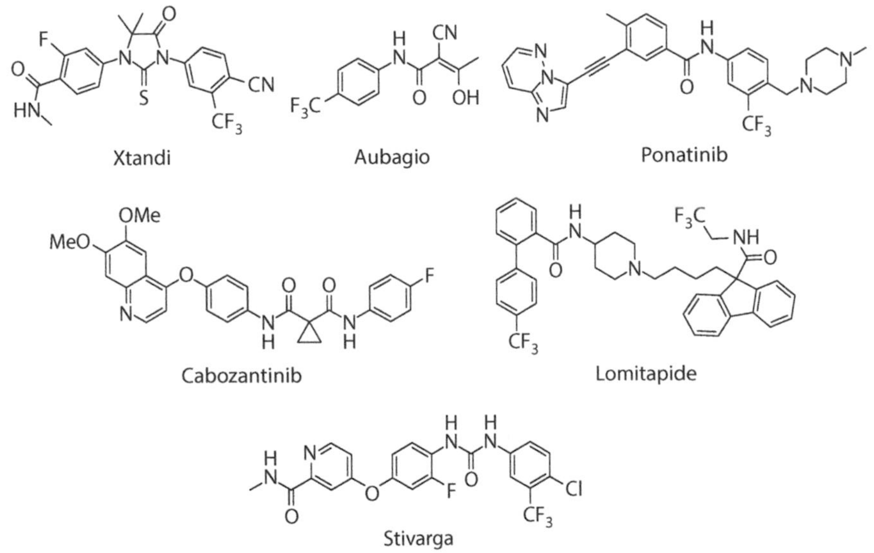

Xtandi

Aubagio

Ponatinib

Cabozantinib

Lomitapide

Stivarga

**SCHEME 9.3** Six molecules approved in 2012 which contain a fluorinated aromatic or trifluoromethyl group.

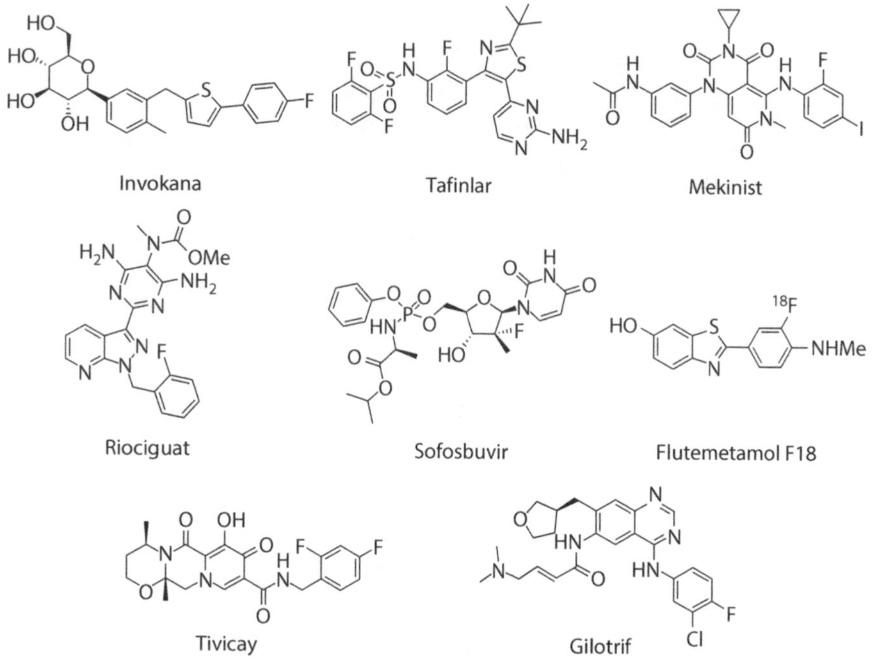

**SCHEME 9.4**   Eight fluorine-containing drugs approved in 2013.

DF-461,[10] which provided better results than the listed HMG-CoA reductase inhibitors. Obviously, the pharmacological benefit of fluorinated compounds intersects with the very exciting development area of organo-fluorine chemistry in which scientific breakthroughs provide a means to accelerate the drug discovery and development process.[11]

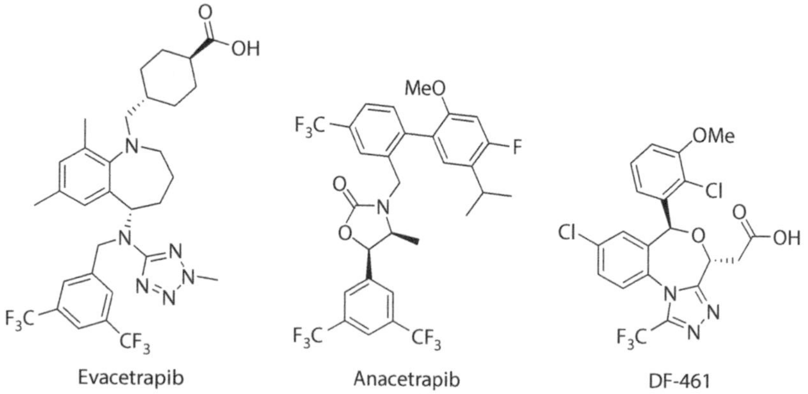

**SCHEME 9.5**   CEPT inhibitors in the clinic.

From the structure of fluorine-containing drugs approved in the last three years, it was found that 12 approved drugs contain fluorinated aromatic hydrocarbons, and six approved drugs contain trifluoromethyl groups. Two drugs have both a fluorine atom and a trifluoromethyl group. Clearly, organic aromatic fluorination and trifluoromethylation reactions are extremely important and useful; therefore, fluorination and trifluoromethylation of arenes and heteroarenes will be analyzed and summarized herein.[12]

## 9.2  HISTORICAL PERSPECTIVES OF FLUORINATION AND TRIFLUOROMETHYLATION

Several new fluorination and trifluoromethylation reactions for facile synthesis of organofluorides have been developed recently.[13] Scheme 9.6 briefly outlines these transformations. However, prior installation of a functional group at the reaction site is typically required. There are only a few reports of direct C—H fluorination or insertion of a trifluoromethyl group into a C—H bond.

Fluorination and trifluoromethylation procedures are conveniently divided into methods that use electrophilic fluorination or trifluoro-methylation and those that proceed by nucleophilic fluorination or trifluoromethylation.[14] Transfer of F⁺ to an electron-rich center is the fundamental process of electrophilic fluorination. However, the species F⁺ is incapable of independent existence, so considerable ingenuity has been required to create F⁺ reagents to allow this process. Some important reagents for fluorination and trifluoromethylation are listed in Schemes 9.7 and 9.8. Excellent reviews provide for a comprehensive discussion.[15] Many of the cited reagents shown

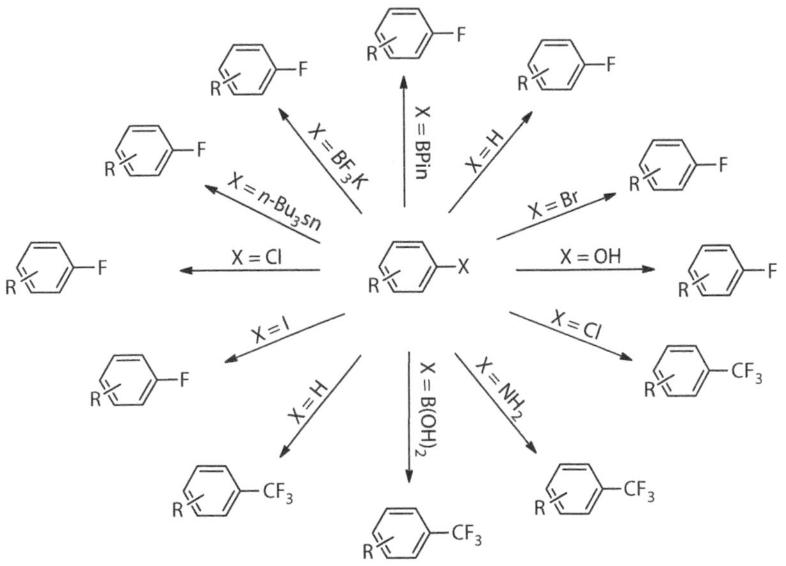

**SCHEME 9.6**  Overview of fluorination and trifluoromethylation of aromatic compounds.

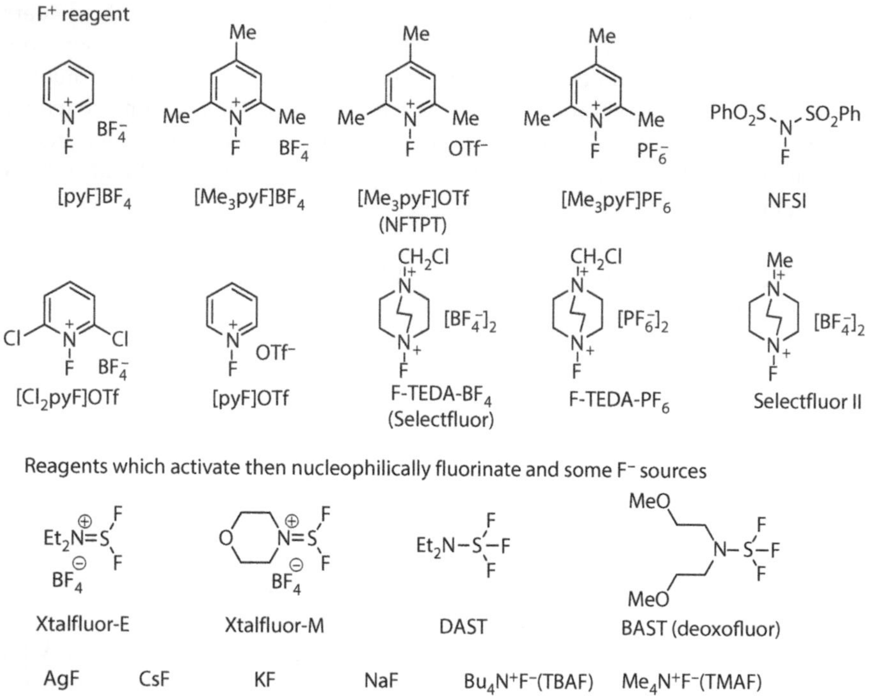

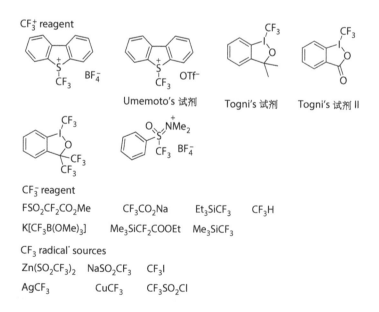

**SCHEME 9.7**  Fluorinating reagents.

**SCHEME 9.8**  Trifluoromethylating reagents.

in Schemes 9.7 and 9.8 have been successfully applied to fluorinate or trifluoromethylate aromatics via C—H activation. Yet the schemes are by no means exhaustive but is hoped to encourage experimentation.

## 9.3  DIRECT FLUORINATION VIA C—H ACTIVATION

Direct C—H bond activation followed by functionalization reactions is limited by two basic challenges: (1) the inert nature of most C—H bonds; and (2) the requirement to control site selectivity in molecules that contain diverse C—H groups. If the lack of reactivity was not a large enough challenge, achieving selective functionalization of a single C—H bond within a complex molecule is a significant test. However, selectivity has been achieved by several different strategies. One strategy is the stoichiometric, ligand-directed C—H activation reaction (also known as cyclo-metalation),[16] whereby appropriate ligands or directing groups bind to the metal and selectively deliver the catalyst to a proximal C—H bond. High selectivity has been achieved in relatively simple systems.

In 2006, Sanford developed the first effective fluorination reaction (Scheme 9.9)[17] for biphenyl derivatives of nitrogen-containing heterocycles such as phenylpyridine, using 10 mol% Pd(OAc)$_2$ in the presence of electrophilic fluorinating agents (oxidants) for the selective introduction of a fluorine atom. This is an example of palladium(II)-catalyzed oxidation under mild conditions using an electrophilic fluorinating reagent via C—H activation/oxidative fluorination.

In 2009, Scripps's Yu and coworkers used a palladium(II) catalyzed *ortho*-directing group-induced C—H activation strategy with F$^+$ reagents in NMP, which led to an efficient preparation of *ortho*-fluorinated aromatic compounds (Scheme 9.10).[18] In this instance, NMP played an important role to promote the reaction.

The use of benzamides as the directing group has been successfully implemented in Yu's group, wherein Pd-catalyzed *ortho* C—H activation and selective

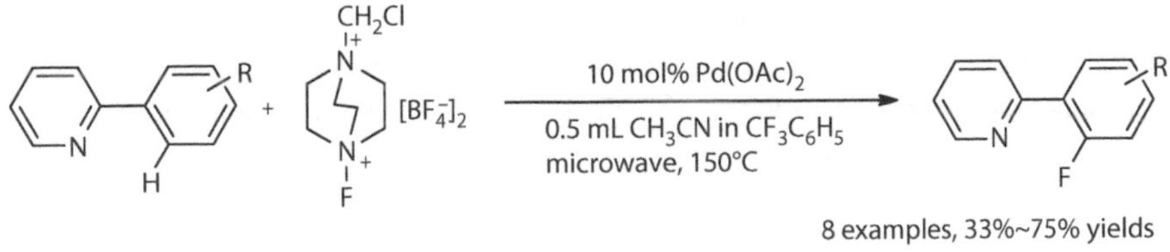

8 examples, 33%~75% yields

**SCHEME 9.9**  C—H activation and fluorination using pyridine as director.

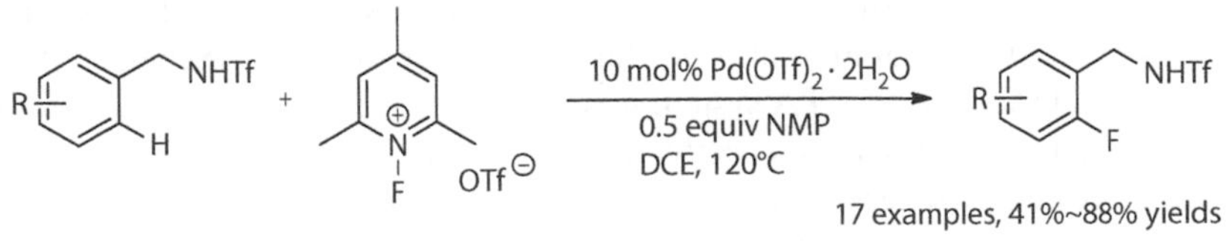

17 examples, 41%~88% yields

**SCHEME 9.10**  C—H activation and fluorination using an amide as director.

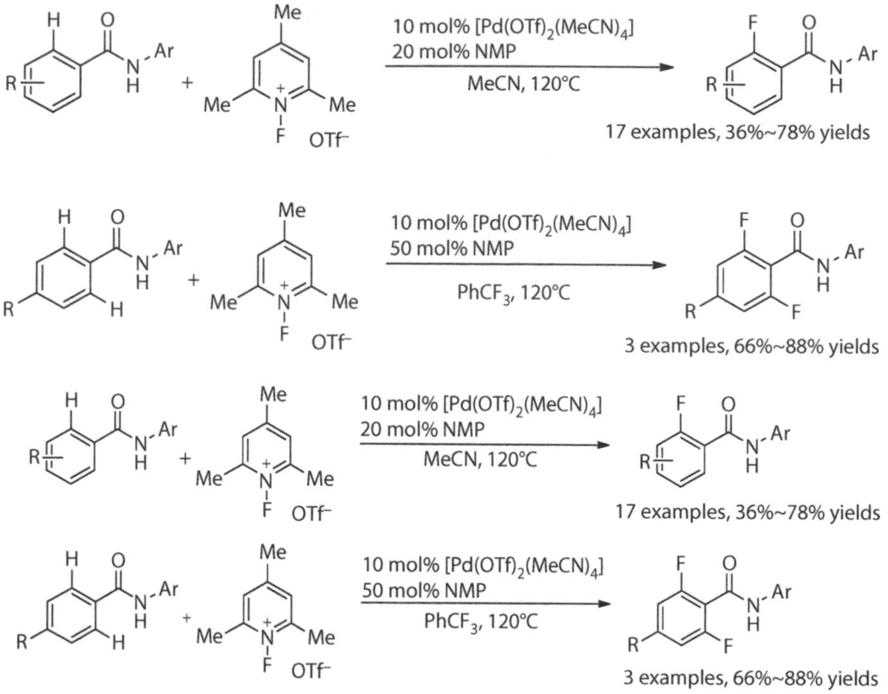

**SCHEME 9.11**  Mono- and difluorination of benzoic amides.

monofluorination of benzoic acids have been successfully exploited. Both mono-fluorinated and difluorinated benzamides were prepared in good conversion and yield (Scheme 9.11).[19]

The above three reactions employed *ortho*-directing groups for assisted C—H activation in the presence of palladium(II) catalysts. The electrophilic fluorination reaction involved cyclopalladation and two electrophilic fluorination processes. Subtly, differential nitrogen atoms (pyridine, substituted amines, and amides) in the molecules were proposed to act as the electron donor, which is required for intramolecular Pd participation in the process. In this case, formation of a five-membered palladium (II) complex (Schemes 9.12 and 9.13) has been proposed.[20]

By using 8-aminoquinoline benzamide substrates and picolinic acid auxiliaries as the directing groups, the Daugulis group reported a novel method for selective mono- or di-fluorination of arene and heteroarene C—H bonds. AgF was used as fluoride source and DMF, pyridine, or DMPU as solvent. This Cu-promoted C—H activation method shows excellent functional group tolerance and provides a straightforward way for the preparation of *ortho*-fluorinated benzoic acids (Scheme 9.14).[21]

A novel Pd(OAc)$_2$–NFSI–TFA system was developed recently by Xu and coworkers demonstrating highly selective, *ortho*-monofluorination. In this case Pd coordination was directed by a diverse range of aryl-*N*-heterocyclic directing groups (Scheme 9.15),[22] such as quinoxaline, pyrazole, benzo[*d*]oxazole, and pyrazine derivatives. A Pd(II/IV) catalytic cycle was proposed.

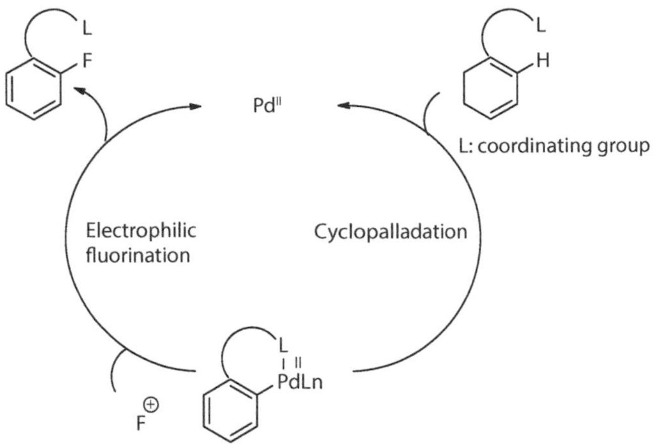

**SCHEME 9.12**  Proposed coordination for C—H activation.

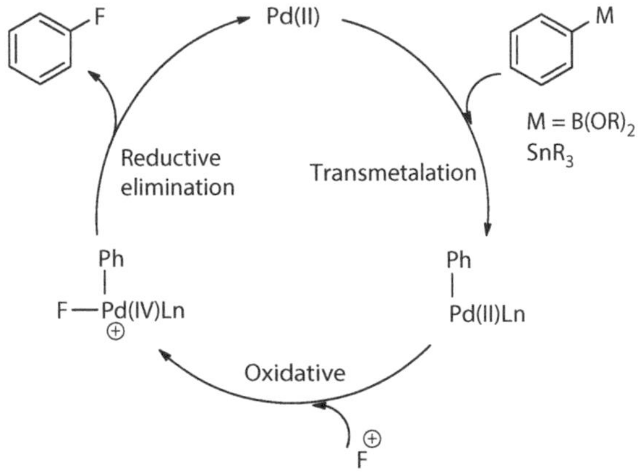

**SCHEME 9.13**  Proposed mechanism for fluorination of an aryl-Pd complex.

## 9.4  DIRECT TRIFLUOROMETHYLATION VIA C—H ACTIVATION

Aromatic nitrogen-containing heterocycles are especially preferred structural units which are often used in drug development. Many drug candidates, containing heterocycles often have poor solubility, offer poor absorption and prove difficult to metabolize. Selectively introducing a trifluoromethyl group as a substituent on the heterocyclic ring can effectively improve the physical and chemical properties of the drug, for example, in the Merck's DPPIV inhibitor, sitagliptin, the trifluoromethyl triazole group is very important.[23]

Using a directing group strategy, Yu reported a new method (Scheme 9.16)[24] for Ar–CF$_3$ bond formation directly via C—H activation, catalyzed by Pd(OAc)$_2$ and with an electrophilic trifluoromethylating reagent and using TFA as a promoter. It was determined that Cu(OAc)$_2$ was effective for enhancing the catalytic turnover.

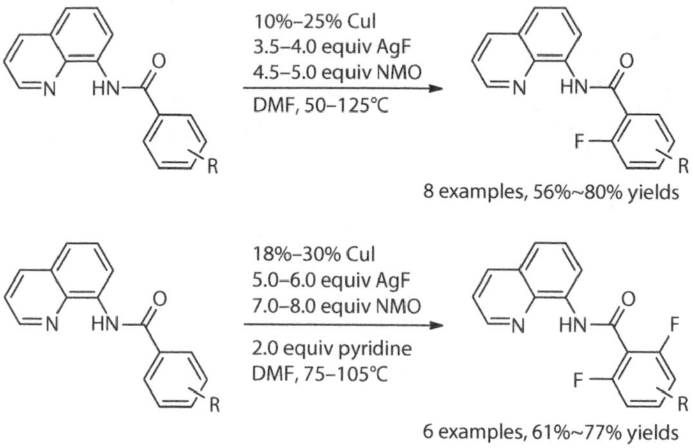

**SCHEME 9.14** Cu promoted mono- and difluorination of benzoic amides.

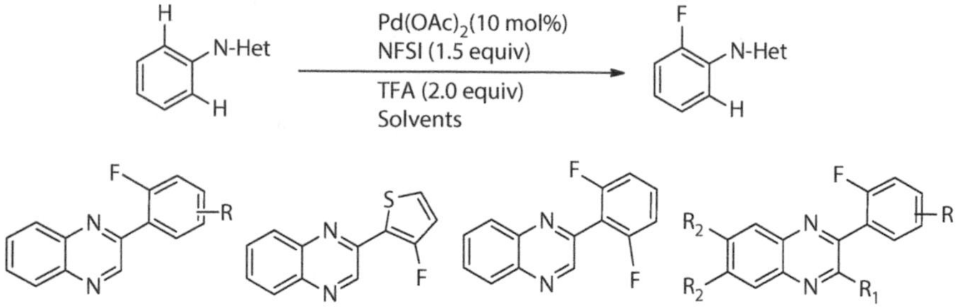

**SCHEME 9.15** *N*-Heterocyclic directing groups for Pd catalyzed fluorination.

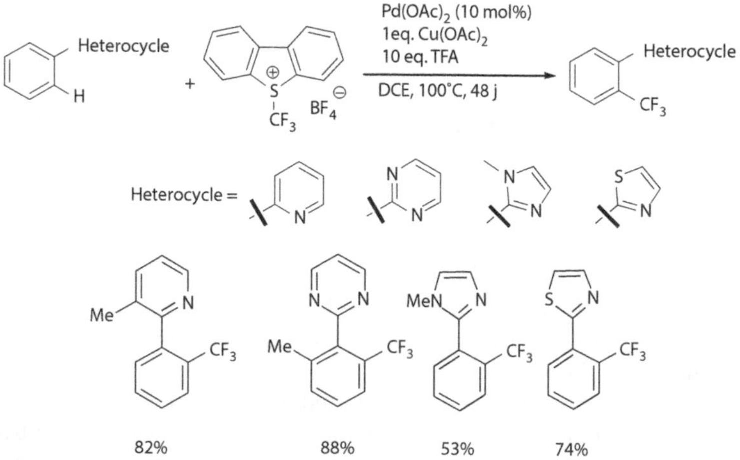

**SCHEME 9.16** *N*-Heterocyclic directing groups for Pd-catalyzed trifluoromethylation.

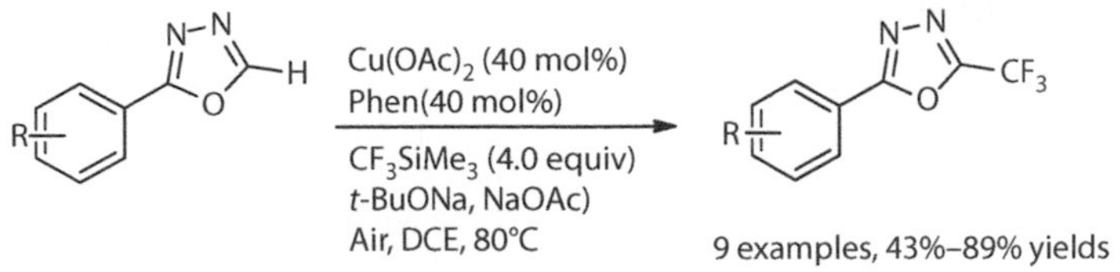

**SCHEME 9.17**    Cu-promoted trifluoromethylation.

By use of direct C—H activation strategy, Qing's group first discovered that copper acetate catalyzed direct C—H oxidative trifluoromethylation of heteroarenes in 2011. Using 1,10-phenanthroline as ligand and $t$-BuONa and NaOAc as cobases, air or $tert$-butyl peroxide as the oxidant, 1,3,4-oxadiazoles or 1,3-azoles reacted smoothly with TMSCF$_3$ providing the trifluoromethylated product (Scheme 9.17).[25] This transformation provided high yield under mild reaction conditions, and could be further exploited as a good method to introduce the trifluoromethyl group into the heterocyclic aromatic compounds in potential, drug candidate molecules.

Qing extended the method described above and successfully substituted a range of benzo-fused heterocyclic systems. Benzoxazole, benzimidazole, benzothiazole, indole and substituted heterocyclic ring systems, and an electron-deficient aromatic system underwent trifluoromethylation achieving fair to good yields (Scheme 9.18).[25]

Shen in 2012 reported an Ir/Cu synergistic catalytic C—H bond activation for the one-pot trifluoromethylation reaction (Scheme 9.19).[26] This catalytic process involves sequential iridium-catalyzed C—H activation borylation and copper-catalyzed trifluoromethylation of arenes with a variety of functional groups. This tandem procedure was successfully applied to the 2,6-disubstituted pyridine, benzofuran, benzothiazole, indole, and substituted quinoline systems.

Trifluoromethylated heteroarenes are widely applicable in the synthesis of pharmaceuticals. Using 10 mol% palladium acetate as catalyst, TMSCF$_3$ as trifluoromethylating reagent, bidentate nitrogen-containing ligand L, and PhI(OAc)$_2$ as oxidant, Liu and coworkers developed a novel palladium catalyzed oxidative trifluoromethylation of indoles at room temperature (Scheme 9.20).[27] This reaction likely involves a palladium (II/IV) mechanism for the formation of the sp$^2$ C—CF$_3$ bond.

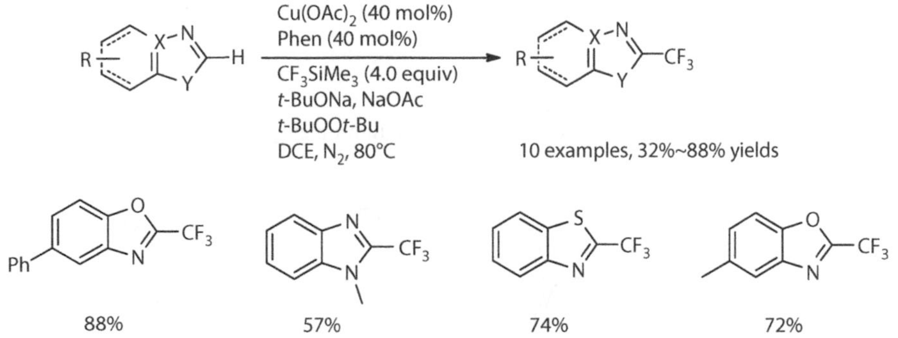

**SCHEME 9.18**    C—H activation and trifluoromethylation of electron deficient heterocycles.

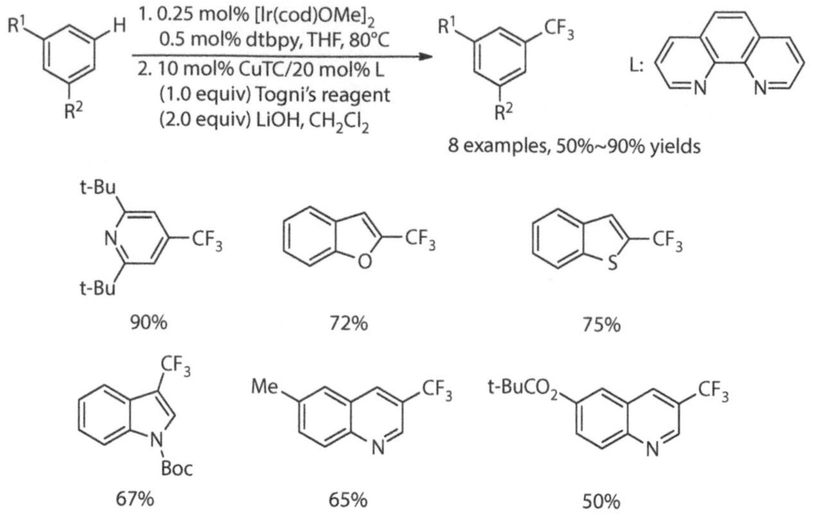

**SCHEME 9.19**  Ir—C—H activation with Cu promoted trifluoromethylation.

A novel methyltrioxorhenium (MTO) catalyzed methodology for the direct trifluoromethylation of both activated and nonactivated arenes and heteroarenes using an easily accessible, shelf-stable hypervalent iodine trifluoromethylating reagent has been developed by Togni and coworkers (Scheme 9.21).[28] This reagent system exhibited a broad substrate scope in direct aromatic trifluoromethylation.

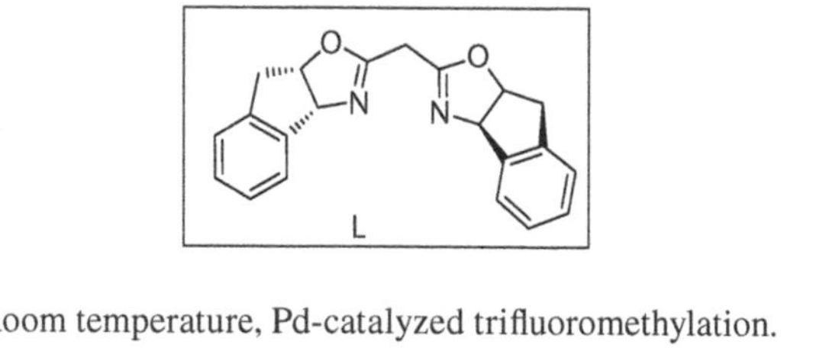

**SCHEME 9.20**  Room temperature, Pd-catalyzed trifluoromethylation.

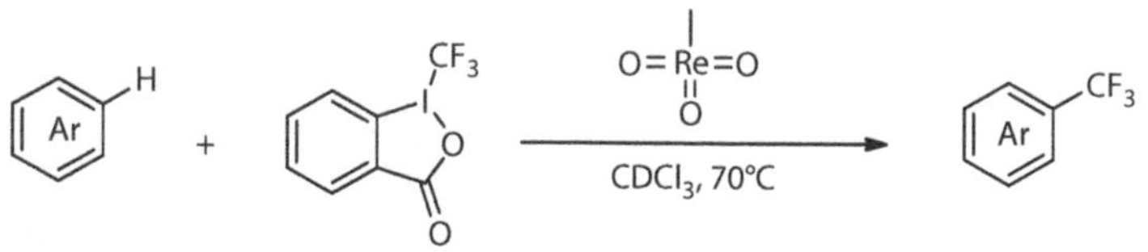

**SCHEME 9.21**  Re-promoted trifluoromethylation.

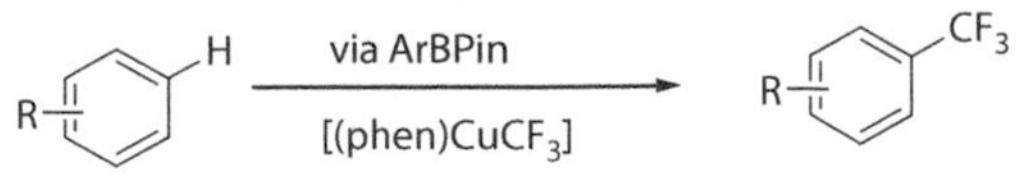

**SCHEME 9.22**   One-pot Cu promoted trifluoromethylation.

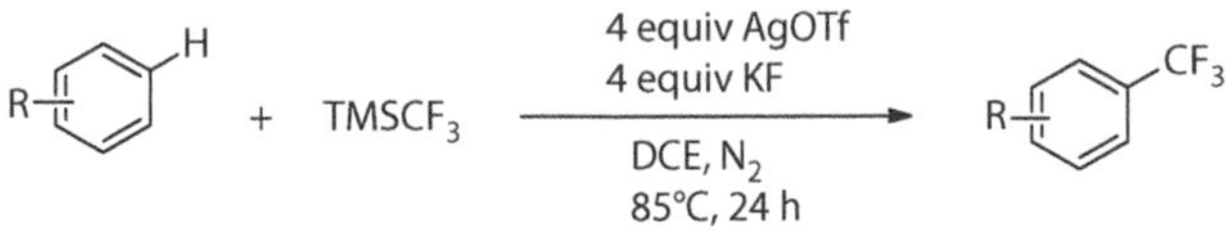

**SCHEME 9.23**   Silver promoted trifluoromethylation.

Ir-catalyzed C—H borylation developed by Hartwig can be used in tandem with reactions of the resulting ArBPin as a two-step, one-pot route to the fluorination of arenes (Scheme 9.22).[29] An aryl boronate ester obtained *in situ* from substituted arenes was shown to readily undergo trifluoromethylation using [(phen)CuCF$_3$] in the presence of air. While the inherent selectivity is derived from the selectivity in the borylation, reagent compatability allows an expedient access to trifluoromethylated arenes.

Using TMSCF$_3$/AgOTf/KF, Sanford developed a new method for direct trifluoromethylation of arenes (Scheme 9.23).[30] The reaction is proposed to proceed via a AgCF$_3$ intermediate, and preliminary studies suggest against free CF$_3^-$ as an intermediate. Importantly, the Ag-mediated reactions proceed with complementary reactivity to analogous transformations of CuCF$_3$ reagents.

## 9.5   C—H ACTIVATION FOR FLUORINATION AND TRIFLUOROMETHYLATION VIA RADICAL PATHWAY

In 2011, Baran developed a new method for innate C—H trifluoromethylation of heterocycles (Scheme 9.24).[31] Using a bench-top stable trifluoromethyl radical source, the Langlois reagent (CF$_3$SO$_2^-$Na$^+$), reacts broadly on a variety of electron deficient and rich heteroaromatic systems. It demonstrates high functional group tolerance. This C—H trifluoromethylation protocol is operationally simple, scalable, proceeds at ambient temperature, and can be used directly on unprotected molecules. Finally, the regioselectivity of C—H trifluoromethylation can be fine-tuned simply by judicious solvent choice.

Photoredox catalysis with a redox-active ruthenium (II) catalyst and CF$_3$SO$_2$Cl developed by MacMillan has been used to generate the electrophilic CF$_3^\bullet$ and leads to the formation of trifluoromethylated products of heteroarenes (Scheme 9.25).[32]

In 2012, Baran reported a new reagent TFMS (Zn(SO$_2$CF$_3$)$_2$) for direct trifluoromethylation of heterocycles (Scheme 9.26).[33] This transformation has several

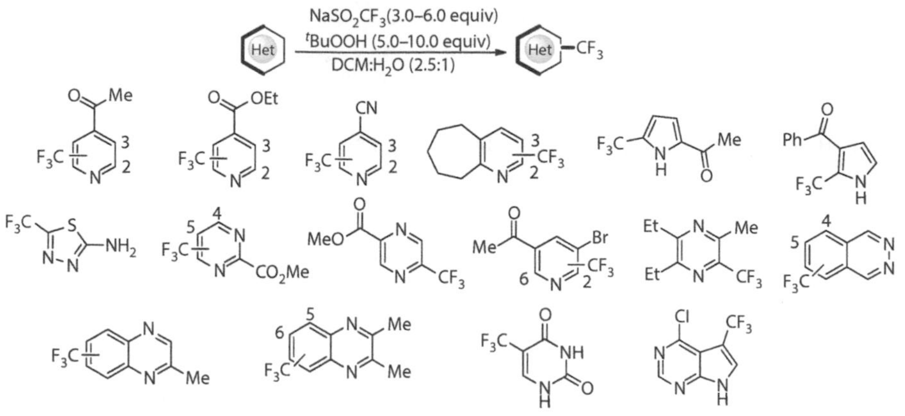

**SCHEME 9.24**   Trifluoromethylation using Langlois' reagent.

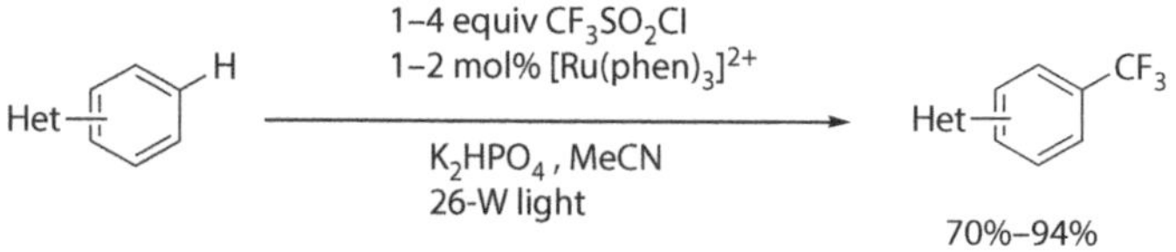

**SCHEME 9.25**   Ru-promoted trifluoromethylation.

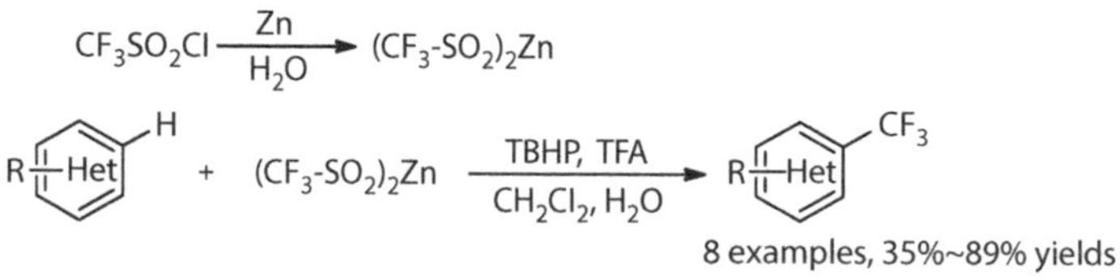

**SCHEME 9.26**   Trifluoromethylation with Zn(OTf)$_2$.

advantages: (1) displays high regioselectivity, (2) proceeds at ambient temperature, (3) runs in aqueous conditions in the presence of air, (4) proves compatible with halogen functionalities, and (5) both electron-rich and electron-deficient heterocycles work well.

Chen recently reported using visible light to promote metal-free C—H activation for selective benzylic difluorination (Scheme 9.27).[34] C—H difluorination is a challenging and underdeveloped area of research. By screening the catalyst and fluorine donor, it was found that xanthone is a better radical generating catalyst and Selectfluor II yields a smooth difluorination. The scope and efficiency of this new C—H activation/fluorination method are significantly better than those of the existing methods.

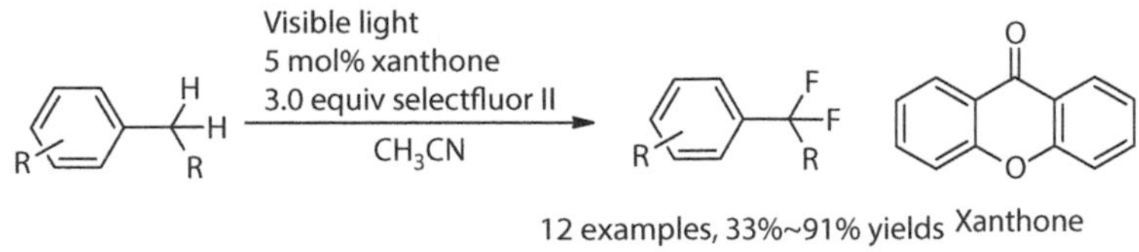

12 examples, 33%~91% yields  Xanthone

**SCHEME 9.27**  Benzylic difluorination.

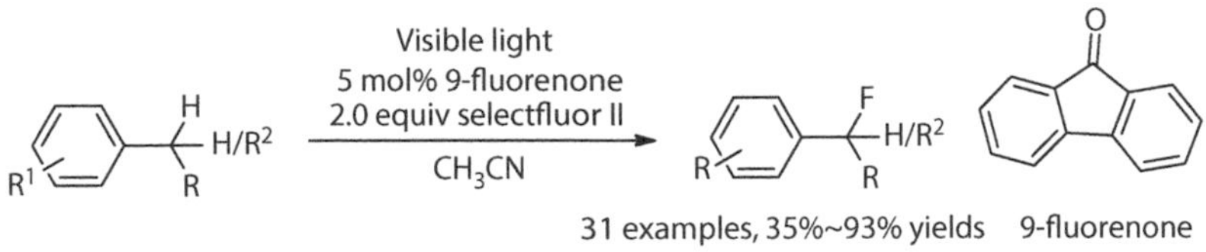

31 examples, 35%~93% yields   9-fluorenone

**SCHEME 9.28**  Benzylic fluorination.

10 examples, 30%~70% yields

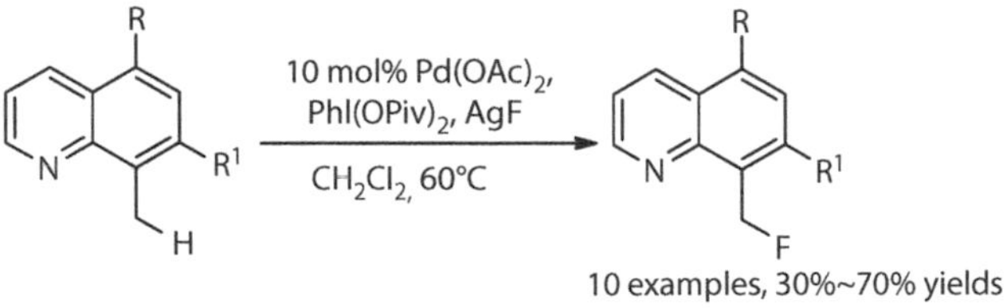

**SCHEME 9.29**  C—H activation benzylic fluorination.

Using 9-fluorenone as the radical initiator for the generation of benzylic radical, monofluorination was found occur with both electron-rich and electron deficient substrates providing the desired product in good-to-excellent yield (Scheme 9.28).[34]

Sanford also found the 8-methylquinoline will undergo facile quinoline-directed C—H activation at Pd(II) to generate a δ-benzyl Pd species and to serve as an excellent substrate for the related Pd-catalyzed C—H activation/oxidative functionalization reactions (Scheme 9.29).[17]

## REFERENCES

1. (a) Ojima, I. *Fluorine in Medicinal Chemistry and Chemical Biology.* Wiley: Hoboken, NJ, **2009.** (b) *Fluorinated Heterocyclic Compounds: Synthesis, Chemistry, and Applications.* Wiley: Hoboken, NJ, **2009.**
2. Roesky, H. W.; Sharpless, K. B. *Efficient Preparation of Fluorine Compounds.* Wiley: Hoboken, NJ, **2012.**
3. Ilardi, E. A.; Vitaku, E.; Njardarson, J. T. *J. Med. Chem.* **2014**, *57*, 2832.
4. Wang, J.; Sanchez, M.; Acena, J.; Pozo, C.; Sorochinsky, A.; Fustero, S.; Soloshonok, V.; Liu, H. *Chem. Rev.* **2014**, *114*, 2432.
5. Zhang, J.; Jin, C.-F.; Zhang, Y.-J. *Chin. J. Org. Chem.* **2014**, *34*, 662.
6. (a) Lu, H.; Silverman, R. B. *J. Med. Chem.* **2006**, 49, 7404. (b) Ojima, I. *J. Org. Chem.* **2013**, *78*, 6358.

7. For reviews on C—H activation, see (a) Kalyani, D.; Sanford, M. S. In *Topic in Organometallic Chemistry: Directed Metalation*; Chatani, N., Ed.; Springer: Berlin, Heidelberg, New York, 2007; Vol. 24, pp 85–116. (b) Bergman, R. G. *Nature* **2007**, 466, 391. (c) Hartwig, J. F. *Nature* **2008**, *455*, 314.

8. Hagmann, W. K. *J. Med. Chem.* **2008**, 51, 4359.

9. Meanwell, N. A. *J. Med. Chem.* **2011**, *54*, 2529.

10. Ichikawa, M.; Ohtsuka, M.; Ohki, H.; Ota, M.; Haginoya, N.; Itoh, M.; Shibata, Y. et al. *ACS Med. Chem. Lett.* **2013**, *4*, 932.

11. Kirsch, P. *Modern Fluoroorganic Chemistry: Synthesis, Reactivity, Applications*, 2nd, Completely Revised and Enlarged Edition. Wiley-VCH Verlag GmbH, Weinheim, Germany, **2013**.

12. (a) Krik, K. L. *Org. Process Res. Dev.* **2008**, *12*, 305 (b) Tomashenko, O. A.; Grushin, V. V. *Chem. Rev.* **2011**, *111*, 4475.

13. (a) Ye, Y.; Sanford, M. S. *J. Am. Chem. Soc.* **2013**, *135*, 4648. (b) Fier, P. S.; Luo, J.; Hartwig, J. F. *J. Am. Chem. Soc.* **2013**, *135*, 2559. (c) Furuya, T.; Ritter, T. *J. Am. Chem. Soc.* **2008**, *130*, 10060. (d) Mazzotti, A. R.; Campbell, M. G.; Tang, P.; Murphy, J. M.; Ritter, T. *J. Am. Chem. Soc.* **2013**, *135*, 14012.(e) Ye, Y.; Schimler, S. D.; Hanley, P. S.; Sanford, M. S. *J. Am. Chem. Soc.* **2013**, *135*, 16292. (f) Cazorla, C.; Metay, E.; Andrioletti, B.; Lemaire, M. *Tetrahedron Lett.* **2009**, *50*, 3936. (g) Dai, J. J.; Fang, C.; Xiao, B.; Yi, J.; Xu, J.; Liu, Z. J.; Lu, X.;Liu, L.; Fu. Y. *J. Am. Chem. Soc.* **2013**, *135*, 8436. (h) Danoun, G.; Bayarmagnai, B.; Grunberg, M. F.; Gooβen, L. J. *Angew. Chem. Int. Ed.* **2013**, *52*, 7972. (i) Wang, X.; Xu, Y.; Mo, F.; Ji, G.; Qiu, D.; Feng, J.; Ye, Y.; Zhang, S.; Zhang, Y.; Wang, J. *J. Am. Chem. Soc.* **2013**, *135*, 10330.

14. Taylor, S. D.; Kotoris, C. C.; Hum, G. *Tetrahedron* **1999**, *55*, 12431.

15. (a) Liang, T.; Neumann, C. N.; Ritter, T. *Angew. Chem. Int. Ed.* **2013**, *52*, 8214. (b) Studer, A. *Angew. Chem. Int. Ed.* **2012**, *51*, 8950. (c) Grushin, V. V. *Acc. Chem. Res.* **2010**, *43*, 160.

16. Dupont, J.; Consorti, C. S.; Spencer, J. *Chem. Rev.* **2005**, *105*, 2527.

17. Hull, K. L.; Anani, W. Q.; Sanford, M. S. *J. Am. Chem. Soc.* **2006**, *128*, 7134.

18. Wang, X.; Mei, T. S.; Yu, J. Q. *J. Am. Chem. Soc.* **2009**, *131*, 7520.

19. Chan, K. S. L.; Wasa, M.; Wang, X.; Yu, J. Q. *Angew. Chem. Int. Ed.* **2011**, *50*, 9081.

20. Lyons, T. W.; Sanford, M. S. *Chem. Rev.* **2010**, *110*, 1147.

21. Truong, T.; Klimovica, K.; Daugulis, O. *J. Am. Chem. Soc.* **2013**, *135*, 9342.

22. Lou, S-J.; Xu, D-Q.; Xia, A-B.; Wang, Y-F.; Liu, Y-K.; Du, X-H.; Xu, Z-Y. *Chem. Commun.* **2013**, *49*, 6218.

23. Hansen, K. B.; Hsiao, Y.; Xu, F.; Rivera, N.; Clausen, A.; Kubryk, M.; Krska, S. et al. *J. Am. Chem. Soc.* **2009**, *131*, 8798.

24. Wang, X.; Truesdale, L.; Yu, J-Q. *J. Am. Chem. Soc.*, **2010**, *132*, 3648.

25. Chu, L.; Qing, F. L. *J. Am. Chem. Soc.* **2012**, *134*, 1298.

26. Liu, T.; Shao, X.; Wu, Y.; Shen, Q. *Angew. Chem. Int. Ed.* **2012**, *51*, 540.

27. Mu, X.; Chen, S.; Zhen, X.; Liu, G. *Chem. Eur. J.* **2011**, *17*, 6039.

28. Mejia, E.; Togni, A. *ACS Catal.* **2012**, *2*, 521.

29. Litvinas, N. D.; Fier, P. S.; Hartwig, J. F. *Angew. Chem. Int. Ed.* **2012**, 51, 536.

30. Ye, Y.; Lee, S. H.; Sanford, M. S. *Org. Lett.* **2011**, *13*, 5464.

31. Ji, Y.; Brueckl, T.; Baxter, R. D.; Fujiwara, Y.; Seiple, I. B.; Su, S.; Blackmond, D. G.; Baran, P. S. *PANS.* **2011**, *108*, 14411.

32. Nagib, D. A.; Macmillan, D. W. C. *Nature* **2011**, *480*, 224.

33. Fujiwara, Y.; Dixon, J. A.; Hara, F.; Funder, E. D.; Dixon, D. D.; Rodrigue, R. A.; Baxter, R. D. et al. *Nature* **2012**, *492*, 95.

34. Xia, J. B.; Zhu, C.; Chen, C. *J. Am. Chem. Soc.* **2013**, *135*, 17494.

# 10 C−H Activation of Heteroaromatics

*Donna A. A. Wilton*

## CONTENTS

## 10.1 INTRODUCTION

Heterocycles are ubiquitous fragments that feature heavily in naturally occurring molecular scaffolds and are of vital interest to the pharmaceutical, agrochemical, and electronic industries.[1] These key building blocks can be found in most small molecule drugs on the market today.[2]

The interactions of aromatic heterocycles with transition metal complexes perturb the molecular orbitals of the heterocycle, resulting in a modified reactivity profile relative to the isolated heterocycle. Frontier molecular orbital theory at the interface of organic and inorganic chemistry has been an indispensable tool in rationalizing feasible reaction pathways for these extended systems and for explaining regiochemical reactivity patterns. This rapidly-growing body of mechanistic knowledge has enabled the rational design of new catalysts which, combined with serendipitous discoveries, has made possible what is now a broad repertoire of valuable transformations. Therefore, the controlled and selective activation of heterocyclic carbon-hydrogen (C—H) bonds by transition metals, followed by mild functionalization, serves as a formidable tool for the synthesis of differentially-substituted heteroaromatic compounds.

In contrast to traditional metal-catalyzed cross-coupling reactions of organometallic nucleophiles with organic electrophiles, C—H activation methods avoid the need for the "pre-activation" of one or both coupling components; not only does this minimize the amount of wasteful by-products generated,[3] but it also dramatically simplifies a synthetic route. In addition, C—H activation methods are often milder than conventional methods, and can overcome the inherent reactivity patterns of the heteroaromatic system and allow access to products that could not otherwise be obtained.

Most of the mechanisms for heteroaromatic C—H bond activation by a transition-metal catalyst fall into one of these four categories: (i) electrophilic aromatic metalation, (ii) carboxylate-ligand-promoted concerted metalation–deprotonation (CMD), (iii) base-assisted metalation and (iv) oxidative addition of C—H to the metal center. The type of mechanism operating in the cleavage of the C—H bond depends on the electronics of the heterocycle (and therefore its substituents) and the reaction conditions being employed.

The field of heteroaromatic C—H activation research is rapidly expanding and new developments appear in the literature on a daily basis. This chapter is divided into sections by heterocycle and aims to mainly cover examples of C—H activation via reactions involving metal insertions into C—H bonds; the description of methods in which a directing group has been used to facilitate C—H bond activation has been kept to a minimum. In order to avoid overlapping with previous publications, and also in consideration of space limitation, this chapter is restricted to the most recent and key developments in this field and particular attention has been paid to unusual reactivity.

## 10.2   HETEROAROMATICS CONTAINING ONE HETEROATOM

### 10.2.1   PYRROLE

Reactivity patterns of pyrrole carbons are determined by delocalization of the nitrogen lone pair. The electron-rich nature of this heterocycle means that reactions with electrophiles are facile and preferentially occur at the C2 (or C5) position unless it is already substituted. This innate reactivity means that C—H activation by metal insertion preferentially occurs at the position adjacent to the heteroatom.

The direct phenylation of 1$H$-pyrrole (**1**) using a variety of metal catalysts has been reported and key examples will be discussed herein (Scheme 10.1). Sames et al. employed a rhodium catalyst to obtain 2-phenyl pyrrole (**2**) in 78% yield.[4] The choice of CsOPiv proved to be crucial in this site-selective transformation as the carbonates and phosphates of alkali metals were ineffective. The authors believe that the mildly basic pivalate ligand catalyzes the rhodium-mediated C—H activation via a CMD mechanism (transition state **3**). A similar carboxylate-assisted mechanism was

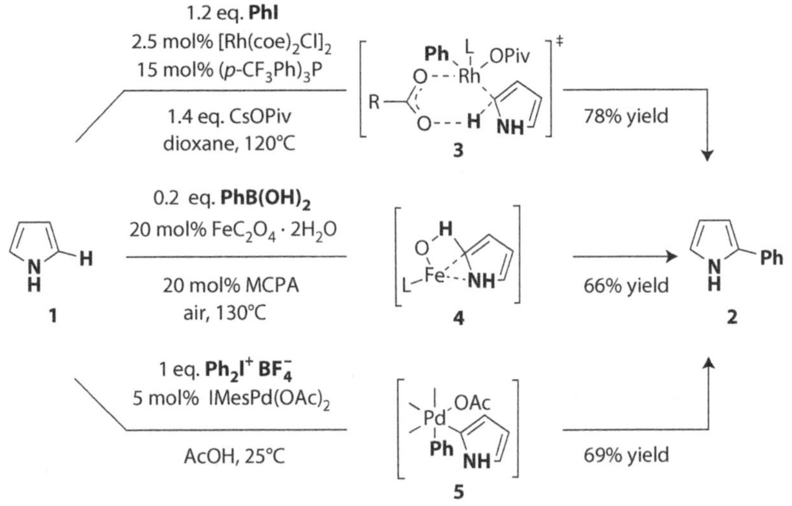

**SCHEME 10.1**  Methods for C2-phenylation of 1*H*-pyrrole.

proposed for pyrrole C2-arylation with a ruthenium(I) catalyst and a bulky carbox-ylic acid cocatalyst; however, this method required a 2-pyridyl substituent directing group at the pyrrole C3 position in order to facilitate the C—H activation reaction.[5]

Hu and Yu disclosed an iron-mediated direct Suzuki–Miyaura coupling[6] between 1*H*-pyrrole (**1**) and arylboronic acids.[7] Macrocyclic polyamines (MCPA) were studied in this process; under the reaction conditions, a tertiary pyrrole–MCPA–oxoiron complex **4** is formed which undergoes transmetallation with the arylboronic acid. After elimination of the iron complex, 2-phenyl pyrrole (**2**) was isolated in 66% yield with only trace amounts of phenol.

The third example from Scheme 10.1 illustrates methodology from the Sanford laboratory whereby a palladium(II) catalyst was used to oxidatively phenylate 1*H*-pyrrole (**1**) with Ph$_2$I$^+$BF$_4^-$ at room temperature.[8] The iodine(III) arylating reagent could either be synthesized independently in 1–2 steps, or generated *in situ* via the reaction of ArI(OAc)$_2$ with the corresponding arylboronic acid. A thorough mechanistic investigation led the authors to suggest that a key catalytic intermediate in the reaction mixture was an unusual palladium(II)/palladium(IV) ("palladium(III)") dimer.[9] Sanford later used these hypervalent iodine arylating reagents in a palladium-catalyzed monoarylation and sequential diarylation of 2,5-substituted pyrroles,[10] thereby offering an alternative approach for accessing densely-substituted pyrroles than the more common Knorr, Paal–Knorr and Hantzch cyclization reactions.[11] Although not depicted here, other palladium-catalyst systems have been used to effect this transformation including Jafarpour's triethanolamine-mediated Pd(OH)$_2$/C catalyst system[12] and Shi's cross-coupling with arylboronic acids[13] where 2-phenyl pyrrole (**2**) was isolated in 58% and 56% yields, respectively.

One simple way to prevent preferred C2-substitution is to override the natural reactivity of pyrrole by appending a bulky substituent on the nitrogen atom. Use of the sterically demanding triisopropylsilyl (TIPS) protecting group biases C—H

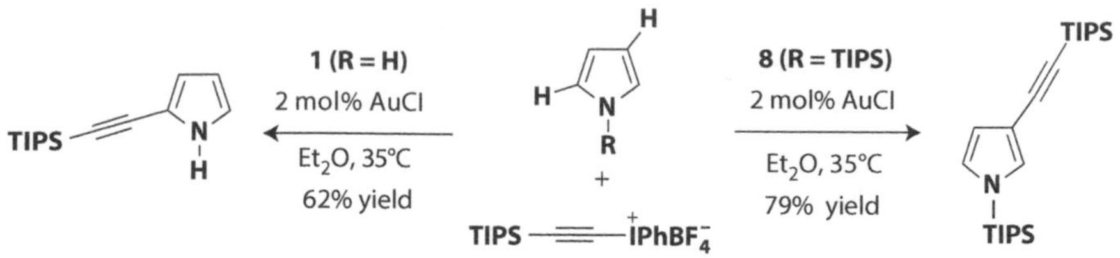

**SCHEME 10.2** Waser's intermolecular regioselective alkynylation of pyrroles.

activation toward the C3 position as the C2 position is effectively blocked from reaction. The *tert*-butyloxycarbonyl (Boc) protecting group has also been shown to guide substitution to the C3 position but this depends on the reaction conditions being employed. The ease with which these protecting groups can be removed contributes to the attractiveness of employing such methods for regioselective control.

Pyrroles almost exclusively add to acetylenes as *N*-centered nucleophiles; however, when the acetylene is activated by a strong electron-withdrawing substituent, pyrrole can add to the triple bond as a *C*-centered nucleophile.[14] In their gold-catalyzed alkynylation of pyrrole with silyl-protected alkynyl benziodoxolone hypervalent iodine reagent **6**,[15] members of the Waser laboratory only saw *C*-alkynylated products (Scheme 10.2). Using this method, 1*H*-pyrrole (**1**) could be C2-funtionalized to afford **7** in 62% yield, whereas *N*-TIPS pyrrole (**8**) was selectively transformed into C3-product **9** in 79% yield. In both cases, auration of the heterocycle led to an organogold(III) intermediate that reductively eliminated to afford the alkynylated products.

Nevado and de Haro later reported a gold-catalyzed arene ethynylation reaction in which both the *sp* C—H and *sp*$^2$ C—H were activated by the gold catalyst.[16] Heterocycles such as *N*-benzyl pyrrole and *N*-benzyl indole were suitable substrates for this reaction but in the former case, ethynylated regioisomers were obtained.

A protecting group strategy was employed by Gaunt and co-workers to control the regioselectivity of oxidative palladium(II)-catalyzed pyrrole alkenylation (Scheme 10.3).[17] When pyrrole was substituted with a Boc group (**10**), its natural reactivity was promoted so that palladation and subsequent Heck reaction with ethyl acrylate

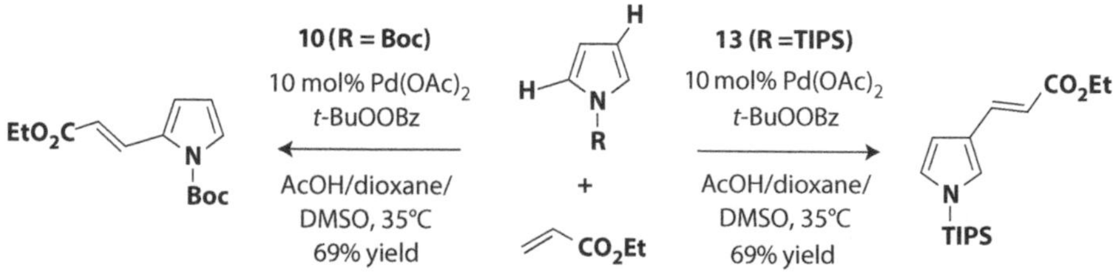

**SCHEME 10.3** Gaunt's intermolecular regioselective alkenylation of pyrroles.

(**11**) occurred at the C2 position to afford product **12** in 69% yield. However, when the *N*-protecting group was switched to TIPS (**13**), palladation took place at the C3 position, leading to formation of pyrrole **14** in 69% yield.

Arrayáyas and Carretero employed a *N*-(2-pyridyl)sulfonyl protecting group to assist in their palladium(II)-catalyzed C2 alkenylation of pyrrole,[18] whereas the research groups of Li and Wang used a *N*,*N*-dimethylcarbamoyl group to affect alkenylation of pyrrole at the C2 position under rhodium-catalyzed conditions.[19]

Stoltz and Ferreira's oxidative indole annulation methodology[20] inspired Oestreich et al. to develop an enantioselective version of the Fujiwara–Moritani reaction (Scheme 10.4).[21] In the presence of Pd(OAc)$_2$ and *tert*-butyl peroxybenzoate as the oxidant, novel NicOx ligand **17** was found to effect 5-*exo-trig* cyclizations on homoallyl pyrroles **15–16** to afford annulation products **19–20** in moderate yields (b.r.s.m) and enantioselectivies; the *Z/E* ratio of the starting material double-bond had a large impact on the stereochemical outcome of the reaction. The ring-closure mechanism is expected to proceed through an electrophilic palladation pathway (**18**).

The direct coupling of 1*H*-pyrrole (**1**) with unactivated alkyl halides, at carbon, is extremely useful from a synthetic standpoint but has had only limited success thus far. The major challenge in the development of such reactions is that non-activated alkyl halides are resistant to oxidative addition; in cases where they do undergo this process, the metal alkyl intermediates are prone to undergoing undesired β-H elimination reactions.[22]

In the late 1990s, Catellani et al. reported that norbornene-mediated palladium-catalyzed domino reactions resulted in *ortho* C—H functionalization.[23] Bach and Jiao later observed that pyrrole-2-carboxylates **21** underwent alkylation at the C5 position with alkyl bromides when a similar palladium–norbornene catalytic manifold was employed (via intermediate **22**, Scheme 10.5).[24] Alkylation at the C5 position is of note as these pyrrole derivatives normally undergo electrophilic substitution at the C4 position due to the directing effect of the electron-withdrawing ester group.[25] The reaction was tolerant of many different functional groups on the primary alkyl bromide, affording the 2,5-disubstituted pyrrole derivatives **23A–D** in good to high yields. 2,3-Disubstituted pyrroles also underwent smooth alkylation reaction to provide access to multi-functionalized 2,3,5-pyrrole derivatives.

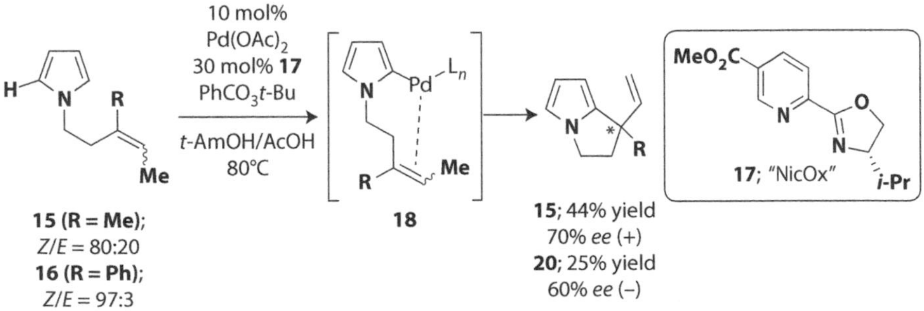

**15 (R = Me);**
*Z/E* = 80:20
**16 (R = Ph);**
*Z/E* = 97:3

**18**

**15**; 44% yield
70% *ee* (+)
**20**; 25% yield
60% *ee* (−)

**17**; "NicOx"

**SCHEME 10.4**  Oestreich's intramolecular annulation of pyrroles.

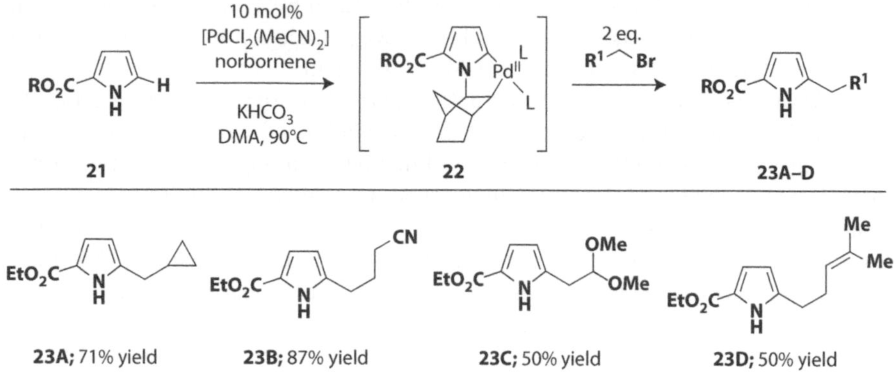
10 mol%
[PdCl$_2$(MeCN)$_2$]
norbornene
KHCO$_3$
DMA, 90°C
2 eq.
R$^1$ Br
21
22
23A–D
23A; 71% yield
23B; 87% yield
23C; 50% yield
23D; 50% yield

**SCHEME 10.5** Bach's alkylation of electron-deficient pyrroles.

Direct oxidative $sp^2$–$sp^3$ couplings between a heterocycle C—H bond and an alkyl C—H bond are extremely rare. As a result, Fagnou's intramolecular cyclization of *N*-pivaloyl pyrroles **24** in air at ambient pressure is of particular interest (Scheme 10.6).[26] Although several combinations of substituted pyrroles and amides were reported, the combination of aromatic and electron-withdrawing pyrrole substituents seemed to facilitate the process (**25A–D**). Kinetic isotope effect studies led to a proposed mechanism beginning with reversible palladation of the arene (intermediate **I**), followed by rate-limiting $sp^3$ C—H activation (intermediate **II**). Reductive elimination led to formation of the coupled product and palladium(0) which could be oxidized to palladium(II) by air. Regioselectivity for initial pyrrole palladation was high for the C—H bond adjacent to the electron-withdrawing acyl/ester substituent.

The iridium-catalyzed borylation of C—H bonds has established itself as a reliable method for heteroaromatic functionalization.[27] The borylation of pyrrole tends to occur at the most acidic C—H bond;[28] treatment of 1*H*-pyrrole (**1**) with $B_2pin_2$

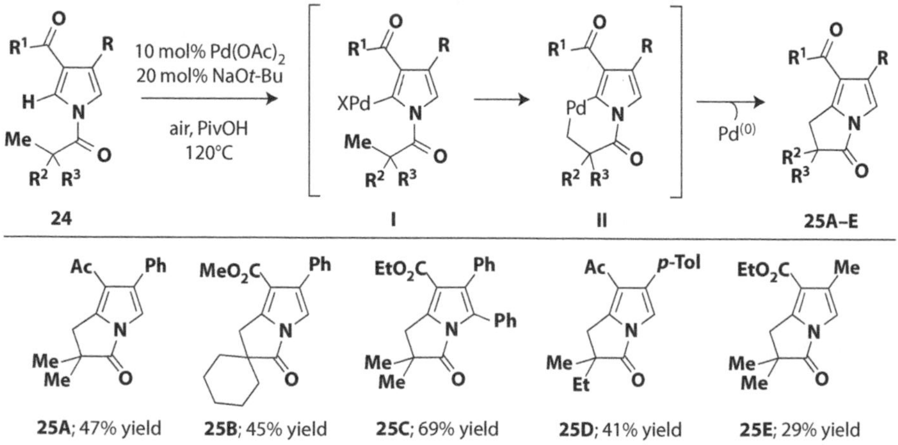
10 mol% Pd(OAc)$_2$
20 mol% NaO*t*-Bu
air, PivOH
120°C
Pd$^{(0)}$
24
I
II
25A–E
25A; 47% yield
25B; 45% yield
25C; 69% yield
25D; 41% yield
25E; 29% yield

**SCHEME 10.6** Fagnou's oxidative intramolecular cyclization of pyrroles.

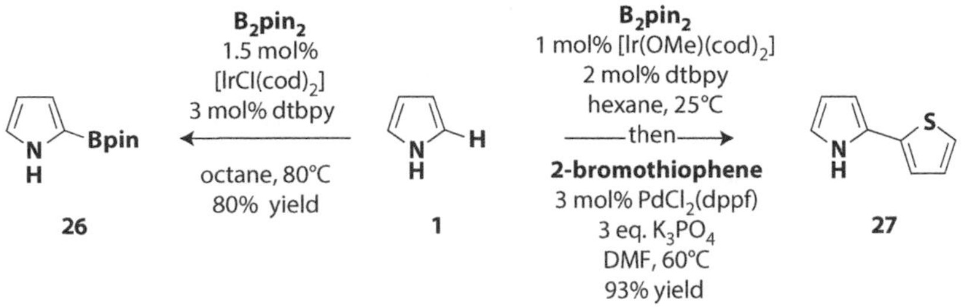

**SCHEME 10.7** C2-borylation of pyrrole and subsequent transformation.

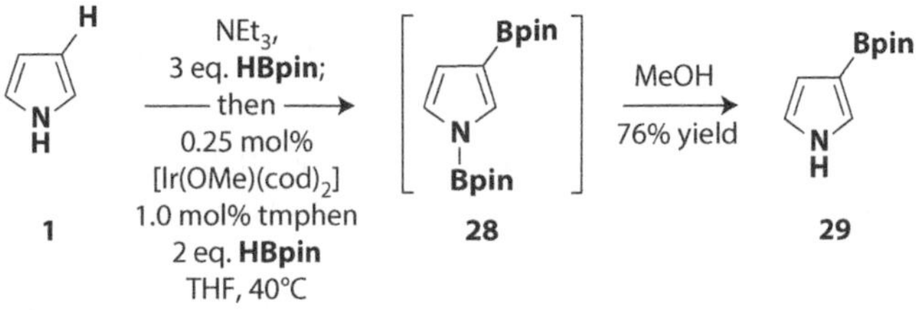

**SCHEME 10.8** Maleczka and Smith's C3-borylation of pyrrole.

(pin = $Me_4C_2O_2$) occurs at the C2 position to afford heteroarylboronate **26** in 80% yield (Scheme 10.7). Traditional cross-coupling methods can then be used to convert the C—B bond into a C—C bond. A one-pot, two-step process for this transformation was realized in 2008 by Miyaura and co-workers; **26** could be prepared *in situ* from reaction of 1*H*-pyrrole (**1**) and an alkoxyborane (either $B_2pin_2$ or HBpin), and subsequently trapped with 2-bromothiophene to allow access to bis-heterocycle **27** in 93% yield.[29]

It has been previously shown that the regioselectivity of pyrrole borylation can be switched from C2 to C3 if the nitrogen is TIPS[27d,30] or Boc[31] protected. More recently, however, Maleczka and Smith described a very useful method that circumvents the installation and removal steps required for pyrrole protection (Scheme 10.8).[32] The researchers demonstrated that 1*H*-pyrrole (**1**) could be exclusively borylated at the C3 position (**29**) when Bpin was employed as a traceless directing group. As the *N*H of pyrrole is not particularly acidic, triethylamine was added to the reaction mixture to make the B—H bond in HBpin more hydridic and ensure successful *N*-borylation.

### 10.2.2 FURAN

The classical reactivity of furans toward electrophilic reagents mirrors that of pyrroles owing to the electron-rich nature of both heterocycles, and the arrangement of heteroatom lone pair completing the aromaticity. Much like pyrrole, furan (**30**) has been shown to undergo regioselective arylation via C—H activation pathways. Dauglis and co-workers employed a palladium(II) catalyst, a Buchwald phosphine ligand and an inorganic base to arylate furan with aryl and heteroaryl chlorides and

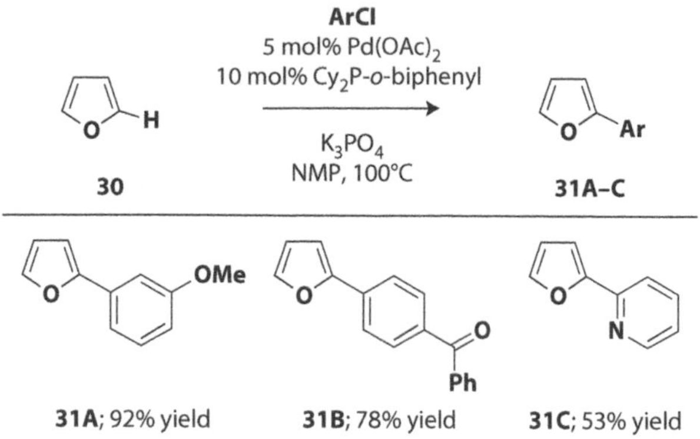

**SCHEME 10.9**   Daugulis' C2-arylation of furan with activated aryl chlorides.

afford products **31A–C** in good to high yields (Scheme 10.9). Although electron-rich and electron-poor aryl chlorides proved to be good reacting partners, unactivated aryl chlorides were not.[33]

Lee and Ghosh addressed this problem when they demonstrated that the highly electron-rich palladium complex **33** could be used to accomplish furan arylation with unreactive and sterically congested aryl chlorides (Scheme 10.10).[34] This method allowed for the synthesis of 2,5-substituted furans **34A–C** that had been previously difficult to access using other methods.[35] This arylation is thought to proceed via a CMD pathway.

Sharp et al. afforded different substitution patterns when they affected arylation of ethyl 3-furoate at the C2 or C5 positions depending on whether Pd(PPh₃)₄ or Pd/C was used as the catalyst.[36] The Fagnou laboratory also demonstrated a method for C5-arylation from C2-substituted furans.[37]

Rovis and co-workers investigated the oxidative cycloadditions of heteroaryl carboxamides using a rhodium(III) catalyst in the presence of a copper(II) oxidant to afford some potentially interesting medicinal chemistry scaffolds (Scheme 10.11).[38] For example, in the presence of a rhodium monomeric species, 3-furyl carboxamide (**35**)

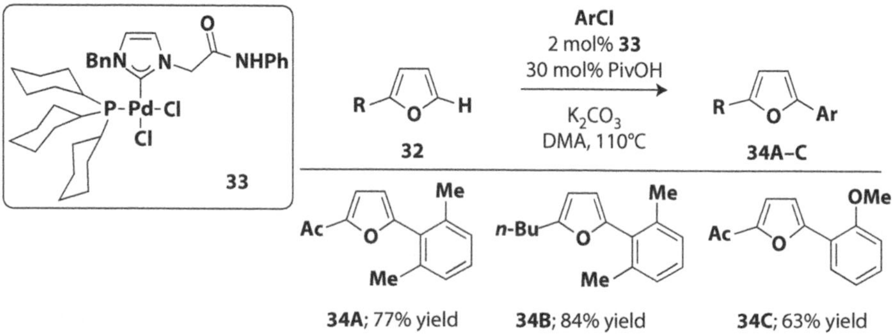

**SCHEME 10.10**   Lee's furan arylation with unactivated, hindered aryl chlorides.

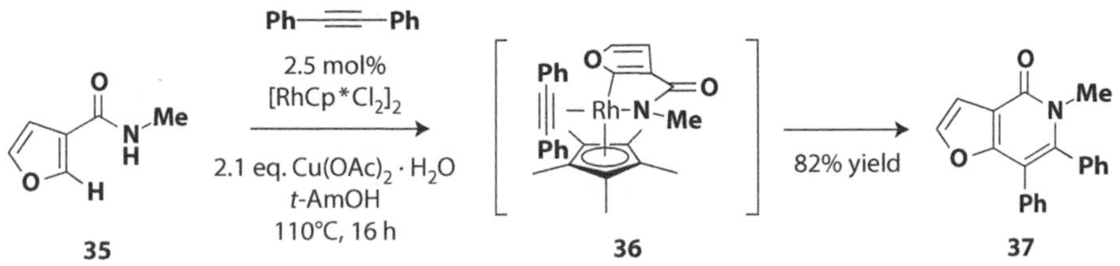

**SCHEME 10.11** Rovis' oxidative furopyridone synthesis.

underwent *ortho* NH metalation and C—H activation at the more reactive 2-position to form the five-membered rhodacycle **36**. Insertion of the alkyne into the open coordination site of the rhodacycle, followed by reductive elimination, generated furopyridone **37** as a single regioisomer in 82% yield. Thiophene, pyrrole, and indolyl carboxamides were all compatible with this methodology, however the less reactive 2-heteroaryl carboxamides required use of a more activated cationic rhodium precatalyst. A broad range of symmetrical and unsymmetrical alkynes were tolerated under these reaction conditions, however terminal alkynes were not.

Hashmi et al. reported the hydroarylation of α,β-unsaturated alkenes by furans (Scheme 10.12).[39] Although the mechanistic rationale for this transformation remains unclear, formation of intermediate **39** could be the result of either: (1) gold activation of methyl vinyl ketone (MVK) followed by electrophilic aromatic substitution at the C5 position of furan **38** or (2) direct auration of furan **38** followed by a 1,4 addition to MVK. Upon protodeauration of the intermediate, unsaturated product **40** would be obtained. Support for the C—H activation pathway can be garnered from literature reports of stoichiometric direct aurations[40] and direct auration of electron-rich arenes.[41]

A ruthenium-catalyzed carbonylative coupling of furfural imine **41** and ethylene was described by Murai and co-workers (Scheme 10.13).[42] The imine directing group was found to be necessary to attain high efficiency and regioselectivity for cleavage of the C—H bond *ortho* to the imine.[43] Product formation probably occurs via metallation of the furfural imine (**I**), reaction with ethylene and CO insertion (ruthenacycle **II** or **III**), followed by reductive elimination to afford propionylated furan **42** in 62% yield.

Like the other five-membered heterocycles, furan can also be regioselectively borylated in high yield to afford the C2-borylated product using the Hartwig–Ishiyama protocol.[27d]

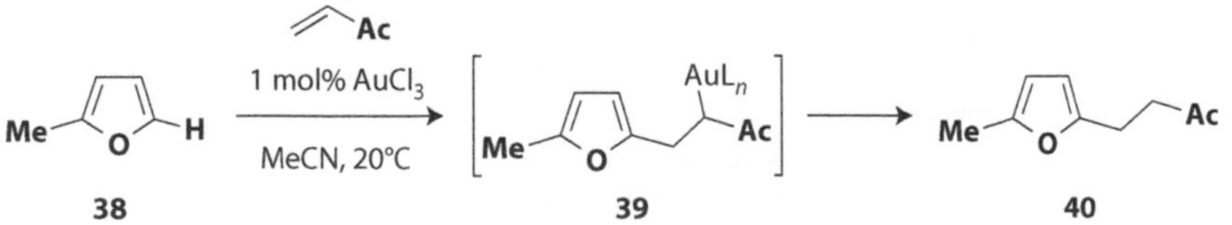

**SCHEME 10.12** Hashmi's hydroarylation of MVK by furan.

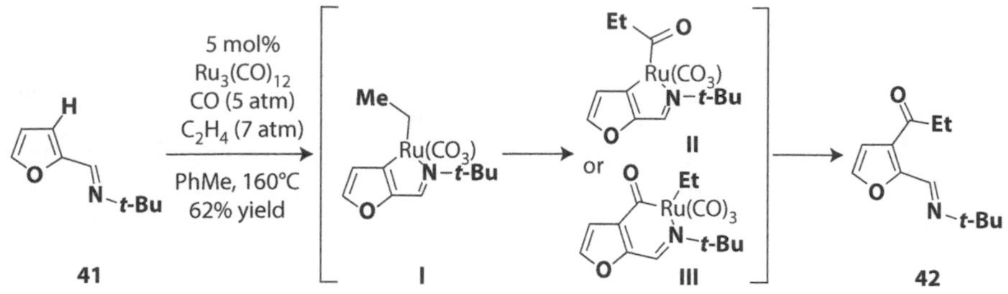

**SCHEME 10.13**   Murai's carbonylative coupling of furan and ethylene.

### 10.2.3   THIOPHENE

Synthetic transformations predominantly take place at the most acidic α-positions of thiophene. Whilst the reactivity patterns of pyrrole and furan also map to this electron-rich heterocycle, examples of selective C—H activation reactions occurring on unsubstituted thiophene are scarce. The most reactive position of a C3-substituted thiophene is at C2, however these substrates undergo direct arylation reactions to give mixtures of C2/C5 monoarylated products as well as the C2,C5-diarylated products. Itami et al. addressed this regioselectivity problem by employing rhodium complex **44** (prepared from mixing [RhCl(CO)$_2$]$_2$ and P[OCH-(CF$_3$)$_2$]$_3$ in toluene) to facilitate direct C—H arylations between five-membered heteroarenes and simply-substituted aryl iodides (Scheme 10.14).[44] For thiophene **43**, a methoxy directing group was used to facilitate arylation at the 2-position, affording products **45** and **46** in good yields of 73% and 94%, respectively.

This biaryl coupling reaction can be viewed to proceed via oxidative addition of the iodoarene followed by electrophilic C—H activation of the thiophene **43** by the cationic arylrhodium(III) intermediate; this pathway was found to be facilitated by the increased π-accepting nature of the phosphine ligands of catalyst **44**.[45] Further optimization of reaction parameters led to a coupling protocol with broader scope.[46]

The Itami group described two protocols for the selective arylation of thiophene derivatives **47** at the least reactive C4 position; this feat was first achieved with iodoarenes[47] (Scheme 10.15) and then with boronic acids[48] (in conjunction with the Studer group). The first report of thiophene functionalization with iodoarenes in the presence of a PdCl$_2$/P[OCH(CF$_3$)$_2$]$_3$/Ag$_2$CO$_3$ catalyst system was shown to be quite general and catalyst-based as the observed regioselectivity seen in products **49A–E**

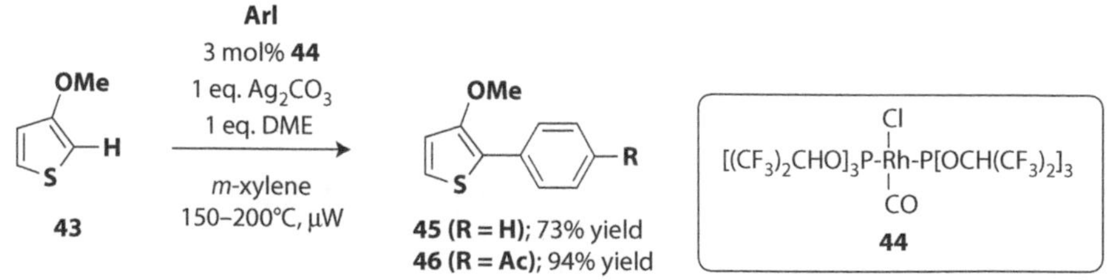

**SCHEME 10.14**   Itami's arylation of a 3-methoxy thiophene.

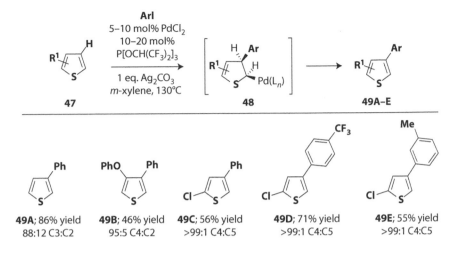

**SCHEME 10.15**  Itami's C4-arylation of furan.

overrode the inherent influence of *ortho* or *meta* directing groups as well as the thiophene ring itself. The nature of the reaction mechanism in this catalyst-controlled arylation has yet to be determined; preliminary experiments by the authors indicated that a Heck-like concerted arylpalladation across the double bond of thiophene was likely occurring (**48**).[49] A computational study by Fu et al. also determined that the C4-selectivity could be explained by a non-traditional Heck-type arylation mechanism that proceeds through an anti-β-hydride elimination process.[50]

When the supporting neutral ligand is switched from P[OCH(CF$_3$)$_2$]$_3$ to 2,2′-bipyridyl (bipy), arylation is expected to occur at the most acidic α-position via a metallation-deprotonation mechanism. However, when arylboronic acids were employed in excess as the arene source, arylation took place at the position β to the heteroatom. The authors rationalized that use of this arylating reagent allowed the deprotonation event to be slowed down sufficiently to allow the aryl group on Pd to irreversibly migrate to the thiophene C4 position. This concept was further elaborated to allow for coupling of thiophene **50** with the hindered arylboronic acid **51** under a Pd(OAc)$_2$/bisoxazoline catalyst manifold and enabled the first enantioselective biaryl coupling via catalytic C—H bond functionalization (Scheme 10.16).[51]

These methodologies set the stage for the synthesis of tetraarylthiophenes (Scheme 10.17).[52] In the presence of the catalyst PdCl$_2$/P[OCH(CF$_3$)$_2$]$_3$ and Ag$_2$CO$_3$ in

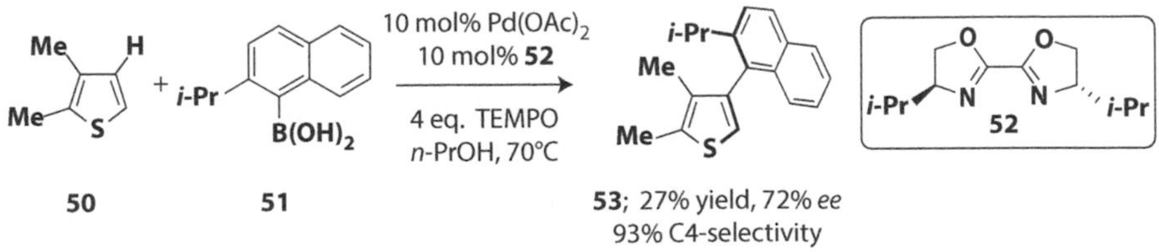

**SCHEME 10.16**  Itami and Studer's enantioselective thiophene arylation.

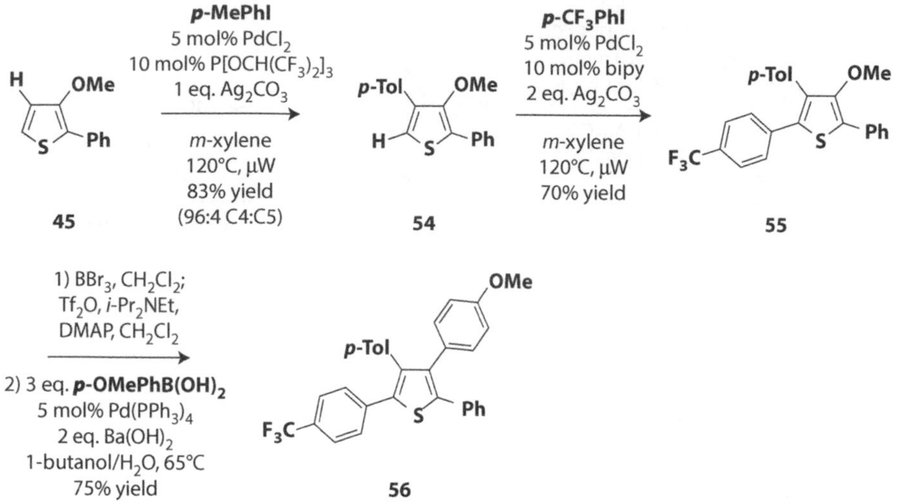

**SCHEME 10.17**   Itami's synthesis of tetraarylthiophenes.

*m*-xylene, thiophene **45** underwent difficult C4-arylation with *p*-tolyliodide to afford **54** in 83% yield and with high regioselectivity (96:4). The PdCl$_2$/bipy catalyst system promoted subsequent C5-arylation with an iodoarene to afford triarylthiophene **55** in 70% yield. Transformation of the methoxy group to the triflate, followed by a Suzuki–Miyaura coupling reaction,[6] afforded tetraarylthiophene **56** in 75% yield.

In 2010, Fagnou and Liégault also addressed the problem of regioselectivity in heteroaromatic arylation reactions by using a chlorine atom on the heterocycle; not only did this blocking group improve the reactivity of the substrate, but it also guided the arylation event to take place at previously inaccessible positions on the heteroarene.[53] Using 3-(*n*-hexyl)thiophene (**57**) as an example, the chlorine atom could be introduced at the C5 position via deprotonation and trapping to afford **58**, or at the C3 position (**59**) via electrophilic chlorination (Scheme 10.18). Upon subjection to their previously established conditions, arylation of the two substrates either took place adjacent (**60**) or remote to the hexyl group (**61**), but always adjacent to the sulfur atom. The chlorine atom could then be efficiently removed with catalytic Pd/C under a hydrogen atmosphere, or used as a handle for subsequent transformations.

A similar temporary ester C2-blocking group strategy was later used by Doucet and co-workers to selectively obtain C5-arylated thiophenes.[54] The authors hypothesized that in instances where other halogens are present on the thiophene, decarboxylation of the blocking group would be an easier reaction than the dehalogenation required in Fagnou's method.[53]

The groups of Hu and You reported a remarkable palladium-catalyzed oxidative cross-coupling of *N*-containing heteroarenes **62** with diversely substituted thiophenes **47** to afford products **63A–D** (Scheme 10.19).[55] The *N*-containing heteroarenes included electron-rich heterocycles such as xanthines and azoles as well as electron-poor heterocycles such as pyridine *N*-oxides. In cases where heteroarenes demonstrated sluggish reactivity, CuBr was used as an additive to assist C—H bond

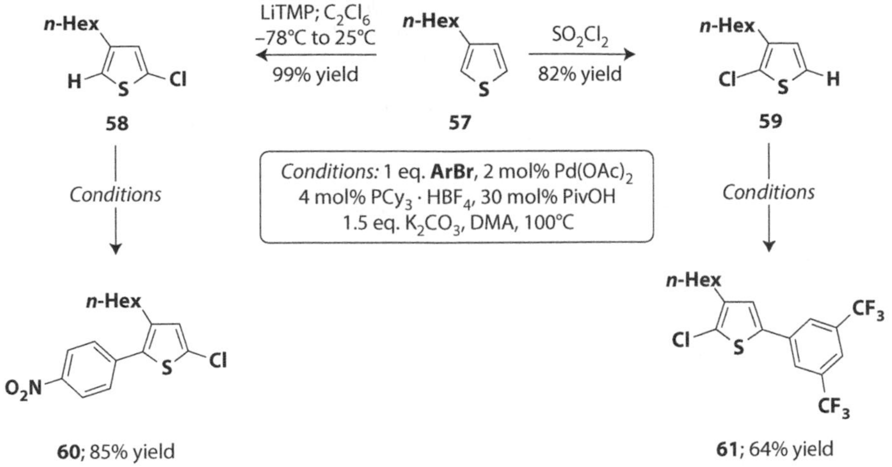

**SCHEME 10.18** Fagnou and Liégault's chloride-directed thiophene arylation.

activation. A computational study provided support for a two-fold C—H activation pathway via a CMD mechanism.

Although there have been reports of thiophene vinylation using palladium catalysts, the requirement of acetic acid in the reaction mixture precludes the inclusion of many functional groups.[56] In order to mitigate this, Satoh and Miura investigated a palladium-catalyzed oxidative coupling reaction under weakly basic conditions (Scheme 10.20).[57] 2-Substituted thiophenes **64** underwent the oxidative coupling reaction with various alkenes **65** to give the corresponding vinylated products **66A–E** in moderate yields; as hoped, acid-sensitive functional groups were compatible with the reaction conditions.

A plausible mechanism would be electrophilic attack of Pd(OAc)$_2$ followed by alkene insertion and subsequent β-H elimination to afford the vinylated product and HPdOAc. It is believed that lithium acetate prevents deactivation of palladium(0), which is formed when HPdOAc releases acetic acid.[58]

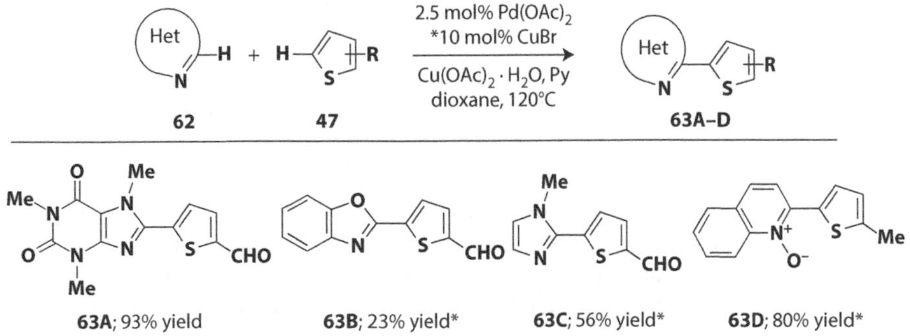

**SCHEME 10.19** You's oxidative coupling with *N*-heteroarenes.

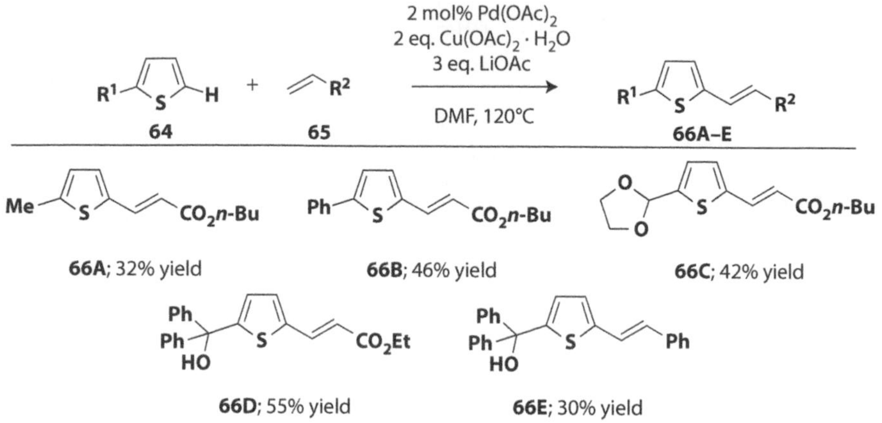

SCHEME 10.20    Miura's oxidative vinylation of thiophene.

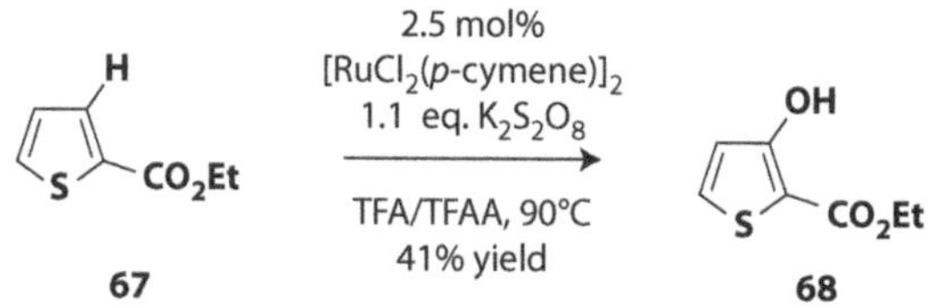

SCHEME 10.21    Rao's hydroxylation of thiophene.

In their paper describing the direct hydroxylation of arene $sp^2$ C–H bonds with a ruthenium catalyst, Rao and co-workers demonstrated that a simple thiophene was also compatible with these reaction conditions (Scheme 10.21).[59] It is proposed that under acidic conditions, $[RuCl_2(p\text{-cymene})]_2$ facilitates C–H bond cleavage of **67** via an orthometalation process through chelation with the ester carbonyl group (kinetic isotope experiments also support a kinetically relevant C–H metalation step). Subsequent reductive elimination afforded the hydroxylated thiophene **68** in 41% yield.

## 10.2.4    PYRIDINE

Pyridine is described as an electron deficient $\pi$-system and this characteristic attenuates its reactivity with electrophiles relative to the electron-rich heterocycles described thus far. Through the use of transition metal catalysis, regioselective C–H activation of pyridine has been described at the $\alpha$-, $\beta$-, and $\gamma$-positions.

The arylation of pyridine is a commonly recognized synthetic challenge because the majority of direct arylation reactions depend upon electrophilic aromatic substitution or CMD mechanisms that are unsuitable for such an electron-poor heteroarene.[60] Moreover, pyridine has a tendency to undergo non-productive binding of the metal catalyst to the lone pair of electrons on the $sp^2$-hybridized nitrogen. A partial solution to this issue has been provided by converting pyridine to its more reactive

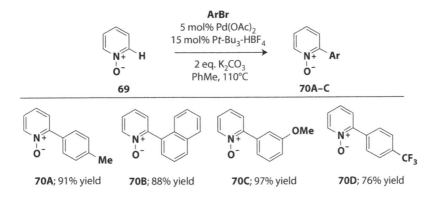

**SCHEME 10.22**  Fagnou's C2-arylation of pyridine *N*-oxide.

pyridine *N*-oxide analog[61] (i.e. with the MeReO$_3$/H$_2$O$_2$ system first described by Sharpless)[62] which has enabled a large breadth of C2-functionalization methods.[63] After the given transformation has taken place, the *N*-oxide product can be reduced back to pyridine using conditions such as palladium-catalyzed hydrogenolysis. The earliest example of this came from the Fagnou laboratory where it was shown that pyridine *N*-oxides **69** undergo regioselective and high-yielding C2-arylation (**70A–C**) in the presence of a Pd(OAc)$_2$/P$t$-Bu$_3$ · HBF$_4$ catalytic system (Scheme 10.22). This was initially demonstrated with aryl bromides[63(a)] and later with aryl triflates;[64] C3- and C4-substituted pyridine *N*-oxides were also shown to participate in the arylation reaction.[65]

Although no mechanistic rationale for the arylation reaction was put forward, kinetic isotope effect studies suggested that this transformation was not occurring via a S$_E$Ar mechanism. This seminal publication led to other C—H activation methodologies for the synthesis of C2-heteroarylated products,[55,66] C2-alkenylated products,[67] and C2-alkylated pyridine derivatives.[68]

Whilst this arylation approach is elegant, it requires two additional synthetic steps: one to activate the pyridine ring system and another to reduce the *N*-oxide at the end of the reaction. Bergman and Ellman reported a rhodium(I)-catalyzed direct arylation reaction that obviated these additional steps; however, the scope was limited to the functionalization of 2-substituted pyridines such as **71** with bulky 3,5-dimethylbromobenzene (**72**) to afford **73** in 53% yield (Scheme 10.23).[69]

The C3-arylation of pyridine (**74**) was achieved by Yu and co-workers using a Pd(OAc)$_2$/1,10-phenanthroline (phen) catalytic system; under these conditions,

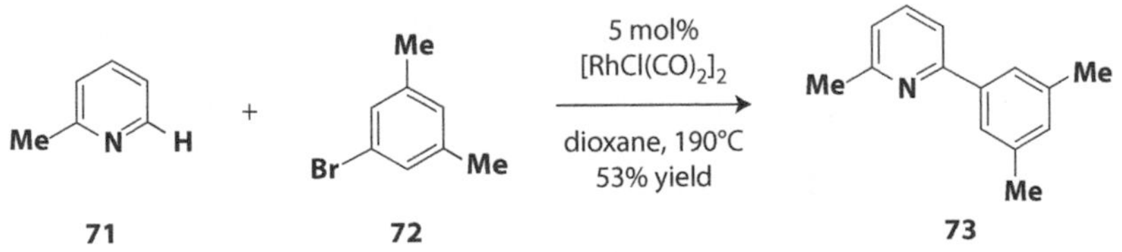

**SCHEME 10.23**  Bergman and Ellman's pyridine arylation.

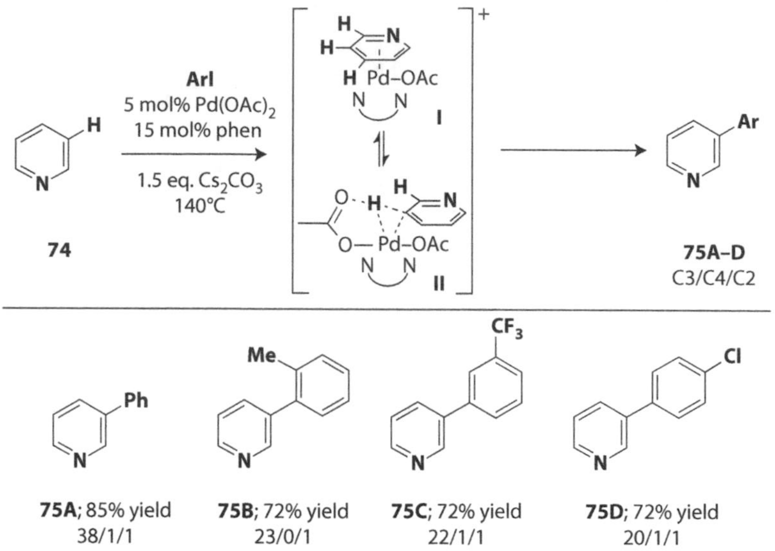

**SCHEME 10.24** Yu's C3-arylation of pyridine.

products **75A–D** were obtained in good yields (Scheme 10.24).[70] In marked contrast to previous protocols, no additional substituents on the pyridine were required to influence reactivity or guide selectivity. Mechanistic investigations led the authors to hypothesize that after initial coordination of pyridine to the metal center via the nitrogen atom, ligand-assisted dissociation causes it to reorient itself and bind the palladium through the π-system, which leads to C3—H bond activation. Oxidative addition of the iodoarene, followed by reductive elimination afforded the C3 product in good yield and high levels of selectivity.

The Yu laboratory described the first selective pyridine C3-olefination reaction (Scheme 10.25).[71] Using a similar rationale to that used in their arylation method,[70]

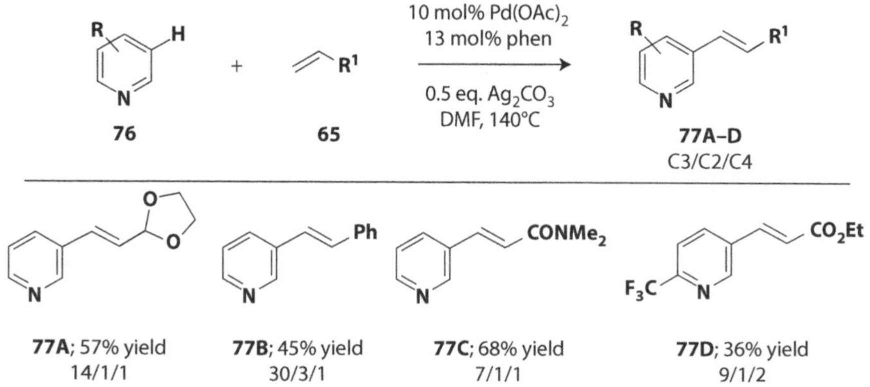

**SCHEME 10.25** Yu's C3-olefination of pyridine.

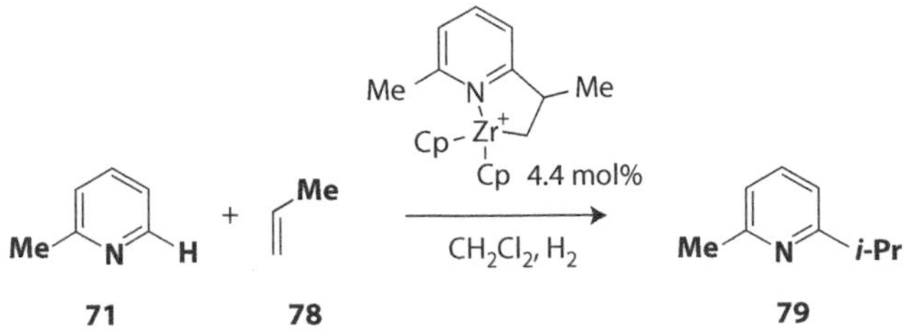

**SCHEME 10.26**   Jordan and Taylor's coupling of 2-picoline and propene.

it was hypothesized that the success of this method was due to unique ability of the bidentate phen ligand to weaken the coordination of the palladium catalyst with the pyridyl *N* atom through the *trans* effect.

In one of the earliest examples of a catalytic direct C—H activation process, Jordan and Taylor described the zirconium-catalyzed coupling of α-picoline (**71**) and propene (**78**) under an atmosphere of hydrogen (Scheme 10.26).[72] It is presumed that the catalytically active zirconocene species is formed upon hydrogenolysis. Coordination of the active catalyst by the 2-picoline nitrogen promotes C—H insertion at the α-position.[60]

Bergman and Ellman later investigated a rhodium-catalyzed version of this alkylation reaction (Scheme 10.27).[73] Following alkylation of 2-TIPS pyridine (**80**) with 3,3-dimethylbutene (**81**) in 64% yield, the silyl group could be removed with aqueous hydrogen fluoride.

The iridium- and rhodium-catalyzed borylation of pyridines via C—H activation can be conducted under mild reaction conditions and is atom-efficient.[27] In general, the borylation of mono- and di-substituted pyridines occurs at the C—H bond(s) located β or γ to the basic nitrogen to afford C3 and C4 products in a statistical mixture.[27i] 2,6-Disubstituted pyridines, however, undergo borylation at the C4 position.[74] Hartwig and Larsen have shown that the lack of borylation of C—H bonds adjacent to the basic nitrogen is not the result of coordination to a bulky Lewis acid prior to C—H activation, but is rather due to the higher-energy pathway required for borylation, coupled with the instability of the resulting product which has a propensity to undergo protodeboronation.[75,76] The iridium-catalyzed borylation of 3-picoline (**83**) yields **84** as a single product; a kinetic investigation of this reaction revealed a first-order dependence of the rate on the concentration of the catalyst, zero-order on the concentration of $B_2pin_2$ and zero-order on the concentration of substrate. This data led the authors to propose a catalytic cycle in which the resting state of the

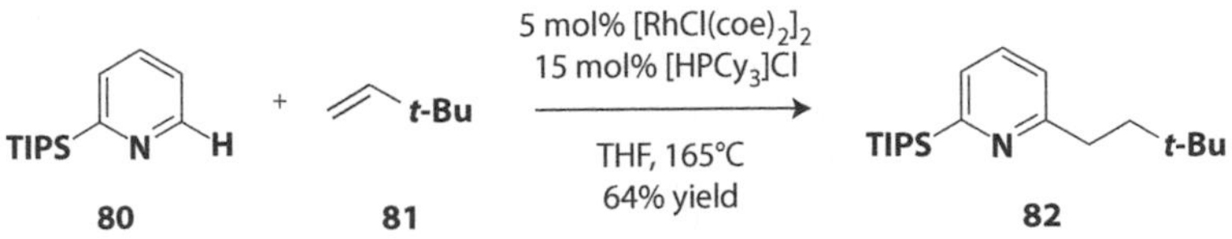

**SCHEME 10.27**   Bergman and Ellman's alkylation of pyridine.

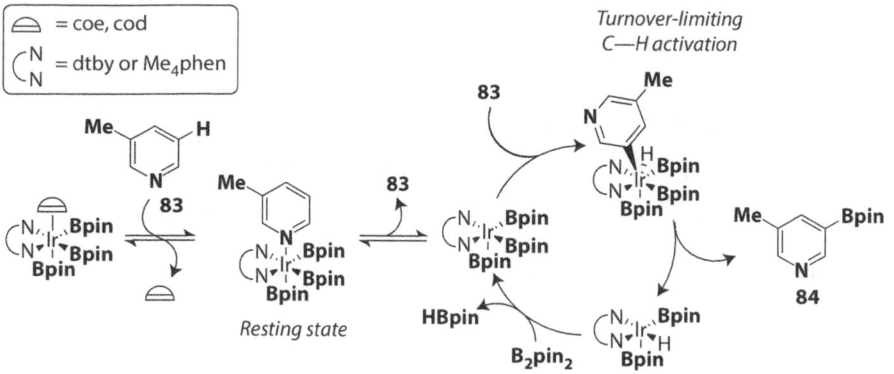

**SCHEME 10.28**  Hartwig's proposed mechanism for the borylation of 3-picoline.

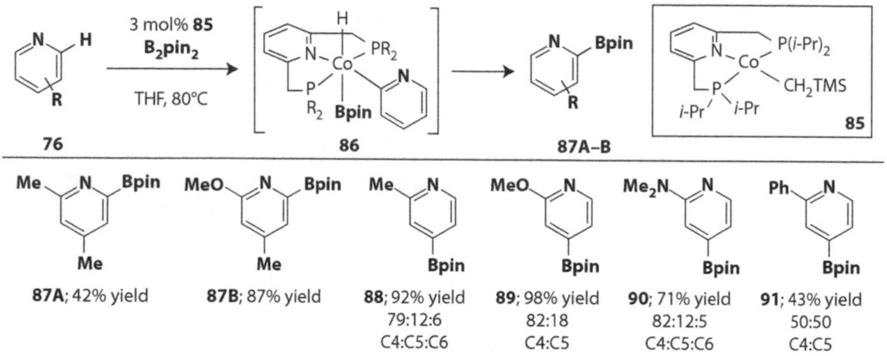

**SCHEME 10.29**  Chirik's borylation of pyridine derivatives.

catalyst contains the heteroarene instead of the cyclooctene (coe)/1,5-cyclooctadiene (cod) alkene precursor (Scheme 10.28).

Recently Chirik et al. reported the synthesis of cobalt pincer complexes (such as **85**) that were shown to be effective catalysts for the borylation of substituted pyridines **76** with $B_2pin_2$ (Scheme 10.29).[77] Use of this catalyst allowed access to novel boronate esters **87A** and **87B**. 2-Substituted pyridines underwent smooth monoborylation to afford the C4 product as the major positional isomer (**88–91**).

## 10.2.5  INDOLE

Of all the heterocycles featured in nature and in man-made compounds, indole is the most abundant. This electron-rich heteroaromatic compound reacts with a range of electrophiles, predominantly at the C3 position via an electrophilic aromatic substitution pathway. The C2 position of indole is the most reactive site for metalation[27d] however C—H activation reactions can proceed at either the C2 or C3 positions; therefore, the control of site-selectivity is the major challenge.

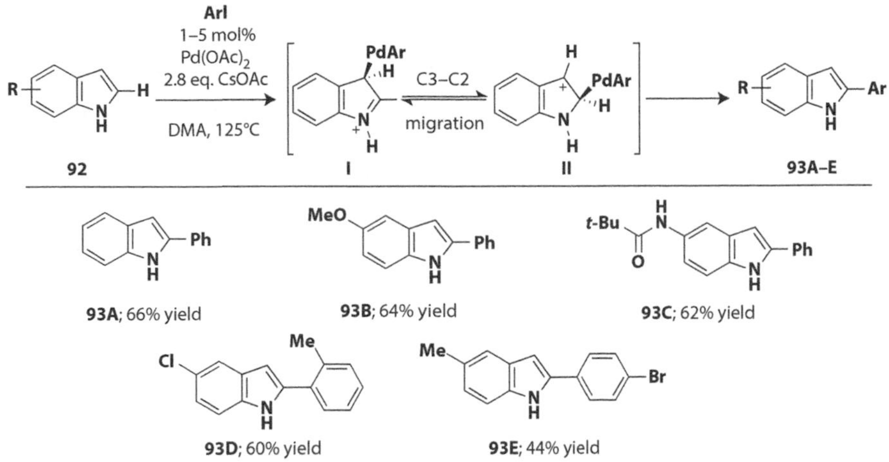

**SCHEME 10.30**   Sames' C2-arylation of indole with aryl iodides.

The arylindole motif can be found in a huge array of biologically active compounds.[78] The first palladium-catalyzed direct C2 indole arylation with 2-chloropyrazines was reported by Ohta and co-workers in 1985.[79] Two decades later, the Sames group expanded on this work by arylating a variety of *N*-methyl and *N*-[2-(trimethylsilyl) ethoxy]methyl protected indoles.[80] A later report by the same authors described the arylation of unprotected indoles **92** with aryliodides in the presence of Pd(OAc)$_2$ as the catalyst and CsOAc as the base (Scheme 10.30).[81] Under these phosphine-free reaction conditions, C2-arylated products **93A–E** were obtained in good yields via a proposed mechanism whereby electrophilic addition of palladium(II) initially takes place at the more electron-rich C3 position (intermediate **I**) before a C3–C2 metal migration (intermediate **II**) which is thought to be driven by stabilization of the C— Pd bond by the adjacent nitrogen atom.

A number of other research groups utilized aryl halides in their indole aryla-tion methodologies[82] including Bellina et al. who arylated indole at the C2 posi-tion under base-free conditions.[83] Shi and co-workers were able to demonstrate that boronic acids were also competent arylating reagents for this transformation.[13] The first mild room temperature protocol for C2-arylation came from the Sanford labora-tory (Scheme 10.31).[8] Differentially substituted indoles **92** could be functionalized in good yields (**93F–I**) using an electron-deficient catalyst and preformed iodine(III) arylating reagents via a palladium(II)/(IV) pathway.

Gaunt et al. found that the same diaryliodonium reagents could effect ambient temperature Cu(OTf)$_2$-catalyzed regioselective arylation of indole at either the C2 (when indole was *N*-methylated) or C3 position (when indole was *N*-acetylated).[84] Oxidative addition of the diaryliodonium coupling partner by a copper(I) salt gener-ates a highly electrophilic copper(III)-aryl species capable of facilitating electro-philic metalation of indole. The authors suggested that C2-arylation was the result of *N*-acetyl-assisted migration of the copper(III)-aryl species from the initial C3 position to C2.

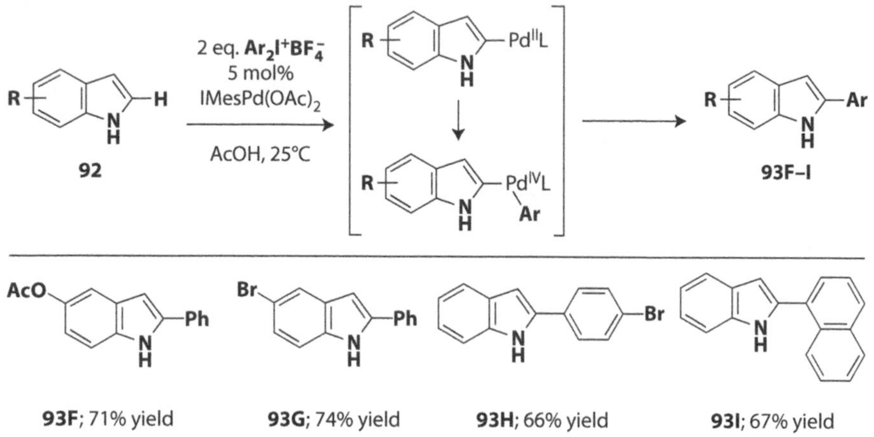

**SCHEME 10.31**    Sanford's mild indole C2-arylation.

Methodologies for the selective arylation of indoles at the C3 position are largely limited to couplings of free indole with bromoarenes.[85] Larrosa et al. employed a decarboxylative functionalization approach to selectively C3-arylate *N*-pivaloylindole (**94**) with electron-poor benzoic acids to afford **95A–C** in good yields;[86] one of the reasons that this method is attractive is because $CO_2$ is the sole waste by-product (Scheme 10.32). The authors proposed a mechanism based on two catalytic cycles

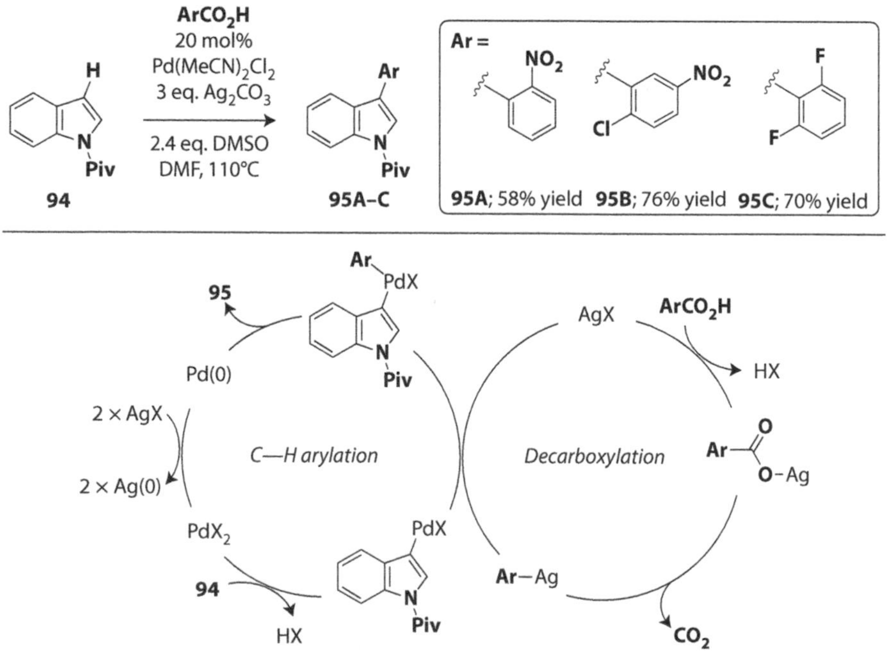

**SCHEME 10.32**    Larrosa's decarboxylative indole C3-arylation.

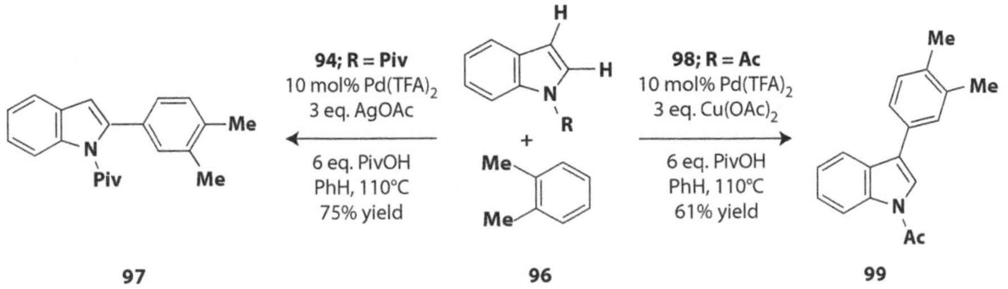

**SCHEME 10.33** Fagnou's intermolecular oxidative indole arylation.

linked by the transmetallation of an arylsilver species to palladium in which the metal catalyst is responsible for the C—H activation and reductive elimination steps and the silver salts perform the decarboxylative activation step.

In their report for palladium-catalyzed oxidative cross-couplings, the Fagnou group were able to access both C2- (**97**) and C3-arylated (**99**) indoles simply by changing the oxidant and the indole *N*-protecting group (Scheme 10.33).[87] Side reactions owing to homocoupling of the indole were avoided by using an excess of the benzene derivatives.

The direct palladium-catalyzed C3-alkynylation of free indoles with bromoacetylenes was first described by Gu and Wang in 2009.[88] The C2-selective alkynylation of indole proved to be especially challenging, and this was not realized until Waser et al. described a mild protocol using TIPS-protected hypervalent iodine reagent **101** and a palladium(II) catalyst (Scheme 10.34).[89] Under optimized conditions, a variety of *N*-alkylated indoles **100** could undergo the alkynylation reaction via a palladium(II)/palladium(IV) mechanism to afford products **102–106** in moderate to good yields. It is unknown whether the C2-palladated intermediate is formed as result of a CMD mechanism, or via a pathway of electrophilic palladation at C3 followed by metal migration to C2.

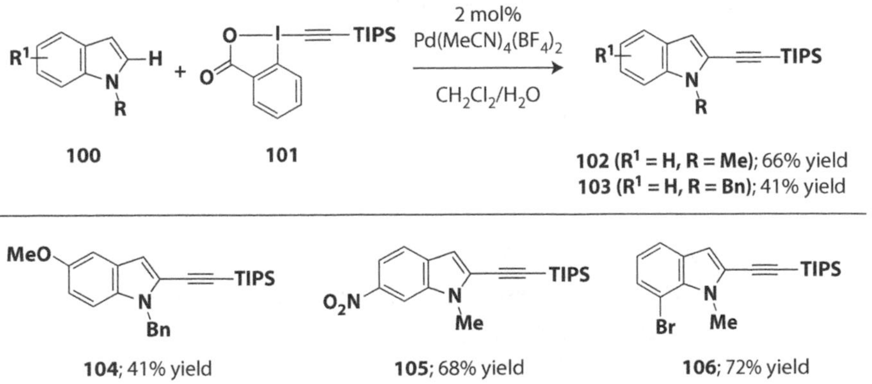

**SCHEME 10.34** Waser's C2-alkynylation of *N*-alkyl indoles.

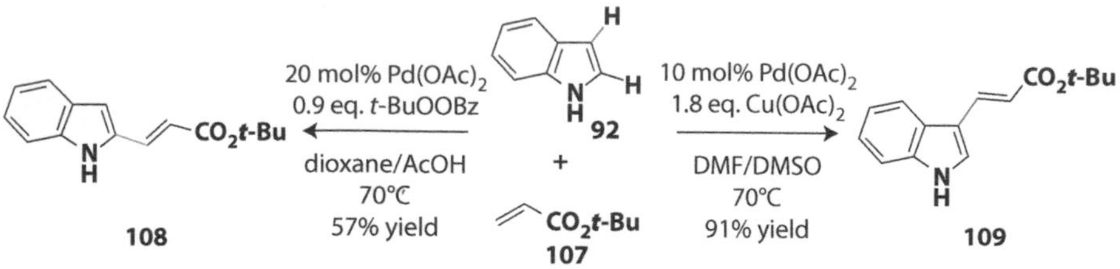

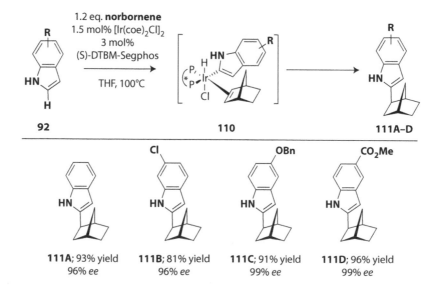

**SCHEME 10.35**  Gaunt's regioselective intermolecular alkenylation of indole.

Gaunt and coworkers were able to affect a solvent-controlled regioselective inter-molecular oxidative Heck reaction between indole and a range of alkenes (Scheme 10.35).[90] Under neutral reaction conditions, the C3-alkenylated product **109** was obtained predominantly. Regioselectivity favored the C2-alkenylated product **108** when acidic medium was employed, presumably due to retardation of the deprot-onation step, thereby allowing the palladium to migrate from the C3 position to C2.

In 2008, Shibata et al. described an asymmetric iridium-catalyzed arene C—H bond alkylation with norbornene.[91] This was expanded upon by Hartwig and Sevov with their report of asymmetric additions of indoles **92** to bicycloalkenes (Scheme 10.36).[92] Indoles, substituted with halides, protected alcohols, and esters, were tol-erated under the reaction conditions. Remarkably, reaction occurred at the indole C2 position (without the need for a *N* protecting or directing group) to afford the *exo*-substituted isomer exclusively (**111A–D**). The hydroheteroarylation reaction was also found to proceed with norbornadiene. The authors propose that the C—H bond addition reaction occurs prior to insertion of the alkene into the Ir–C bond (**110**).

**SCHEME 10.36**  Hartwig's asymmetric addition of indoles to bicycloalkenes.

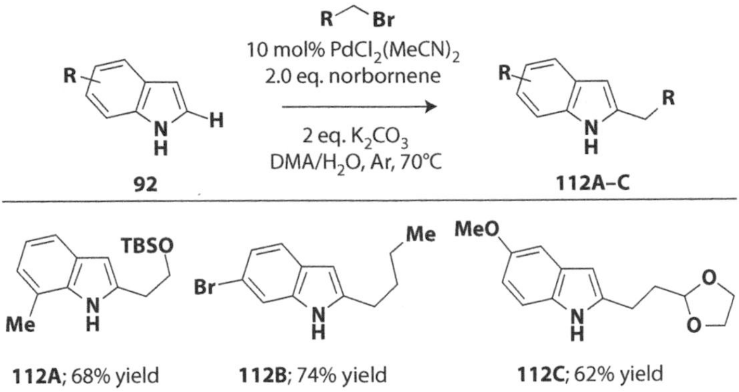

**SCHEME 10.37**  Bach's intermolecular alkylation of indoles.

Prior to his work on pyrrole C—H activation,[24] Bach applied his palladium-catalyzed norbornene-mediated methodology to the indole ring system (Scheme 10.37).[93] The method demonstrated that free *N*H indoles (**92**) could be regioselectively alkylated at the C2 position with simple and functionalized alkyl bromides to afford product **112A–C** in good yields.

A cobalt-catalyzed C2-alkylation of indole was shown to be possible with alkyl chlorides; however, the indole *N* had to be functionalized with a pyridyl or pyrimidyl directing group (both protecting groups are easily installable and readily removable).[94]

In their "green" catalytic method for the C—H alkylation of alkenes with alcohols, Yi and co-workers employed ruthenium complex **114** to alkylate *N*-methylindole (**113**) in almost quantitative yields (**116–117**) whilst only generating water as the by-product (Scheme 10.38).[95] Mechanistic experiments led the authors to propose a cationic Ru–indole–alkyl intermediate (**115**) that would arise from C2—H activation; the exact mechanism of the C—O cleavage step is unknown. A related example with *N*-methoxycarbonyl-L-tryptophan methyl ester demonstrated that an indole protecting group was not necessary.

As one might expect from its inherent reactivity for the metallation of indole, the selective iridium-catalyzed monoborylation of indole at the C2 position proceeded efficiently to afford the corresponding boronate ester in 92% yield.[27d] A later report

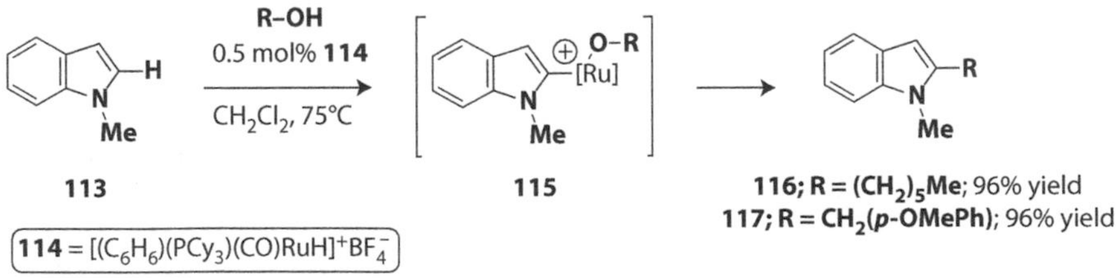

**SCHEME 10.38**  Yi's "green" alkylation of indole.

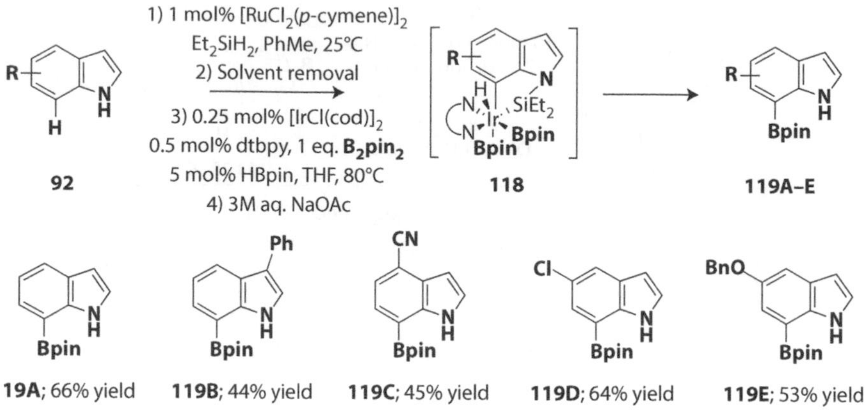

**SCHEME 10.39**    Hartwig's borylation of indole at the C7 position.

outlined the C7-borylation of indole in systems where the C2 position had been blocked with an alkyl or ester substituent.[96] The need for a C2 substituent to achieve exclusive C7 borylation was obviated when a landmark report from the Hartwig laboratory detailed that indoles **92** could undergo C7-borylation when a silyl-directed method was employed (Scheme 10.39).[97] The *N*-hydrosilylindole formed in the first part of the reaction can undergo a reversible reaction with the iridium catalyst, thereby generating HBpin and an intermediate silyl complex. The authors proposed that the C7—H bond is selectively cleaved instead of the C2—H bond due to the formation of the five-membered metallocycle **118** rather than a four-membered metallocycle. The reaction was tolerant of a variety of indole substituents at the 3-, 4-, and 5-positions and afforded the C7-boronate esters **119A–E** in good yields.

## 10.2.6    BENZOXAZOLE

DeBoef and co-workers described the aerobic oxidative inter- and intra-molecular coupling of benzoxazole (**120**) with benzene derivatives in the presence of catalytic amounts of Pd(OAc)$_2$ and heteropolymolybdovanadic acid (HPMV) (Scheme 10.40).[98] Under these conditions, the monoarylated C2 products **121–123** were obtained in moderate to good yields in less than 2 hours; extended reaction times led

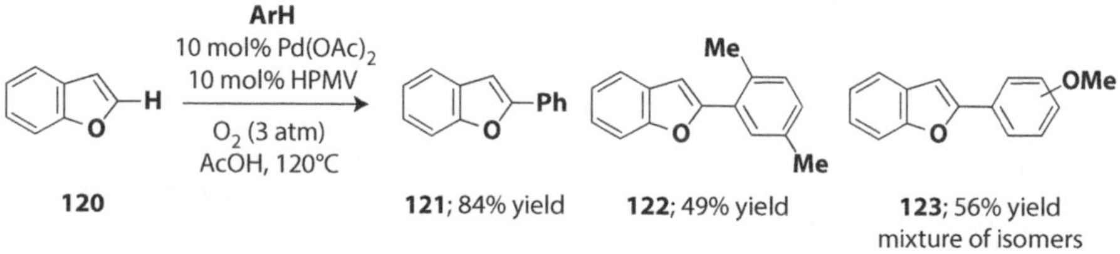

**SCHEME 10.40**    DeBoef's benzoxazole arylation.

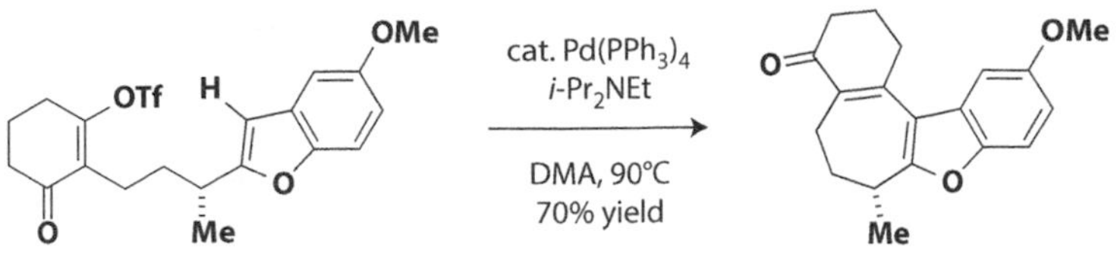

**SCHEME 10.41** Key cyclization step in Trauner's synthesis of (–)-frondosin B.

to formation of the undesired 2,3-diarylated products. Problematic coupling partners included anisole (which afforded regiomeric mixtures of the C2 product) and electron-poor or acidic arenes which led to no product formation. Palladation is thought to occur at the more nucleophilic C3 position before migration to the C2 position and biaryl product formation.

In the total synthesis of the marine terpenoid (–)-frondosin B, Trauner and Hughes described an intramolecular palladium-catalyzed alkenylation reaction between a benzofuran and an enol triflate (**124** to **125**, Scheme 10.41).[99] Although the mechanism to form the key seven-membered ring is still unclear, a reasonable hypothesis would involve oxidative addition of palladium(0) to the C—OTf bond, C3-palladation of the benzofuran and reductive elimination to form the new C—C bond. This work is notable as it was the first example of heteroaromatic C—H activation in a complex molecule setting.

## 10.2.7 QUINOLINE

A number of methods for the derivitazion of quinoline have been described in the literature; many of them are complementary (Scheme 10.42). One example for regioselective C2-arylation came from laboratories of Bergman and Ellman.[69,100] In this method, a rhodium(I) complex was shown to catalyze the direct arylation of quinolines **126** with electron-deficient and electron-rich aryl bromides to afford products **127A–D**. The authors proposed a mechanism in which the quinoline $N$ coordinates the [RhCl(CO)$_2$]$_2$ catalyst to provide an adduct which can undergo *ortho*-C—H activation prior to oxidative addition of the aryl bromide. Shown in the same scheme is a method for site-selective C8-arylation, first described by Chang and co-workers.[101] A rhodium(NHC) active catalyst effectively coupled a broad range of quinolines **126** and aryl bromides via a base-assisted concerted proton abstraction and metallation pathway to afford products **128A–E**. *Ortho*-substituted aryl bromides were found to be too sluggish to undergo the coupling reaction; the authors suggested steric hindrance as a possible explanation.

Also worthy of mention is Yu's palladium(0)-catalyzed arylation protocol whereby an amido directing group enables functionalization at the C3 or C4 positions of quinoline.[102]

## 10.2.8 BENZO[*h*]QUINOLINE

In 2007 Sanford and Hull described the palladium-catalyzed oxidative cross-coupling of benzo[*h*]quinoline (**129**) with arenes in the presence of benzoquinone

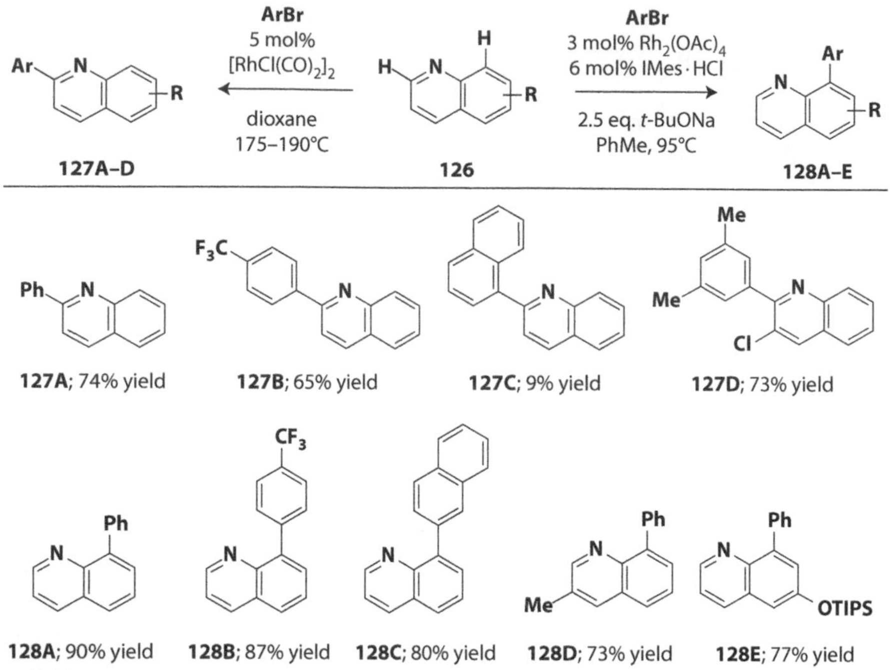

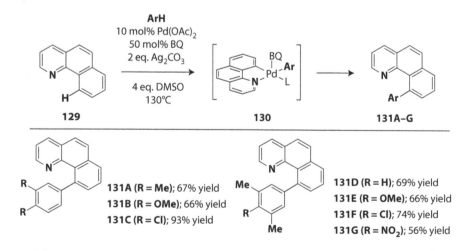

**SCHEME 10.42** Site selective quinoline arylation.

(BQ) and a silver(I) oxidant (Scheme 10.43).[103] Various di- and tri-substituted arenes underwent selective coupling with the heterocycle at the least hindered position of the arene; for 1,2-disubstituted arenes this was at the 4-position (**131A–C**) and for 1,3-di- and 1,2,3-trisubstituted arenes this was at the 5-position (**131D–G**). A later communication by the same authors showed that this

**SCHEME 10.43** Sanford's arylation of benzo[h]quinoline.

regioisomeric preference could be altered when a carbonate ligand was included in the reaction mixture.[104]

Detailed mechanistic studies led the authors to implicate a reaction pathway involving cyclopalladation of the heterocycle, reversible activation of the arene (**130**) and BQ-binding/BQ-promoted C—C bond-forming reductive elimination.[105]

## 10.2.9 INDOLIZINE

Gevorgyan et al. were the first to develop a synthetically useful protocol for the C3-arylation of indolizines with aryl and heteroaryl bromides.[106] In the presence of a $PdCl_2(PPh_3)_2$–KOAc catalyst system and phenyl bromide, indolizine (**132**) was C3-phenylated to afford **133** in 71% yield (Scheme 10.44). The arylation study provided the groundwork for a subsequent $sp^2$–$sp$ coupling method of indolizine with differentially substituted alkynyl bromides whereby product **135** could be isolated in 62% yield.[107]

In both methodologies, the authors implicate an electrophilic substitution mechanism analogous to that proposed by Miura in the arylation of azole compounds:[108] the most electron-rich C3 position of the indolizine attacks an aryl/alkynylpalladium species to form an iminium intermediate which can subsequently be deprotonated (**134**). Reductive elimination generates the products **133** and **135**.

In a later study, Zhao reported the arylation of indolizines with aryltrifluoroborate salts and used this same $Pd(OAc)_2$/AgOAc catalyst system to affect C3-alkynylation with phenylpropiolic acids via a decarboxylative pathway.[109]

Zhang et al. effected palladium-catalyzed oxidative C3-vinylations of indolizine-1-carboxylates **136** with various styrenes (Scheme 10.45).[110] Use of the bidentate nitrogen ligand bipy was key to formation of the branched α-product regioisomer **137** over the linear β-product. The authors later demonstrated that the same indolizine-1-carboxylates could be C3-acylated with a variety of α,β-unsaturated carboxylic acids via C—H and C—C double bond cleavage under oxidative conditions (**138**).[111]

In this novel reaction, potassium chromate plays a dual role in the oxidation of the tertiary C—H bond and the final oxidative cleavage step (Scheme 10.46). Interestingly, when the oxidant was switched to BQ, a decarboxylation event took place and an annulation product was isolated instead of **138**.

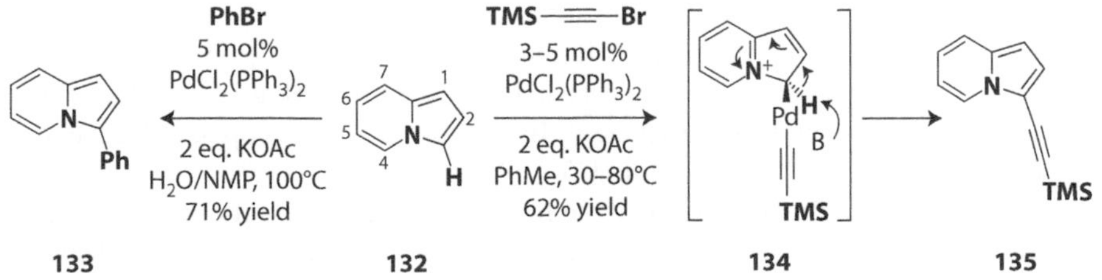

**SCHEME 10.44**    Gevorgyan's indolizine C3-functionalization.

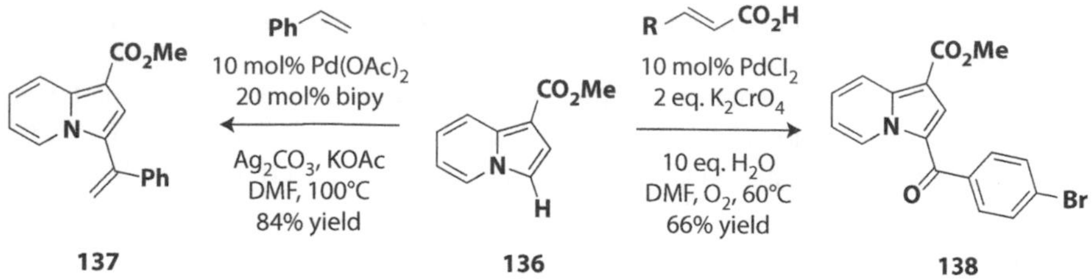

**SCHEME 10.45**  Zhang's indolizine C3-functionalization.

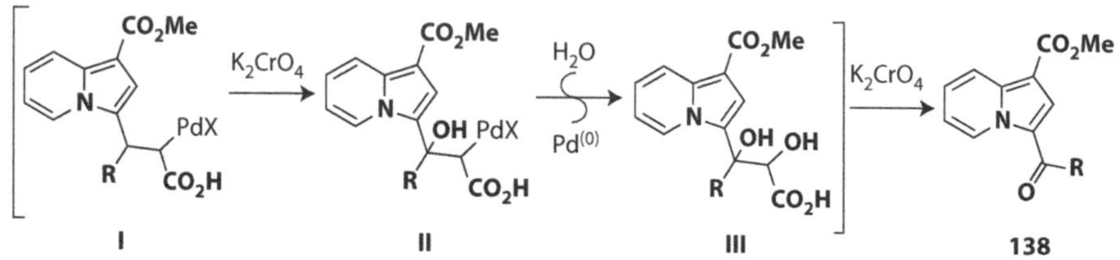

**SCHEME 10.46**  Proposed mechanism for C3 acylation.

## 10.3  HETEROAROMATICS CONTAINING TWO HETEROATOMS

### 10.3.1  AZOLES

The three C—H bonds of thiazole are chemically nonequivalent: the C5 site is considered to be the most nucleophilic, and the C2 site the most acidic.[112] In transition metal catalyzed arylation reactions, the C2—H and/or C5—H bonds are preferentially activated over the least reactive C4—H bond.

Miura et al. first described the arylation of imidazole and thiazole (**139**) C—H bonds with palladium acetate in the presence of a base.[108] The method allowed 2-substituted azoles such as **140** to be arylated at the C5 position in good yields (Chang later developed a method for thiazole C5-arylation that did not require a C2-blocking group).[113] When Mori established a method for C2-azole arylation (i.e. **139** to **140**)[114] it was found to be orthogonal to Miura's and the two could be coupled to iteratively functionalize thiazole and obtain unsymmetrical 2,5-diarylthiazoles like **141** in good overall yields (Scheme 10.47).

The azole arylation findings can be rationalized by Gorelsky's observations that azole $N$ coordination to copper(I) imparts an increase in C2—H acidity and makes it the most reactive site on the heterocycle.[115] In the absence of the copper additive, the reactivity switches back to preferential C5-arylation.

In 2011, Itami and Studer described the first regioselective C4-arylation of 2-phenyl thiazole (**142**) with various arylboronic acids (Scheme 10.48).[48] This breakthrough finding laid the groundwork for Itami's powerful catalyst-controlled programmed

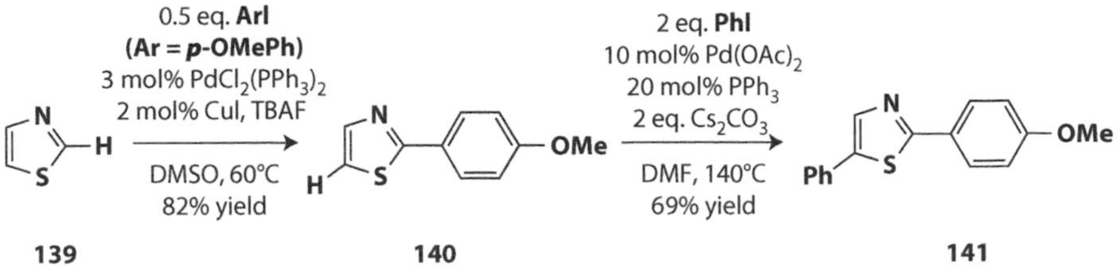

**SCHEME 10.47** Iterative thiazole arylation.

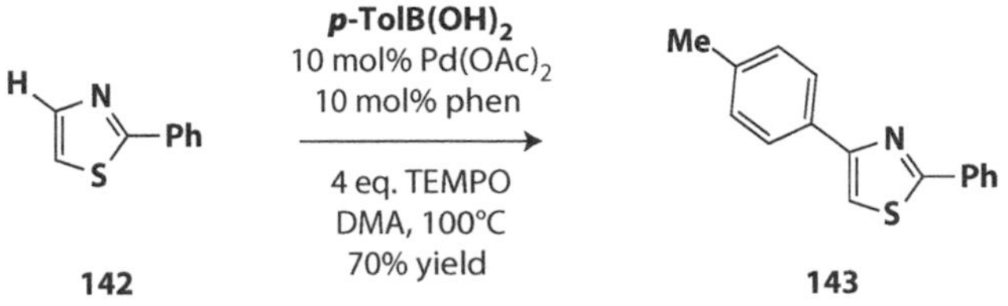

**SCHEME 10.48** Itami and Studer's thiazole C4-arylation.

synthesis of all possible arylthiazole substitution patterns (2-aryl, 4-aryl, 5-aryl, 2,4-diaryl, 2,5-diaryl, 4,5-diaryl, and 2,4,5-triaryl) to afford triarylthiazole **144** from thiazole (**139**) (Scheme 10.49).[116]

The direct alkenylation of azoles has been shown to be possible with alkenyl halides[117] and was later made possible with alkenes via an oxidative Fujiwara–Moritani reaction.[118] In the presence of palladium acetate as the catalyst and silver acetate as the oxidant, Miura showed that substituted thiazoles and oxazoles **145** could be alkenylated to afford products **146A–D** in good yields (Scheme 10.50).

Intramolecular azole alkylation via C—H activation was first reported by Bergman and Ellman in the presence of [RhCl(coe)₂]₂ as the catalyst precursor.[119] Activation of the imidazole **147** C2—H bond led to isolation of product **148** in 71% yield (Scheme 10.51). A subsequent study demonstrated that the same derivatives

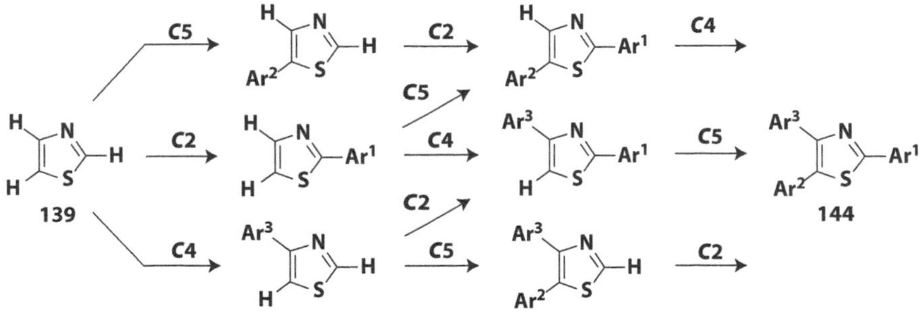

**SCHEME 10.49** Itami's synthesis of triarylthiazoles via iterative C—H activation.

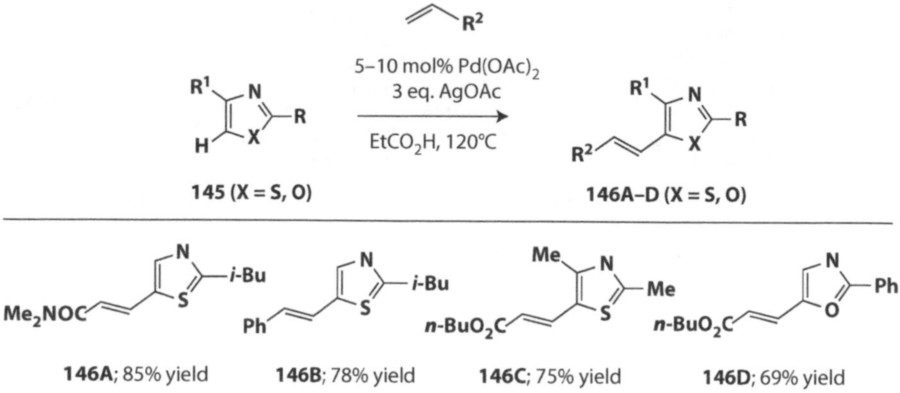

**145 (X = S, O)**                                          **146A–D (X = S, O)**

**146A**; 85% yield       **146B**; 78% yield       **146C**; 75% yield       **146D**; 69% yield

**SCHEME 10.50**   Miura's oxidative alkenylation of azoles.

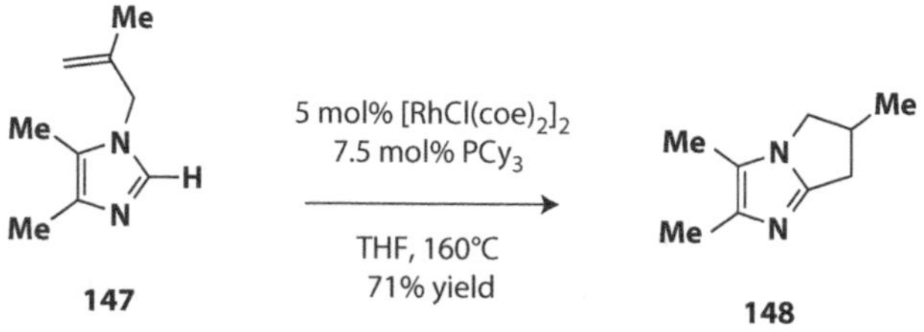

**SCHEME 10.51**   Bergman and Ellman's intramolecular azole cyclization.

could be accessed in a more efficient manner when the reaction was conducted in a microwave, and the ligand was switched to [HPCy₃]Cl which acts as a Brønsted acid source.[120]

An enantioselective version of this rhodium-catalyzed cyclization reaction with a TangPhos ligand was developed by the authors at a later date.[121] It was hypothesized that the highly electron-rich ligand was so efficient at promoting asymmetric induction because of its ability to create a vacant coordinating site on the metal center through partial dissociation.[122]

## 10.3.2   BENZAZOLES

Building upon methodology discussed herein,[100] Bergman and Ellman described the arylation of azoles with aryl bromides in a microwave reactor (Scheme 10.52).[123] The combination of phosphephine ligand **150** and [RhCl(coe)₂]₂ provides a highly active dimer complex which dissociates and coordinates benzazole **149** to afford intermediate **I**. Carbene **II** is proposed to be the result of a C—H activation/tautomerization process and this intermediate undergoes oxidative addition of 3-bromothiophene to generate intermediate **III**. Reductive elimination then occurs to generate biaryl **151** in 64% yield. The hindered amine base is thought to minimize protodehalogenation of the aryl halide and assist in elimination of HBr.

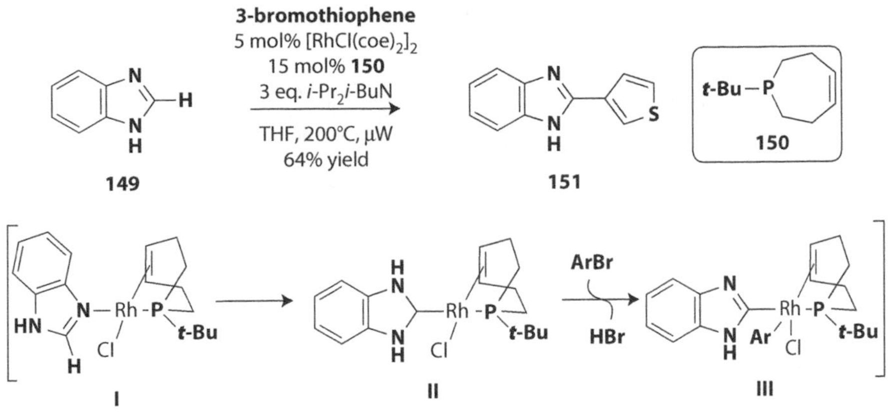

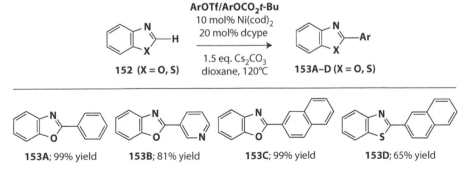

SCHEME 10.52   Bergman and Ellman's azole arylation.

Itami et al. observed that an aryl ester could replace an aryl halide in their azole arylation protocol; such a finding led to the first method for Ar–H/Ar–O couplings and dramatically increased its synthetic utility simply due to the availability of phenols. The Ni(cod)$_2$/dcype catalyst system was active for the coupling of azoles **152** with phenol derivatives such as carbamates, carbonates, mesylates, triflates, and tosylates to afford 2-arylazole products **153A–D** in high yields (Scheme 10.53).[124]

The same authors used the same Ni(cod)$_2$/dcype catalytic manifold in their decarbonylative C—H biaryl coupling reaction of azoles and aryl esters.[125] This finding expands the substrate scope of azole arylation as many aryl coupling partners are substituted with an ester.

Bergman and Ellman developed a rhodium-catalyzed intermolecular reaction between a broad number of heterocycles and a variety of alkenes.[120,126] Benzimidazole, benzothiazole, and benzoxazole **(154)** underwent facile intermolecular alkylation with neohexene to afford heterocyclic products **155–157** in high yields (Scheme 10.54). Substitution of either coupling partner with electron-rich or electron-deficient functional groups did not affect the efficiency of the reaction.

SCHEME 10.53   Itami's C—H/C—O biaryl coupling.

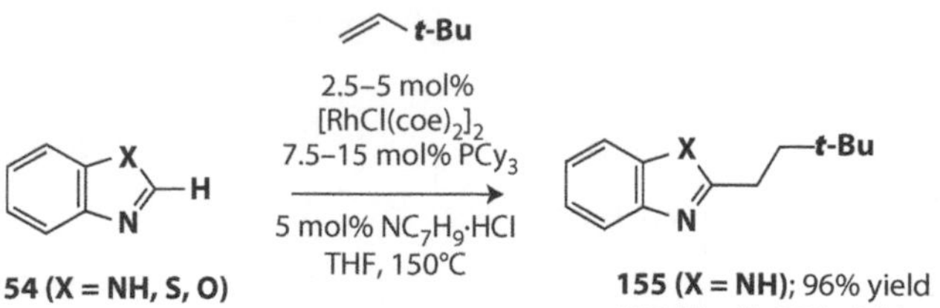

154 (X = NH, S, O)

155 (X = NH); 96% yield
156 (X = S); 97% yield
157 (X = O); 77% yield

**SCHEME 10.54** Bergman and Ellman's alkylation of benzazoles.

The same Rh/[HPCy$_3$]Cl catalyst system was found to facilitate alkylation reactions on notoriously challenging heterocycles such as electron-deficient pyridines and quinolines, thereby displaying its broad utility. Although the exact mechanism for intra- and inter-molecular heterocycle alkylation still remains unclear, activation is thought to proceed via a mechanistically distinct pathway that involves formation of a Rh(I)-*N*-heterocyclic carbene complex.

## 10.4 HETEROAROMATICS CONTAINING THREE HETEROATOMS

### 10.4.1 Triazole

In Daugulis' disclosure of intermolecular electron-rich heterocycle arylation by cheap and commercially available aryl chlorides, C5-arylation of 1-methyl-1,2,4-triazole (**158**) was described (Scheme 10.55).[127] An electrophilic substitution mechanism was identified as the most likely reaction pathway.

Gevorgyan subsequently reported a general and high-yielding method for the C5-arylation of unsymmetrically-substituted 1,2,3-triazoles **161** (Scheme 10.56).[128] Experimental and computational studies strongly support an electrophilic mechanism for this palladium-catalyzed arylation reaction.

### 10.4.2 Imidazopyrimidine

Chemists at Merck reported an effective palladium-catalyzed arylation protocol for imidazo[1,2-*a*]pyrimidine (**164**) (Scheme 10.57).[129] Using a Pd(OAc)$_2$/PPh$_3$/K$_2$CO$_3$ catalytic manifold, the authors found that arylation with various aryl bromides

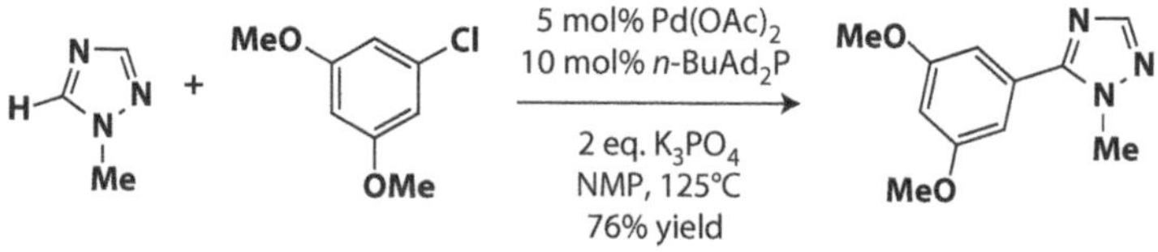

158 159 160

**SCHEME 10.55** Daugulis' C5-arylation of 1-methyl-1,2,4-triazole.

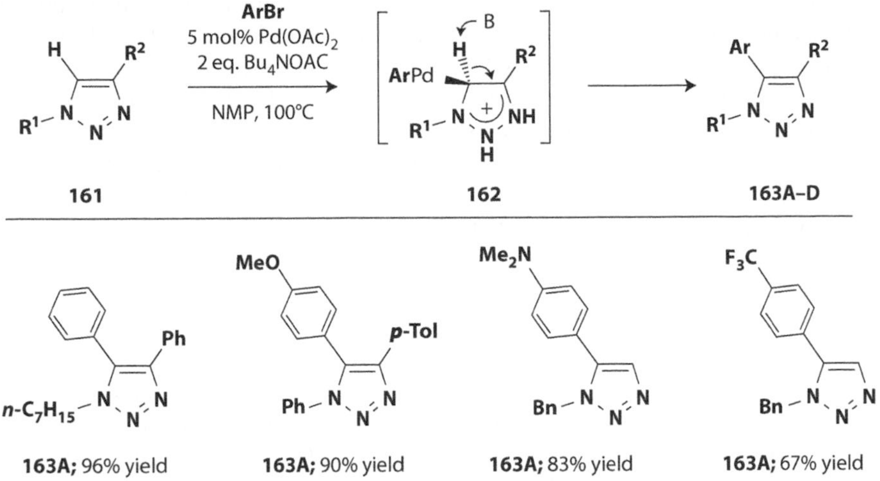

**SCHEME 10.56** Gevorgyan's C5-arylation of unsymmetrical 1,2,3-triazoles.

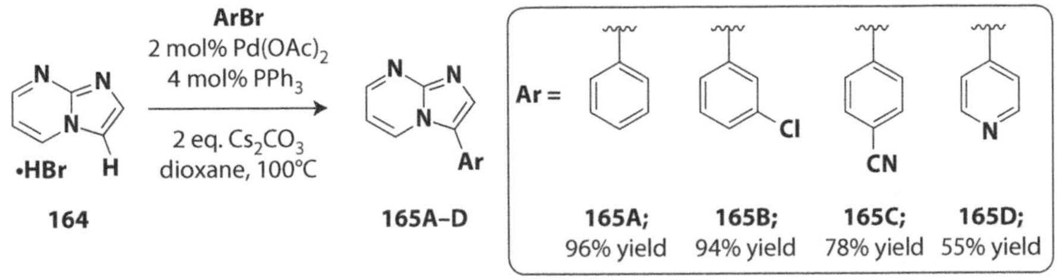

**SCHEME 10.57** C3-arylation of imidazo[1,2-$a$]pyrimidines.

occurred exclusively at the C3 position of the imidazole B-ring to afford products **165A–D** in good to high yields.

This reaction is thought to be electrophilic in character, with arylation occurring at the site most susceptible to electrophilic attack. It was reasoned that attack occurred at the C3 position instead of C2 due to the formation of a more stable arenium ion.

## 10.5 CONCLUSIONS

The area of C—H activation has experienced a flurry of activity over the past two decades and transformations such as those summarized and described herein are fast gaining in their popularity as powerful tools with which the organic chemist can design and construct heteroaromatic molecules. In addition, these metal-catalyzed C—H bond functionalization reactions have had a significant impact on the synthesis of natural products, pharmaceuticals, and functional organic materials.

Future disclosures from the chemistry community will undoubtedly focus on the continued development of highly efficient and active transition metal catalysts that operate at mild temperature and low loadings. More specifically, catalysts that can selectively functionalize unactivated heteroaromatic C—H bonds, without the need for activating or directing groups, are extremely sought-after.

Although tremendous ground has already been broken by researchers in this field, the type of transformations that can be unlocked via the activation of heteroaromatic C—H bonds must still be expanded upon. An ideal scenario would be the disconnection of any carbon–element bond via a C—H activation reaction.

## REFERENCES

1. Eicher, T.; Hauptmann, S. *The Chemistry of Heterocycles*, Wiley-VCH: Weinheim, 2003.
2. (a) McGrath, N. A.; Brichacek, M.; Njardarson, J. T. *J. Chem. Ed.* **2010**, *87*, 1348. (b) Taylor, R. D.; MacCoss, M.; Lawson, A. D. G. *J. Med. Chem.* **2014**, *57*, 5845.
3. (a) Trost, B. M. *Science.* **1991**, *254*, 1471. (b) Anastas, P. T.; Warner, J. C.; *Green Chemistry: Theory and Practice.* Ed.; Oxford University Press: New York, 1998. (c) Wender, P. A.; Verma, V. A.; Paxton, T. J.; Pillow, T. H. *Acc. Chem. Res.* **2008**, *41*, 40.
4. Wang, X.; Lane, B. S.; Sames, D. *J. Am. Chem. Soc.* **2005**, *127*, 4996.
5. Ackermann, L.; Lygin, A. V. *Org. Lett.* **2011**, *13*, 3332.
6. Miyaura, N.; Suzuki, A. *Chem. Rev.* **1995**, *95*, 2457.
7. Wen, J.; Qin, S.; Ma, L.-F.; Dong, L.; Zhang, J.; Liu, S.-S.; Duan, Y.-S.; Chen, S.-Y.; Hu, C.-W.; Yu, X.-Q. *Org. Lett.* **2010**, *12*, 2694.
8. Deprez, N. R.; Kalyani, D.; Krause, A.; Sanford, M. S. *J. Am. Chem. Soc.* **2006**, *128*, 4972.
9. Deprez, N. R.; Sanford, M. S. *J. Am. Chem. Soc.* **2009**, *131*, 11234.
10. Wagner, A. M.; Sanford, M. S. *Org. Lett.* **2011**, *13*, 288.
11. Taylor, E. C.; Jones, R. A. *Pyrroles*; Wiley: New York, 1990.
12. Jafarpour, F.; Rahiminejadan, S.; Hazrati, H. *J. Org. Chem.* **2010**, *75*, 3109.
13. Yang, S.-D.; Sun, C.-L.; Fang, Z.; Li, B.-J.; Li, Y.-Z.; Shi, Z.-J. *Angew. Chem. Int. Ed.* **2008**, *47*, 1473.
14. Trofimov, B. A. In *The Chemistry of Heterocyclic Compounds, Vol. 48: Pyrroles, Part II*; Jones, R. A., Ed.; Wiley: New York, 1992; pp 131–298.
15. Brand, J. P.; Charpentier, J.; Waser, J. *Angew. Chem. Int. Ed.* **2009**, *48*, 9346.
16. De Haro, T.; Nevado, C. *J. Am. Chem. Soc.* **2010**, *132*, 1512.
17. Beck, E. M.; Grimster, N. P.; Hatley, R.; Gaunt, M. J. *J. Am. Chem. Soc.* **2006**, *128*, 2528.
18. (a) García-Rubia, A.; Arrayás, R. G.; Carretero, J. C. *Angew. Chem. Int. Ed.* **2009**, *48*, 6511. (b) García-Rubia, A.; Urones, B.; Arrayás, R. G.; Carretero, J. C. *Chem. Eur. J.* **2010**, *16*, 9676.
19. Li, B.; Ma, J.; Xie, W.; Song, H.; Xu, S.; Wang, B. *Chem. Eur. J.* **2013**, *19*, 11863.
20. (a) Ferreira, E. M.; Stoltz, B. M. *J. Am. Chem. Soc.* **2003**, *125*, 9578. (b) Ferreira, E. M.; Zhang, H.; Stoltz, B. M. *Tetrahedron.* **2008**, *64*, 5987.
21. Schiffner, J. A.; Wöste, T. H.; Oestreich, M. *Eur. J. Org. Chem.* **2010**, 174.
22. (a) Vechorkin, O.; Proust, V.; Hu, X. *J. Am. Chem. Soc.* **2009**, *131*, 9756. (b) Netherton, M. R.; Fu, G. C. *Adv. Synth. Catal.* **2004**, *346*, 1525. (c) Rudolph, A.; Lautens, M. *Angew. Chem. Int. Ed.* **2009**, *48*, 2656.
23. Catellani, M.; Frignani, F.; Rangoni, A. *Angew. Chem. Int. Ed.* **1997**, *36*, 119.
24. Jiao, L.; Bach. T.; *Angew. Chem. Int. Ed.* **2013**, *52*, 6080.

25. (a) Anderson, H. J.; Hopkins, L. C. *Can. J. Chem.* **1964**, *42*, 1279. (b) Tani, M.; Ariyasu, T.; Nishiyama, C.; Hagiwara, H.; Watanabe, T.; Yokoyama, Y.; Murakami, Y. *Chem. Pharm. Bull.* **1996**, *44*, 48.
26. Liégault, B.; Fagnou, K. *Organometallics.* **2008**, *27*, 4841.
27. (a) Iverson, C. N.; Smith, M. R., III. *J. Am. Chem. Soc.* **1999**, *121*, 7696. (b) Ishiyama, T.; Takagi, J.; Hartwig, J. F.; Miyaura, N. *Angew. Chem. Int. Ed.* **2002**, *41*, 3056. (c) Cho, J.-Y.; Tse, M. K.; Holmes, D.; Maleczka, R. E., Jr.; Smith, M. R., III. *Science*, **2002**, *295*, 305. (d) Takagi, J.; Sato, K.; Hartwig, J. F.; Ishiyama, T.; Miyaura, N. *Tetrahedron Lett.* **2002**, *43*, 5649. (e) Ishiyama, T.; Takagi, J.; Ishida, K.; Miyaura, N.; Anastasi, N. R.; Hartwig, J. F. *J. Am. Chem. Soc.* **2002**, *124*, 390. (f) Ishiyama, T.; Takagi, J.; Yonekawa, Y.; Hartwig, J. F.; Miyaura, N. *Adv. Synth. Catal.* **2003**, *345*, 1103. (g) Ishiyama, T.; Miyaura, N. *J. Organomet. Chem.* **2003**, *680*, 3. (h) Ishiyama, T.; Miyaura, N. *Chem. Rec.* **2004**, *3*, 271. (i) Ishiyama, T.; Miyaura, N. *Pure Appl. Chem.* **2006**, *78*, 1369.
28. Vanchura II, B. A.; Preshlock, S. M.; Roosen, P. C.; Kallepalli, V. A.; Staples, R. J.; Maleczka, R. E., Jr.; Singleton, D. A.; Smith, M. R., III. *Chem. Commun.* **2010**, *46*, 7724.
29. Kikuchi, T.; Nobuta, Y.; Umeda, J.; Yamamoto, Y.; Ishiyama, T.; Miyaura, N. *Tetrahedron.* **2008**, *64*, 4967.
30. Tse, M. K.; Cho, J.-Y.; Smith, M. R., III. *Org. Lett.* **2001**, *3*, 2831.
31. Kallepalli, V. A.; Shi, F.; Paul, S.; Onyeozili, E. N.; Maleczka, R. E., Jr.; Smith, M. R., III. *J. Org. Chem.* **2009**, *74*, 9199.
32. Preshlock, S. M.; Plattner, D. L.; Maligres, P. E.; Krska, S. W.; Maleczka, R. E., Jr.; Smith, M. R., III. *Angew. Chem. Int. Ed.* **2013**, *52*, 12915.
33. Nadres, E. T.; Lazareva, A.; Daugulis, O. *J. Org. Chem.* **2011**, *76*, 471.
34. Ghosh, D.; Lee, H. M. *Org. Lett.* **2012**, *14*, 5534.
35. Roy, D.; Mom, S.; Lucas, D.; Cattey, H.; Hierso, J.-C.; Doucet, H. *Chem. Eur. J.* **2011**, *17*, 6453.
36. Glover, B.; Harvey, K. A.; Liu, B.; Sharp, M. J.; Tymoschenko, M. F. *Org. Lett.* **2003**, *5*, 301.
37. Liégault, B.; Lapointe, D.; Caron, L.; Vlassova, A.; Fagnou, K. *J. Org. Chem.* **2009**, *74*, 1826.
38. Hyster, T. K.; Rovis, T. *J. Am. Chem. Soc.* **2010**, *132*, 10565.
39. Hashmi, A. S. K.; Schwarz, L.; Choi, J.-H.; Frost, T. M. *Angew. Chem. Int. Ed.* **2000**, *39*, 2285.
40. (a) Kharasch, M. S.; Beck, T. M. *J. Am. Chem. Soc.* **1934**, *56*, 2057. (b) Fuchita, Y.; Ieda, H.; Yasutake, M. *J. Chem. Soc. Dalton Trans.* **2000**, 271. (c) Fuchita, Y.; Utsunomiya, Y.; Yasutake, M. *J. Chem. Soc. Dalton Trans.* **2001**, 2330.
41. (a) Porter, K. A.; Schier, A.; Schmidbaur, H. *Organometallics.* **2003**, *22*, 4922. (b) Shi, Z.; He, C. *J. Am. Chem. Soc.* **2004**, *126*, 13596.
42. Fukuyama, T.; Chatani, N.; Kakiuchi, F.; Murai, S. *J. Org. Chem.* **1997**, *62*, 5647.
43. Chatani, N.; Ie, Y.; Kakiuchi, F.; Murai, S. *J. Org. Chem.* **1997**, *62*, 2604.
44. Yanagisawa, S.; Sudo, T.; Noyori, R.; Itami, K. *J. Am. Chem. Soc.* **2006**, *128*, 11748.
45. Itami, K. *J. Synth. Org. Chem., Jpn.* **2010**, *28*, 1132.
46. Yanagisawa, S.; Sudo, T.; Noyori, R.; Itami, K. *Tetrahedron.* **2008**, *64*, 6073.
47. Ueda, K.; Yanagisawa, S.; Yamaguchi, J.; Itami, K. *Angew. Chem. Int. Ed.* **2010**, *49*, 8946.
48. Kirchberg, S.; Tani, S.; Ueda, K.; Yamaguchi, J.; Studer, A.; Itami, K. *Angew. Chem. Int. Ed.* **2011**, *50*, 2387.
49. Steinmetz, M.; Ueda, K.; Grimme, S.; Yamaguchi, J.; Kirchberg, S.; Itami, K.; Studer, A. *Chem. Asian J.* **2012**, *7*, 1256.
50. Tang, S.-Y.; Guo, Q.-X.; Fu, Y. *Chem. Eur. J.* **2011**, *17*, 13866.
51. Yamaguchi, K.; Yamaguchi, J.; Studer, A.; Itami, K. *Chem. Sci.* **2012**, *3*, 2165.

52. Yanagisawa, S.; Ueda, K.; Sekizawa, H.; Itami, K. *J. Am. Chem. Soc.* **2009**, *131*, 14622.
53. Liégault, B.; Petrov, I.; Gorelsky, S.I.; Fagnou, K. *J. Org. Chem.* **2010**, *75*, 1047.
54. Chen, L.; Bruneau, C.; Dixneuf, P. H.; Doucet, H. *Green Chem.* **2012**, *14*, 1111.
55. Xi, P.; Yang, F.; Qin, S.; Zhao, D.; Lan, J.; Gao, G.; Hu, C.; You, J. *J. Am. Chem. Soc.* **2010**, *132*, 1822.
56. (a) Fujiwara, Y.; Maruyama, O.; Yoshidomi, M.; Taniguchi, H. *J. Org. Chem.* **1981**, *46*, 851. (b) Tsuji, J.; Nagashima, H. *Tetrahedron.* **1984**, *40*, 2699. (c) Jia, C.; Lu, W.; Kitamura, T.; Fujiwara, Y. *Org. Lett.* **1999**, *1*, 2097.
57. Maehara, A.; Satoh, T.; Miura, M. *Tetrahedron.* **2008**, *64*, 5982.
58. Amatore, C.; Jutand, A. *J. Organomet. Chem.* **1999**, *576*, 254.
59. Yang, Y.; Lin, Y.; Rao, Y. *Org. Lett.* **2012**, *14*, 2874.
60. Nakao, Y. *Synthesis.* **2011**, *20*, 3209.
61. For a recent review see: Liu, C.; Luo, J.; Xu, L.; Huo, Z. *Arkivoc.* **2013**,154.
62. Copéret, C.; Adolfsson, H.; Khuong, T.-A.V.; Yudin, A. K.; Sharpless, K. B. *J. Org. Chem.* **1998**, *63*, 1740.
63. Some of which include: (a) Campeau, L.-C.; Rousseaux, S.; Fagnou, K. *J. Am. Chem. Soc.* **2005**, *127*, 18020. (b) Do, H.-Q.; Kashif Khan, R. M.; Daugulis, O. *J. Am. Chem. Soc.* **2008**, *130*, 15185. (c) Ackermann, L; Fenner, S. *Chem. Commun.* **2011**, *47*, 430.
64. Schipper, D. J.; El-Salfiti, M.; Whipp, C. J.; Fagnou, K. *Tetrahedron.* **2009**, *65*, 4977.
65. Campeau, L.-C.; Stuart, D. R.; Leclerc, J.-P.; Bertrand-Laperle, M.; Villemure, E.; Sun, H.-Y.; Lasserre, S.; Guimond, N.; Lecavallier, M.; Fagnou, K. *J. Am. Chem. Soc.* **2009**, *131*, 3291.
66. (a) Gong, X.; Song, G.; Zhang, H.; Li, X. *Org. Lett.* **2011**, *13*, 1766. (b) Yamaguchi, A. D.; Mandal, D.; Yamaguchi, J.; Itami, K. *Chem. Lett.* **2011**, *40*, 555.
67. (a) Kanyiva, K. S.; Nakao, Y.; Hiyama, T. *Angew. Chem. Int. Ed.* **2007**, *46*, 8872. (b) Cho, S. H.; Hwang, S. J.; Chang, S. *J. Am. Chem. Soc.* **2008**, *130*, 9254. (c) Wu, J.; Cui, X.; Chen, L.; Jiang, G.; Wu, Y. *J. Am. Chem. Soc.* **2009**, *131*, 13888.
68. Xiao, B.; Liu, Z.-J.; Liu, L.; Fu, Y. *J. Am. Chem. Soc.* **2013**, *135*, 616.
69. Berman, A. M.; Lewis, J. C.; Bergman, R. G.; Ellman, J. A. *J. Am. Chem. Soc.* **2008**, *130*, 14926.
70. Ye, M.; Gao, G.-L.; Edmunds, A. J. F.; Worthington, P. A.; Morris, J. A.; Yu, J.-Q. *J. Am. Chem. Soc.* **2011**, *133*, 19090.
71. Ye, M.; Gao, G.-L.; Yu, J.-Q. *J. Am. Chem. Soc.* **2011**, *133*, 6964.
72. Jordan, R. F.; Taylor, D. F. *J. Am. Chem. Soc.* **1989**, *111*, 778.
73. Lewis, J. C.; Bergman, R. G.; Ellman, J. A. *J. Am. Chem. Soc.* **2007**, *129*, 5332.
74. Chotana, G. A.; Rak, M. A.; Smith, M. R., III. *J. Am. Chem. Soc.* **2005**, *127*, 10539.
75. (a) Mandolesi, S. D.; Vaillard, S. E.; Podestá, J. C.; Rossi, R.A. *Organometallics.* **2002**, *21*, 4886. (b) Fidelibus, P. M.; Silbestri, G. F.; Lockhart, M. T.; Mandolesi, S. D.; Chopa, A. B.; Podestá, J. C. *Appl. Organomet. Chem.* **2007**, *21*, 682. (c) Schneider, C.; Broda, E.; Snieckus, V. *Org. Lett.* **2011**, *13*, 3588.
76. Larsen, M. A.; Hartwig, J. F. *J. Am. Chem. Soc.* **2014**, *136*, 4287.
77. Obligacion, J. V.; Semproni, S. P.; Chirik, P. J. *J. Am. Chem. Soc.* **2014**, *136*, 4133.
78. Cacchi, S.; Fabrizi, G. *Chem. Rev.* **2005**, *105*, 2873.
79. Akita, Y.; Inoue, A.; Yamamoto, K.; Ohta, A.; Kurihara, T.; Shimizu, M. *Heterocycles.* **1985**, *23*, 2327.
80. (a) Lane, B. S.; Sames, D. *Org. Lett.* **2004**, *6*, 2897. (b) Lane, B. S.; Brown, M. A.; Sames, D. *J. Am. Chem. Soc.* **2005**, *127*, 8050. (c) Touré, B. B.; Lane, B. S.; Sames, D. *Org. Lett.* **2006**, *8*, 1979.
81. Wang, X.; Gribkov, D. V.; Sames, D. *J. Org. Chem.* **2007**, *72*, 1476.
82. (a) Lebrasseur, N.; Larrosa, I. *J. Am. Chem. Soc.* **2008**, *130*, 2926. (b) Nandurkar, N. S.; Bhanushali, M. J.; Bhor, M. D.; Bhanage, B. M. *Tetrahedron Lett.* **2008**, *49*, 1045. (c) Miyasaka, M.; Fukushima, A.; Satoh, T.; Hirano, K.; Miura, M. *Chem. Eur. J.* **2009**,

15, 3674. (d) Joucla, L.; Batail, N.; Djakovitch, L. *Adv. Synth. Catal.* **2010**, *352*, 2929. (e) Ruiz-Rodríguez, J.; Albericio, F.; Lavilla, R. *Chem. Eur. J.* **2010**, *16*, 1124.

83. (a) Bellina, F.; Cauteruccio, S.; Rossi, R. *Eur. J. Org. Chem.* **2006**, *6*, 1379. (b) Bellina, F.; Calandri, C.; Cauteruccio, S.; Rossi, R. *Tetrahedron.* **2007**, *63*, 1970.

84. Phipps, R. J.; Grimster, N. P.; Gaunt, M. J. *J. Am. Chem. Soc.* **2008**, *130*, 8172.

85. (a) Zhang, Z.; Hu, Z.; Yu, Z.; Lei, P.; Chi, H.; Wang, Y.; He, R. *Tetrahedron Lett.* **2007**, *48*, 2415. (b) Ackermann, L.; Barfuesser, S. *Synlett.* **2009**, 808. (c) Cusati, G.; Djakovitch, L. *Tetrahedron Lett.* **2008**, *49*, 2499.

86. Cornella, J.; Lu, P.; Larrosa, I. *Org. Lett.* **2009**, *11*, 5506.

87. (a) Stuart, D. R.; Fagnou, K. *Science.* **2007**, *316*, 1172. (b) Stuart, D. R.; Villemure, E.; Fagnou, K. *J. Am. Chem. Soc.* **2007**, *129*, 12072.

88. Gu, Y.; Wang, X.-M. *Tetrahedron Lett.* **2009**, *50*, 763.

89. Tolnai, G. L.; Ganss, S.; Brand, J. P.; Waser, J. *Org. Lett.* **2013**, *15*, 112.

90. Grimster, N. P.; Gauntlett, C.; Godfrey, C. R. A.; Gaunt, M. J. *Angew. Chem. Int. Ed.* **2005**, *44*, 3125.

91. Tsuchikama, K.; Kasagawa, M.; Hashimoto, Y.-K.; Endo, K.; Shibata, T. *J. Organomet. Chem.* **2008**, *693*, 3939.

92. Sevov, C. S.; Hartwig, J. F. *J. Am. Chem. Soc.* **2013**, *135*, 2116.

93. Jiao, L.; Bach, T. *J. Am. Chem. Soc.* **2011**, *133*, 12990.

94. Punji, B.; Song, W.; Shevchenko, G. A.; Ackermann, L. *Chem. Eur. J.* **2013**, *19*, 10605.

95. Lee, D.-H.; Kwon, K.-H.; Yi, C. S. *Science.* **2011**, *333*, 1613.

96. Paul, S.; Chotana, G. A.; Holmes, D.; Reichle, R. C.; Maleczka, R. E., Jr.; Smith, M. R., III. *J. Am. Chem. Soc.* **2006**, *128*, 15552.

97. Robbins, D. W.; Boebel, T. A.; Hartwig, J. F. *J. Am. Chem. Soc.* **2010**, *132*, 4068.

98. Dwight, T. A.; Rue, N. R.; Charyk, D.; Josselyn, R.; DeBoef, B. *Org. Lett.* **2007**, *9*, 3137.

99. Hughes, C. C.; Trauner, D. *Angew. Chem. Int. Ed.* **2002**, *41*, 1569.

100. (a) Berman, A. M.; Bergman, R. G.; Ellman, J. A. *J. Org. Chem.* **2010**, *75*, 7863; (b) Lewis, J. C.; Wiedemann, S. H.; Bergman, R. G.; Ellman, J. A. *Org. Lett.* **2004**, *6*, 35. (c) Lewis, J. C.; Berman, A. M.; Bergman, R. G.; Ellman, J. A. *J. Am. Chem. Soc.* **2008**, *130*, 2493.

101. Kwak, J.; Kim, M.; Chang, S. *J. Am. Chem. Soc.* **2011**, *133*, 3780.

102. Wasa, M.; Worrell, B. T.; Yu, J.-Q. *Angew. Chem. Int. Ed.* **2010**, *49*, 1275.

103. Hull, K. L.; Sanford, M. S. *J. Am. Chem. Soc.* **2007**, *129*, 11904.

104. Lyons, T. W.; Hull, K. L.; Sanford, M. S. *J. Am. Chem. Soc.* **2011**, *133*, 4455.

105. Hull, K. L.; Sanford, M. S. *J. Am. Chem. Soc.* **2009**, *131*, 9651.

106. Park, C.-H.; Ryabova, V.; Seregin, I. V.; Sromek, A. W.; Gevorgyan, V. *Org. Lett.* **2004**, *6*, 1159.

107. Seregin, I. V.; Ryabova, V.; Gevorgyan, V. *J. Am. Chem. Soc.* **2007**, *129*, 7742.

108. Pivsa-Art, S.; Satoh, T.; Kawamura, Y.; Miura, M.; Nomura, M. *Bull. Chem. Soc. Jpn.* **1998**, *71*, 467.

109. Zhao, B. *Org. Biomol. Chem.* **2012**, *10*, 7108.

110. Yang, Y.; Cheng, K.; Zhang, Y. *Org. Lett.* **2009**, *11*, 5606.

111. Yang, Y.; Chen, L.; Zhang, Z.; Zhang, Y. *Org. Lett.* **2011**, *13*, 1342.

112. (a) Joo, J. M.; Touré, B. B.; Sames, D. *J. Org. Chem.* **2010**, *75*, 4911. (b) Shibahara, F.; Yamaguchi, E.; Murai, T. *J. Org. Chem.* **2011**, *76*, 2680.

113. Liu, X.-W.; Shi, J.-L.; Yan, J.-X.; Wei, J.-B.; Peng, K.; Dai, L.; Li, C.-G.; Wang, B.-Q.; Shi, Z.-J. *Org. Lett.* **2013**, *15*, 5774.

114. Mori, A.; Sekiguchi, A.; Masui, K.; Shimada, T.; Horie, M.; Osakada, K.; Kawamoto, M.; Ikeda, T. *J. Am. Chem. Soc.* **2003**, *125*, 1700.

115. Gorelsky, S. I. *Organometallics.* **2012**, *31*, 794.

116. Tani, S.; Uehara, T. N.; Yamaguchi, J.; Itami, K. *Chem. Sci.* **2014**, *5*, 123.

117. (a) Gottumukkala, A. L.; Derridj, F.; Djebbar, S.; Doucet, H. *Tetrahedron Lett.* **2008**, *49*, 2926. (b) Besselièvre, F.; Piguel, S.; Mahuteau-Betzer, F.; Grierson, D. S. *Org. Lett.* **2008**, *10*, 4029.
118. Miyasaka, M.; Hirano, K.; Satoh, T.; Miura, M. *J. Org. Chem.* **2010**, *75*, 5421.
119. Tan, K. L.; Bergman, R. G.; Ellman, J. A. *J. Am. Chem. Soc.* **2001**, *123*, 2685.
120. Tan, K. L.; Vasudevan, A.; Bergman, R. G.; Ellman, J. A.; Souers, A. J. *Org. Lett.* **2003**, *5*, 2131.
121. Tsai, A. S.; Wilson, R. M.; Harada, H.; Bergman, R. G.; Ellman, J. A. *Chem. Commun.* **2009**, 3910.
122. Zheng, C.; You, S.-L. *RSC Adv.* **2014**, *4*, 6173.
123. Lewis, J. C.; Wu, J. Y.; Bergman, R. G.; Ellman, J. A. *Angew. Chem. Int. Ed.* **2006**, *45*, 1589.
124. Muto, K.; Yamaguchi, J.; Itami, K. *J. Am. Chem. Soc.* **2012**, *134*, 169.
125. Amaike, K.; Muto, K.; Yamaguchi, J.; Itami, K. *J. Am. Chem. Soc.* **2012**, *134*, 13573.
126. (a) Tan, K. L; Bergman, R. G.; Ellman, J. A. *J. Am. Chem. Soc.* **2002**, *124*, 13964. (b) Wiedemann, S. H.; Bergman, R. G.; Ellman, J. A.; *Org. Lett.* **2004**, *6*, 1685. (c) Tan, K. L.; Park, S.; Ellman, J. A.; Bergman, R. G. *J. Org. Chem.* **2004**, *69*, 7329.
127. Chiong, H. A.; Daugulis, O. *Org. Lett.* **2007**, *9*, 1449.
128. Chuprakov, S.; Chernyak, N.; Dudnik, A. S.; Gevorgyan, V. *Org. Lett.* **2007**, *9*, 2333.
129. Li, W.; Nelson, D. P.; Jensen, M. S.; Hoerrner, R. S.; Javadi, G. J.; Cai, D.; Larsen, R. D. *Org. Lett.* **2003**, *5*, 4835.

# Index